Agriculture Automation and Control

The ultimate goal of agricultural research and technology development is to help farmers produce sufficient foods, feeds, fibers, or biofuels while at the same time, minimize the environmental impacts caused by these large scale activities. Automation offers a potential means by which improved productivity, resource optimization, and worker health and safety, can be accomplished. Although research on agricultural automation can be found in the published literature, there lacks a curated source of reference that is devoted to the unique characteristics of the agricultural system. This book series aims to fill the gap by bringing together scientists, engineers, and others working in these areas, and from around the world, to share their success stories and challenges. Individual book volume will have a focused theme and will be guest-edited by researchers/scientists renowned for their work within the respective sub-discipline.

More information about this series at https://link.springer.com/bookseries/15728

Shaochun Ma • Tao Lin
Enrong Mao • Zhenghe Song
Kuan-Chong Ting
Editors

Sensing, Data Managing, and Control Technologies for Agricultural Systems

Springer

Editors
Shaochun Ma
College of Engineering
China Agricultural University
Beijing, China

Tao Lin
College of Biosystems Eng. & Food Sci.
Zhejiang University
Zhejiang, China

Enrong Mao
College of Engineering
China Agricultural University
Beijing, China

Zhenghe Song
College of Engineering
China Agricultural University
Beijing, China

Kuan-Chong Ting
Department of Agricultural and Biological Engineering
University of Illinois at Urbana-Champaign
Urbana, IL, USA

ISSN 2731-3492 ISSN 2731-3506 (electronic)
Agriculture Automation and Control
ISBN 978-3-031-03833-4 ISBN 978-3-031-03834-1 (eBook)
https://doi.org/10.1007/978-3-031-03834-1

This Springer imprint is published by the registered company Springer Nature Switzerland AG
The registered company address is: Gewerbestrasse 11, 6330 Cham, Switzerland

Preface

The global food demand will be doubled by 2050 due to the world's growing population; however, labor shortages continue to negatively affect agri-food productions. To address these challenges, the emerging science and technology should play an important role in improving agricultural food production and reducing labor dependences.

Agricultural automation is the emerging technology which heavily relies on computer-integrated management and advanced control systems. The tedious farming tasks had been taken over by agricultural machines in last century; in the new millennium, computer-aided systems, automation, and robotics have been applied to precisely manage agricultural production system. With agricultural automation technologies, sustainable agriculture is being developed based on efficient use of land, increased conservation of water, fertilizer, and energy resources. The agricultural automation technologies refers to related areas in sensing and perception, reasoning and learning, data communication, and task planning and execution. Since the literature on this diverse subject is widely scattered, it is necessary to review the current status and capture the future challenges through a comprehensive monograph.

Tremendous progress has been made in the past decades toward successful application of automation technologies in field crops such as corn, wheat, soybean, etc. For example, tractor guidance and steering control technologies have been successfully commercialized to increase steering accuracy and reduce driver fatigue with minimum interaction. In future, driverless tractors will change the role of driver to a fleet manager who oversees robot fleet working in the field 24/7. However, serious technical challenges in automating operations have simultaneously emerged especially in specialty crops such as tree fruit crops. For instance, it is difficult to achieve automated harvesting of fresh fruits and vegetables due to the high bruising rate. In future, agricultural robots need to undergo significant changes to be suited to harvest delicate produce.

In this book, we have limited the scope to agricultural automation and invited a number of internationally recognized experts to provide critical reviews of advanced control technologies, their merits and limitations, application areas, and research opportunities for further development. Such a monograph will serve as an authoritative treatise that can help researchers, engineers, educators, and students in the field of sensing, control, and automation technologies for agriculture.

Beijing, China Shaochun Ma

Contents

Chapter 1
Overview of Sensing, Data Management, and Control Technologies for Agricultural Systems

Peng Huo, Kuan-Chong Ting, and Shaochun Ma

1.1 Introduction

Agriculture is a complex industry, which includes not only production activities, but also processing, storage, transportation, and marketing of agricultural products. The development of agricultural production is affected by the natural environment and social conditions. In the past, to solve the problem of insufficient food supply, the development of agricultural technologies has been focusing on increasing production. However, for the modern society, the provision of basic food and clothing has been mostly achieved. Since entering the twenty-first century, agricultural labor force is increasingly in short supply; therefore, "Who will farm and how to farm" has become an important social problem.

In the human history, there was much evidence that agriculture was critical to the well-being of a country. Shang Yang in the Qin Dynasty of China once said: "A country with 100 farmers out of every 100 people will be a dominating kingdom; with 10 farmers out of every 11 people will be a strong nation, and with only 50% of the population being farmers will be insecure. Therefore, the ruler of a country encourages all his/her people to work in agriculture." In other words, if agricultural production is well performed, a country's food and clothing supply can be adequately provided. With the development of the current agricultural process continues to accelerate, the rural labor force and farmers' concept of labor have also

P. Huo · S. Ma (✉)
College of Engineering, China Agricultural University, Beijing, China
e-mail: shaochun2004@cau.edu.cn

K.-C. Ting
Department of Agricultural and Biological Engineering,
University of Illinois at Urbana-Champaign, Urbana, IL, USA
e-mail: kcting@intl.zju.edu.cn; kcting@illinois.edu

S. Ma et al. (eds.), *Sensing, Data Managing, and Control Technologies for Agricultural Systems*, Agriculture Automation and Control,
https://doi.org/10.1007/978-3-031-03834-1_1

undergone certain changes. The recent agricultural development trend shows that farmers welcome new ways of enhancing labor productivity in agricultural production. Therefore, their demand for agricultural mechanization has been gradually increasing, and they also have a great dependence on the application of mechanized equipment in agricultural production. It is clear that agricultural mechanization has had a great impact on agricultural development. The productivity of every farmer has become more important than the proportion of population engaged in farming.

Building on the success of mechanization that has dramatically reduced the strong dependence of agricultural production on labor force, new development, such as integrated Electronic Information, Internet of Things, Cloud Computing, and other technologies, will facilitate the realization of automation, intelligence, and roboticization in agricultural production processes. The development of agricultural science, engineering, and technology has advanced rapidly. The field of agricultural engineering is no exception. It has made very impressive impact to agricultural production by adding to mechanization with the use of information technology. The research and application of human-like machine capabilities, namely automation to agricultural production has been one of the most active research and development efforts in the agricultural engineering discipline.

As an emerging and fast growing subject area, agricultural automation relies on computer-assisted management and advanced control systems, involving perception (i.e., sensors and sensing), reasoning and learning, communication, task planning and execution, and systems integration. Agricultural automation is also expected to enable sustainably intensified agriculture by improving the use of land, water, fertilizer, and energy, as well as increasing productivity.

Literature on the subject of agricultural automation is abundant but widely scattered; therefore, it is necessary and beneficial to capture the past development, current activities, and future challenges in a comprehensive monograph. This monograph introduces the sensing technology (part A), data processing (Part B), and control technology (Part C) as related to agricultural automation. Specifically, the content of this monograph include: application of agricultural sensor technology in the agricultural Internet of Things (Minzan Li); application of machine vision technology in the production of special crops (Manoj Karkee); imaging technology of plant phenotypes (Jianfeng Zhou); analysis of agricultural production using the automation-culture-environment oriented system approach (ACESYS) (Tao Lin; Kuan-Chong (KC) Ting); application of deep learning algorithms in agricultural image recognition (Luis F Rodriguez; Tao Lin); application of big data in complex agricultural systems (Shih-Fang Chen); research direction of intelligent agricultural management (Du Chen); introduction to the opportunities and challenges of control technology in agricultural automation technology through emerging tractor automation technology (Enrong Mao); application of agricultural machinery time–frequency control (Xiuheng Wu; Zhenghe Song; Steve Suh); application of agricultural drone technology (Yongjun Zheng); and fruit harvesting robot (Long He).

1.2 Part A: Sensing and Perception

The world's agriculture is currently in the transition period from traditional agriculture to modern agriculture. The rapid acquisition of agricultural information is the prerequisite and basis for automated agricultural production (Earl et al., 1996). New technologies, such as agricultural Internet of Things and smart sensors, will play unique and important roles and also provide unprecedented opportunities for the development of modern agriculture. Generally, a wireless sensor node consists of a processing module (usually a low-power microcontroller unit, MCU), one or more sensor modules (embedded, external analog, or digital sensor equipment), and an RF communication module. The RF communication module is usually supported by a low-power wireless communication technology (Tzounis et al., 2017; Fig. 1.1). A sensor in the agricultural Internet of Things system integrates agricultural information perception, data transmission, and intelligent information processing technologies. The specific performance is based on the need of agricultural information to be detected and gathered, such as soil fertility (Mouazen & Ramon, 2009), moisture (Tian et al., 2016), diseases, pests and weeds (Zhao et al., 2015), cultivated layer status information (Andrade-Sanchez et al., 2008), nutrients information (Hashimoto et al., 2007), crop growth and seedling information (Hoshi et al., 2007), farmland ecological environment information (Liu et al., 2014), and crop yield (Zhao et al., 2011) and quality information (Wang et al., 2012). These varieties of agricultural information can be measured using different types of sensing technologies, such as field-based electronic sensors (Ruiz-Garcia et al., 2009), spectrometers (Mouazen et al., 2005), machine vision (Aquino et al., 2015), airborne multispectral and hyperspectral remote sensing (Yang et al., 2008), satellite imagery (Metternicht

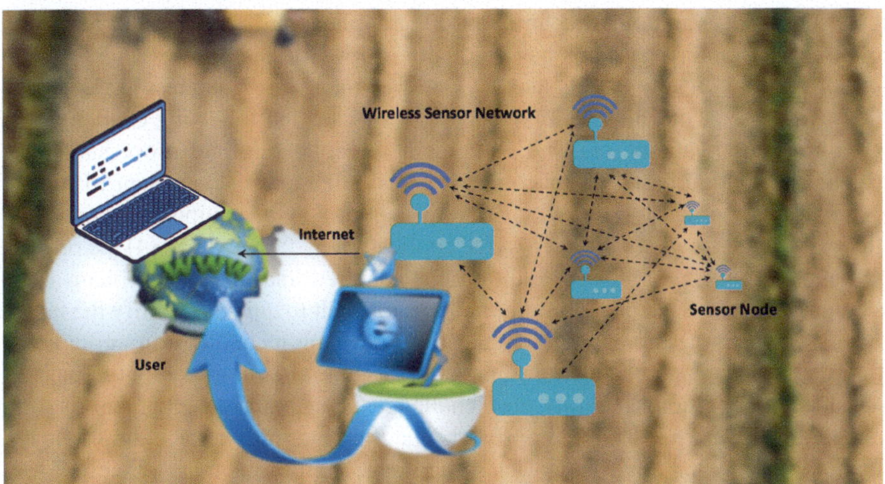

Fig. 1.1 The architecture of a typical wireless sensor node

& Zinck, 2003), thermal imaging (Ondimu & Murase, 2008), radiofrequency identification (Vellidis et al., 2008), machine olfactory system (Benedetti et al., 2008), and so on. The sensing technologies for crop biomass detection, weed detection, soil characteristics, and nutrients are the most advanced and can provide the data required for management in specific locations. On the other hand, sensing technologies for disease detection and characterization and crop moisture status are based on more complex interactions between plants and sensors, making them more difficult to implement on a large scale in the field and more complex to handle (Lee et al., 2010).

In the modern agricultural technology development, the application of agricultural sensor technology in the agricultural Internet of Things, machine vision technology in the production of specialty crops, and imaging technology in plant phenotyping are widely studied.

As the "eyes and ears" of agricultural information perception, the development of new agricultural sensors will be an indispensable and important enabler for the agricultural Internet of Things industry. At the same time, the wide deployment of sensor networks in agriculture and field monitoring requires the following features: low cost, easy operation, durability, remote communication, and scalability of a large number of sensor nodes. At present, these goals have been basically achieved in addition to "low cost." Although mass production can realize low-cost sensor networks, mass production is still difficult in the initial stage of introducing sensor networks into agriculture. To accurately obtain large-scale crop and growth environment information, it is necessary to increase the deployment of sensors, which will greatly increase the cost. Thus, low-cost informatization has become a development direction of sensor technology in the agricultural Internet of Things (Wang et al., 2016; Lee et al., 2010).

With the continuous development of image processing, pattern recognition, artificial intelligence, and other technologies, the application of machine vision technology in agricultural systems has gradually deepened, has been extended to the field of agricultural automation, and has achieved many important results (He et al., 2002). At present, machine vision technology still has the following challenges in agricultural applications: (1) The equipment for collecting and processing crop growth information based on machine vision is bulky and inconvenient to supply power; (2) the portability of field operations is poor; (3) the equipment is not cost-effective, which is not conducive to wide application by farmers; and (4) the operating environment is complex. In addition, the acquired images have problems such as complex and varying backgrounds, and it is difficult to quickly identify the target object, which limits the practical application of vision technology in agriculture. High-definition image acquisition, image processing algorithm computing speed, and object recognition capability pose challenges; hence, the development of software and hardware needs to be improved. The illumination in agricultural fields is affected by time and environment, and the image quality fluctuates greatly. Most existing researches are based on specific light source environments. Varying light, mechanical vibrations, and varying backgrounds make machine vision technology moving from the laboratory to the field more challenging. At present, the speed of image processing by embedded machine vision technology is relatively slow. For occasions with high real-time requirements, image processing algorithms and system hardware need to be further optimized (Zhang, 2016).

High-throughput, automated, high-resolution plant phenotypic information collection platform and analysis technology are essential for accelerating plant improvement and breeding to increase yield and resistance to diseases and insect pests. The platform is used to analyze genomic information and quantitatively study complex traits related to growth, yield, and adaptation to biotic or abiotic stresses. Phenotypic plasticity is an important way to build plant growth models and collect crop data, rich phenotypic data sets that can fill the gap between genomic information and plant phenotypic plasticity (Zhang et al., 2020). It cannot be ignored that the above-mentioned techniques have advantages in analyzing the above-ground parts of plants under natural conditions; however, they have limitations on underground root systems. The cost of high-precision imaging sensors (such as hyperspectral, chlorophyll fluorescence, etc.) limits its broad field deployment. How to convert the phenotypic data generated by spectral imaging into an effective reference for planting and breeding is also a major challenge that imaging spectroscopy technology faces in the analysis of plant abiotic stress phenotypes. Therefore, the fusion of imaging spectroscopy technology can be the directions for future research such as nuclear magnetic resonance, terahertz imaging, comprehensive analysis of plant phenotypes, and efficient algorithms for tabular big data mining and high-throughput analysis framework (Cao et al., 2020).

At present, some major crop yield estimates have adopted commercial yield monitoring systems. Although the yield cannot be predicted accurately during the growing season, the yield patterns and field management areas identified from the airborne multi-spectral and hyper-spectral images are important for intra-season and seasonal production. Post management is very useful. Due to the complexity and nonstructural nature of the agricultural environment, computer vision still needs to improve efficiency in future research. The hyperspectral imaging image processing method needs more research in the field of image automatic registration, and the properties of the monitored objects should be considered when building the model to solve the problem of image fusion skillfully. The development of thermal infrared imaging not only has the high-precision advantages of thermal imaging but also compensate for the relatively low spatial resolution of the camera. Future research needs to develop better acquisition and processing technologies. RFID technology is used to track pesticide and irrigation sensor node information and is expected to gain wide applications. Electronic noses have also shown application potential in specialty crops, and further development is needed in terms of selectivity, sensitivity, and repeatability. Infinite sensor networks have been developed, and the applications of sensor technology are too numerous to enumerate. There are many advanced sensing technologies in various sectors, which can be applied in the agricultural sector in the future, including but not limited to yield estimation, soil detection, water regime monitoring, pest detection, etc. The future of agricultural sensor technology is full of opportunities and challenges, and it is hopeful to improve agricultural operating conditions. In the near future, the development and application of sensor technology will strongly facilitate the development of agricultural automation.

1.3 Part B: Data Utilization and Decision-Making for Automation

Smart agriculture emphasizes the application of information and communication technology in the information–physical farm management cycle. With the emergence of smart farming machines and sensors and the increasing number and scope of farm data, sowing, farming, and harvesting processes will increasingly rely on data and management. The rapid development of the Internet of Things and cloud computing has promoted the so-called smart agriculture (Sundmaeker et al., 2016). Although precision agriculture only considers the variability of the field, smart agriculture goes beyond this emphasis. It is not only based on geographic location but also based on data. It enhances management tasks through real-time event-triggered operations and situational awareness. The real-time auxiliary reconstruction function is necessary to perform agile operations, especially in the case of sudden changes in operating or other conditions (such as weather or disease warnings), including implementation, maintenance, and technology utilizations. Figure 1.2 summarizes the concept of smart agriculture in the management cycle as a cyber–physical system, which involves engineered computing and communicating systems interfacing the farming system (Wolfert et al., 2014).

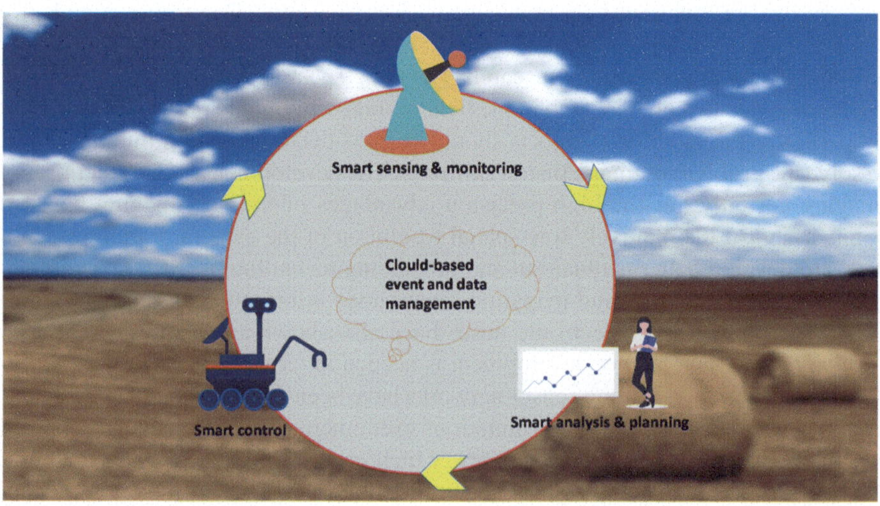

Fig. 1.2 The cyber–physical management cycle of smart farming enhanced by cloud-based event and data management

With the rapid development of network information technology, the agricultural Internet of Things has become an important tool in boosting productivity and preserving energy. The traditional extensive agricultural production model is gradually shifting towards intensification, intelligence, and data. The role of big data in agricultural applications is of great practical significance to the integration of agricultural informatization and modernization (Han & Ma, 2021). New technologies such as the Internet of Things and cloud computing are expected to take advantage of this development, which incorporate the application of big data for decision-making. The scope of application of big data in smart agriculture is not limited to primary production, but also affecting the entire food supply chain, playing a vital role in the development of the agricultural Internet of Things: the machines are equipped with a variety of sensors to capture data in the machines' operating environment. This ranges from relatively simple feedback mechanisms (Nandurkar et al., 2014) (such as thermostats to adjust the temperature) to deep learning algorithms (Chen et al., 2020) (such as implementing correct crop protection strategies). This is achieved by combining with other external big data sources (Elgendy & Elragal, 2014), such as weather or market data, or with benchmark data from other farms. Due to the rapid development of this area, it is difficult to give a unified definition of big data, which usually refers to such a large or complex data set that traditional data processing applications cannot meet the requirements (Wikipedia, 2016).

In the data management of agricultural systems, field agricultural production system analysis for automation-culture-environment oriented system (ACESys), application of deep learning algorithms in agricultural image recognition, application of big data in complex agricultural systems, and smart agricultural management become more cutting-edge and popular; meanwhile, there exists many problems and challenges.

Plant-based engineering systems have evolved from simple structure for plant protection to sophisticated forms for optimizing the productivity of plants and human labor. In developing decision support to aid in the analysis of such complex systems, an automation-culture-environment oriented systems (ACESys) analysis concept has been perceived. The ACESys concept has been used in guiding the process of object-oriented analysis, design, and programming in the development of computerized systems analysis tools. The implementation of these system analysis tools accompanied by information gathering interfaces on the internet which enables the decision supports functions available in a real-time fashion. This concurrent science, engineering, and technology (ConSEnT) platform is expected to encourage broad user participation and effective information integration within the scientific and engineering communities (Ting et al., 2003, 2016; Ting, 1997).

In agricultural systems, big data can be captured, analyzed, and used for decision-making. There are multiple linked stages in the supply chains of agriculture from production to sales. The relevant data is collected, analyzed, and processed through the agricultural big data analysis platform, and then applied to agricultural production. It can provide more accurate data analysis reports for agricultural production throughout the supply chains (Li & Zhang, 2020). The current big data in agricultural economic management (Zilberman, 2019), agricultural e-commerce (Yin &

Liu, 2019), agricultural water conservancy (Sun et al., 2017; Kamienski et al., 2019), mountain agriculture (Kyere et al., 2020), agricultural machinery operations (Li et al., 2019), and agricultural Internet of Things (Wolfert et al., 2017), as well as other fields of applied research are cutting-edge. However, there are some problems and challenges in the research of agricultural system big data. The first is how to ensure privacy and security (Sonka, 2014). Since the data on the farm is usually in the hands of individual entities, it is necessary to invest in public infrastructure to transfer and integrate data, and finally develop applications from the collected data (Schönfeld et al., 2018). Another problem is that the availability and quality of the data may be poor, it needs to be validated before use, and the data needs to be integrated (Yang, 2014). The prospects of big data in agriculture are broad, but to improve the utilization rate of big data applications, the above-mentioned challenges must be dealt with.

Deep learning is an emerging image processing and data analysis technology with broad application prospects and development potential (Kamilaris & Prenafeta-Boldú, 2018). Detecting the main organs (flowers, fruits, stems, leaves, etc.) of plants from plant images is a typical target object recognition capability, which can be used for pest monitoring (Wang et al., 2021) and targeted pesticide application (Tufail et al., 2021), solar greenhouse scene understanding (Jung et al., 2020), and the development of intelligent agricultural machinery (Zhang et al., 2019). The deep convolutional neural network developed in recent years is a new object classification and recognition method, which can realize automatic image feature extraction, integrate with the classification and recognition process, and realize self-learning through data (Zhou et al., 2017). It is different from the shallow architecture of the traditional neural network method. When extracting features of the target, the original image does not need to have a fixed size, so there is no need to trim or compress the original image, and the extracted feature information is more complete and accurate (Yan et al., 2019). Deep learning provides better performance and is superior to other popular image processing techniques. In future work, it will be further used for smarter, more sustainable agriculture, and safer food production.

The future of agriculture is data-centric, more accurate, and smarter than ever before. We call it smart agriculture. It uses advanced modern information technology to intelligently manage agricultural production and operations, thereby promoting precise, visualized, and intelligent agricultural production management models (Ayaz et al., 2019). Enhancing the capabilities of smart agriculture is one of the most important tasks in agricultural modernization. In recent years, the continuous development of information technology has also driven the improvement of cloud computing, big data, and Internet of Things technologies (Channe et al., 2015). The beneficial use of big data in the implementation of smart agriculture can provide a variety of development paths, which will positively promote the transformation and upgrading of agricultural production and management (Kshetri, 2014). Intelligent field control (Popescu et al., 2020), intelligent cloud decision-making (Symeonaki et al., 2020), and mobile terminal monitoring and dispatching are the future direction of intelligent agriculture (Liu et al., 2020a).

The demand for crops continues to increase in quantity and quality, raising the demand for intensification and industrialization of agricultural production. "Agricultural Internet of Things" is a very effective technology that can provide many solutions for agricultural modernization. Scientific teams and research institutions, as well as the industry, are working hard to provide more and more IoT products to the stakeholders of agricultural enterprises, and ultimately lay the foundation for the Internet of Things to play a clear role as a mainstream technology. At the same time, the cloud computing and deep learning provide sufficient resources and solutions to maintain, store, and analyze the massive amounts of data generated by IoT devices. The management and analysis of IoT data ("big data") can be used to automate processes, predict conditions, and improve agricultural activities, and more importantly, the data can be updated in real time. These technologies and foundations have promoted the development of intelligent agriculture, gradually eased the pressure on the rural labor force, and solved the problem of low technology in agricultural development more effectively. In the future, data management technology will be more advanced and smarter, and agriculture can be truly unmanned, precise, visualized, and intelligent.

1.4 Part C: Task Planning and Execution

To reduce the dependence of agricultural production on labor force, the fundamental way is to realize the automation, intelligence, and roboticization of agricultural production processes. Intelligent agricultural equipment integrates electronic information, Internet of things, cloud computing, and other technologies, and the intelligence and automation level has become an important indicator to the development level of modern agriculture (Li et al., 2018). Intelligent agricultural equipment is a complex system integrating machinery, intelligent perception/decision/control, big data/cloud platform/Internet of things, and other technologies, which can independently, efficiently, safely, and reliably complete agricultural tasks (Liu et al., 2020a). These technologies are widely used in various applications of agricultural production, such as common parameter perception of agricultural machinery (Wu et al., 2018), obstacle avoidance of agricultural machinery (Cai et al., 2019), detection of tillage parameters (Yin & Liu, 2019), precision fertilization and sowing (Chen et al., 2018), precision plant protection (Pertot et al., 2017), agricultural UAV (Wang et al., 2019), harvesting robot (Mu et al., 2020), etc. Figure 1.3 shows the integration of small and large sensor networks, UAVs, autonomous vehicles, robots, and agricultural machinery supported by a cloud infrastructure. At present, intelligent agricultural machinery and equipment is a hot research area, which involves positioning system, wireless communication, vehicle communication, data structure, automatic navigation system, automatic turning, and harvest automation. In future, we need to pay attention to problems in these research areas as follows: the responsibility of field machinery, single robot control architecture, multi-robot control architecture,

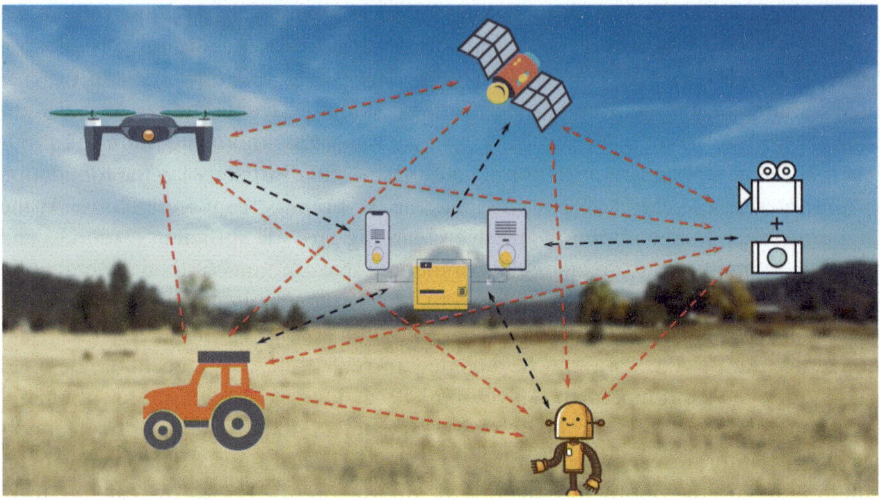

Fig. 1.3 The fusion of small- and large-scale sensor networks, drones, autonomous vehicles, robots, and agrimachinery supported by a cloud infrastructure in open-field cultivation

and intelligent cooperative operation of new-generation agricultural equipment (Shearer et al., 2010).

In the control technology of agricultural system, tractor automation technology is emerging, including application of time frequency control of agricultural machinery, application of agricultural UAV technology, and fruit harvesting robot. Research is cutting-edge, but there are many emerging opportunities and challenges.

The development of agricultural system industrialization puts forward higher requirements for the quantity and performance of agricultural machinery and equipment. As the core of agricultural equipment, tractor technology development reflects the level of national agricultural mechanization and modernization. In recent years, while new technologies are widely used in tractors, new structures and products are constantly emerging, with the continuous improvement of technology content and product performance (Xie et al., 2018). Intelligent tractors for agricultural automation, intelligent perception and decision-making, efficient/autonomous driving, and remote control will play an important role (Baillie et al., 2018). Throughout the development of agricultural tractor technology, it has experienced large-scale breakthroughs in power, transmission, hydraulic, electronic control, and mechatronics. Focusing on the four aspects of "efficiency, intelligence, environmental protection, and information integration," agricultural tractors have made breakthrough progresses in power (Rymaniak et al., 2020), transmission (Ahn et al., 2021), walking (Raheman & Snigdharani, 2020), hydraulic (Roeber et al., 2016), suspension (Yang et al., 2009), driving comfort (Liu et al., 2020b), Internet of Things (Lyle, 2013), and integrated service/management platform (Colezea et al., 2018), as well as many other aspects. As an important node of intelligent agricultural machinery equipment and agricultural production information network, tractors will become new bright spots and new directions in efficient operation, energy saving, environmental protection, informatization, and intelligence.

The application of automated and intellectualized technology in agricultural machine can enhance its working performance, operational safety, economic performance, and environmental friendliness to a great extent. The vital technique to achieve these objectives is the control of actuators, especially under the rough condition which agricultural machine often encounters. A time–frequency control technique which emphasizes that the control algorithm must be implemented in the time domain and frequency domain at the same time. The control error is small in the time domain, and the broadband oscillation spectrum can be suppressed in the frequency domain. This is a broad control concept which has many specific operations on agricultural machines. For example, by applying wavelet analysis technology to improve the performance of traditional controller, the dynamic performance of the controlled system in the whole working frequency domain can be improved (Wu et al., 2017).

With the development of information technology, the Internet of things technology and low-altitude remote sensing technology provided by UAV are widely used in the fields of production practice and environmental monitoring. Agricultural UAV has shown obvious characteristics and advantages in practice and application because it is widely used in agricultural areas where the ground machinery is difficult to cultivate (Lan et al., 2018). In modern agricultural pest monitoring, Internet of things and low altitude remote sensing technology can respectively monitor crop pests and diseases from the ground micro and air macro perspective and analyze their pathogenic environmental factors (Tsouros et al., 2019). Compared with manned aircraft, agricultural UAV has the advantages of not requiring a special runway, better efficiency than manual operation, high quality, low cost, and strong adaptability to the working environment, which is widely recognized by enterprises and the public. However, at present, UAVs are mainly controlled by remote control and pre-programming, which is easy to cause problems, such as poor perception of autonomous environment, slow information processing speed, slow convergence of path planning algorithm, and low crop recognition rate (Radoglou-Grammatikis et al., 2020).

In fruit production, harvesting accounts for about 40% of the total operation. The quality of picking directly affects the storage, processing, and sales of fruits and vegetables, and ultimately affects the market price and economic benefits. Due to the complexity of picking operation, the degree of picking automation is still very low (Song et al., 2006). In the past few decades, many researchers around the world have conducted extensive research on fruit harvesting. However, few robot prototypes have been successfully commercialized (Zhang et al., 2016). The main reasons are that the detection/recognition rate and picking rate of fruits are not high, the average picking cycle is long, and the manufacturing cost of picking robot is high (Jia et al., 2020). Under the current development pattern of environment industrialization, structure standardization, multi-machine co-melting and man–machine co-melting, the robot picking technology has developed from the former picking auxiliary equipment to the automatic picking equipment, and will be intelligent and in the future. (Zhao et al., 2016).

Through the above analysis, we can see that the agricultural IoT is a very promising technology that can provide a variety of solutions for agricultural automation. In addition, the concept of interoperability between heterogeneous devices has inspired the innovation of smart agricultural equipment. With these tools, new applications and services can be created with added values to the data flow generated at the edge of the network.

1.5 Summary

In the past two decades, people have had high expectations for the application of automation technology in agriculture, but they have also seen many challenges. To achieve agricultural automation, it is necessary to learn and apply modern technology in agriculture and integrate intelligent network technologies. Now it is a good time to move toward modern sustainable agriculture. Agriculture can demonstrate the full power of data-driven management to meet the challenges faced by food production in the twenty-first century. In the next 10 years, intelligent agricultural equipment will enable the rapid development of agricultural automation, which can effectively improve agricultural production efficiency, reduce agricultural production costs, improve product quality and safety, and improve agricultural development. Therefore, it is imperative to implement intelligent agricultural automation. In this book, we will introduce the key technologies in agricultural automation, show the main research results, and analyze the opportunities and challenges faced.

References

Ahn, D. V., Shin, I. K., Oh, J., Chung, W. J., Han, H. W., Kim, J. T., & Park, Y. J. (2021). Reduction of torsional vibration in resonance phenomena for tractor power take-off drivelines using torsional damper. *Transactions of the ASABE, 64*(2), 365–376.

Andrade-Sanchez, P., Upadhyaya, S. K., Plouffe, C., & Poutre, B. (2008). Development and field evaluation of a field-ready soil compaction profile sensor for real-time applications. *Applied Engineering in Agriculture, 24*(6), 743–750.

Aquino, A., Millan, B., Gutiérrez, S., & Tardáguila, J. (2015). Grapevine flower estimation by applying artificial vision techniques on images with uncontrolled scene and multi-model analysis. *Computers and Electronics in Agriculture, 119*, 92–104.

Ayaz, M., Ammad-Uddin, M., Sharif, Z., Mansour, A., & Aggoune, E. H. M. (2019). Internet-of-things (IoT)-based smart agriculture: Toward making the fields talk. *IEEE Access, 7*, 129551–129583.

Baillie, C. P., Lobsey, C. R., Antille, D. L., McCarthy, C. L., & Thomasson, J. A. (2018). A review of the state of the art in agricultural automation. Part III: Agricultural machinery navigation systems. In *2018 ASABE Annual International Meeting* (p. 1). American Society of Agricultural and Biological Engineers.

Benedetti, S., Buratti, S., Spinardi, A., Mannino, S., & Mignani, I. (2008). Electronic nose as a non-destructive tool to characterise peach cultivars and to monitor their ripening stage during shelf-life. *Postharvest Biology and Technology, 47*(2), 181–188.

Cai, D. Q., Li, Y. M., Qin, C. J., & Liu, C. L. (2019). Detection method of boundary of paddy fields using support vector machine. *Transactions of the Chinese Society of Agricultural Machinery, 50*(6), 22–27.

Cao, X. F., Yu, K. Q., Zhao, Y. R., & Zhang, H. H. (2020). Current status of high-throughput plant phenotyping for abiotic stress by imaging spectroscopy: A review. *Spectroscopy and Spectral Analysis, 40*(11), 3365–3372.

Channe, H., Kothari, S., & Kadam, D. (2015). Multidisciplinary model for smart agriculture using internet-of-things (IoT), sensors, cloud-computing, mobile-computing & big-data analysis. *International Journal of Computer Technology & Applications, 6*(3), 374–382.

Chen, H., Chen, A., Xu, L., Xie, H., Qiao, H., Lin, Q., & Cai, K. (2020). A deep learning CNN architecture applied in smart near-infrared analysis of water pollution for agricultural irrigation resources. *Agricultural Water Management, 240*, 106303.

Chen, J., Li, Y., Qin, C., & Liu, C. (2018). Design and test of capacitive detection system for wheat seeding quantity. *Transactions of the Chinese Society of Agricultural Engineering, 34*(18), 51–58.

Colezea, M., Musat, G., Pop, F., Negru, C., Dumitrascu, A., & Mocanu, M. (2018). CLUeFARM: Integrated web-service platform for smart farms. *Computers and Electronics in Agriculture, 154*, 134–154.

Earl, R., Wheeler, P. N., Blackmore, B. S., et al. (1996). Precision farming—The management of variability. *Landwards, 51*(4), 18–25.

Elgendy, N., & Elragal, A. (2014, July). Big data analytics: A literature review paper. In *Industrial Conference on Data Mining* (pp. 214–227). Springer.

Han, S., & Ma, D. X. (2021). Analysis on the progress of agricultural big data application research. *Hubei Agricultural Sciences, 60*(02), 15–19.

Hashimoto, A., Ito, R., Iguchi, N., Nakanishi, K., Mishima, T., Hirozumi, T., Hirafuji, M., Ninomiya, S., & Kameoka, T. (2007, January). An integrated field monitoring system for sustainable and high-quality production of agricultural products based on BIX concept with field server, In *2007 International Symposium on Applications and the Internet Workshops* (pp. 76). IEEE.

He, D. J., Zhang, H. L., Ning, J. F., & Long, M. S. (2002). Application of computer vision technique to automatic production in agriculture. *Transactions of the Chinese Society of Agricultural Engineering, 18*(2), 171–175.

Hoshi, T., Shiozawa, E., Shinma, K., Takaichi, M., & Hirafuji, M. (2007). Development of an application program for field servers to acquire and leverage production history information in protected horticulture. *Agricultural Information Research, 16*(1), 1–8.

Jia, W., Zhang, Y., Lian, J., Zheng, Y., Zhao, D., & Li, C. (2020). Apple harvesting robot under information technology: A review. *International Journal of Advanced Robotic Systems, 17*(3), 1729881420925310.

Jung, D. H., Kim, H. S., Jhin, C., Kim, H. J., & Park, S. H. (2020). Time-serial analysis of deep neural network models for prediction of climatic conditions inside a greenhouse. *Computers and Electronics in Agriculture, 173*, 105402.

Kamienski, C., Soininen, J. P., Taumberger, M., Dantas, R., Toscano, A., Salmon Cinotti, T., Maia, R. F., & Torre Neto, A. (2019). Smart water management platform: IoT-based precision irrigation for agriculture. *Sensors, 19*(2), 276.

Kamilaris, A., & Prenafeta-Boldú, F. X. (2018). Deep learning in agriculture: A survey. *Computers and Electronics in Agriculture, 147*, 70–90.

Kshetri, N. (2014). The emerging role of big data in key development issues: Opportunities, challenges, and concerns. *Big Data & Society, 1*(2), 2053951714564227.

Kyere, I., Astor, T., Graß, R., & Wachendorf, M. (2020). Agricultural crop discrimination in a heterogeneous low-mountain range region based on multi-temporal and multi-sensor satellite data. *Computers and Electronics in Agriculture, 179*, 105864.

Lan, Y. B., Wang, L. L., & Zhang, Y. L. (2018). Application and prospect on obstacle avoidance technology for agricultural UAV. *Transactions of the Chinese Society of Agricultural Engineering, 34*(09), 104–113.

Lee, W. S., Alchanatis, V., Yang, C., Hirafuji, M., Moshou, D., & Li, C. (2010). Sensing technologies for precision specialty crop production. *Computers and Electronics in Agriculture, 74*(1), 2–33.

Li, X., Sun, S., & Xiao, M. (2018, December). Agricultural machinery automation and intelligent research and application. In *IOP Conference Series: Materials Science and Engineering* (Vol. 452, No. 4, p. 042077). IOP Publishing.

Li, S., Xu, H., Ji, Y., Cao, R., Zhang, M., & Li, H. (2019). Development of a following agricultural machinery automatic navigation system. *Computers and Electronics in Agriculture, 158*, 335–344.

Li, S., & Zhang, Y. (2020, December). Construction of big data processing platform for intelligent agriculture. In *International Conference on Big Data Analytics for Cyber-Physical-Systems* (pp. 1206–1212). Springer.

Liu, C., Chang, C., & Hu, J. (2020a, April). Development of key components of wheel tractor driving simulator based on desktop virtual reality. In *IOP Conference Series: Earth and Environmental Science* (Vol. 474, No. 3, p. 032019). IOP Publishing.

Liu, C. L., Lin, H. Z., Li, Y. M., Gong, L., & Miao, Z. H. (2020b). Analysis on status and development trend of intelligent control technology for agricultural equipment. *Transactions of the Chinese Society of Agricultural Machinery, 51*(01), 1–18.

Liu, Y., Wang, R. Z., & Yang, Z. L. (2014). Design of environment monitoring system in multispan vegetable greenhouse based on internet of things. *Transactions of the Chinese Society of Agricultural Machinery, 1*, 121–126.

Lyle, S. D. (2013). Experiment to test RTK GPS with satellite "internet to tractor" for precision agriculture. *International Journal of Agricultural and Environmental Information Systems (IJAEIS), 4*(2), 1–13.

Metternicht, G. I., & Zinck, J. A. (2003). Remote sensing of soil salinity: Potentials and constraints. *Remote Sensing of Environment, 85*(1), 1–20.

Mouazen, A. M., De Baerdemaeker, J., & Ramon, H. (2005). Towards development of on-line soil moisture content sensor using a fibre-type NIR spectrophotometer. *Soil and Tillage Research, 80*(1–2), 171–183.

Mouazen, A. M., & Ramon, H. (2009). Expanding implementation of an on-line measurement system of topsoil compaction in loamy sand, loam, silt loam and silt soils. *Soil and Tillage Research, 103*(1), 98–104.

Mu, L., Cui, G., Liu, Y., Cui, Y., Fu, L., & Gejima, Y. (2020). Design and simulation of an integrated end-effector for picking kiwifruit by robot. *Information Processing in Agriculture, 7*(1), 58–71.

Nandurkar, S. R., Thool, V. R., & Thool, R. C. (2014, February). Design and development of precision agriculture system using wireless sensor network. In *2014 First International Conference on Automation, Control, Energy and Systems (ACES)* (pp. 1–6). IEEE.

Ondimu, S., & Murase, H. (2008). Water stress detection in Sunagoke moss (Rhacomitrium canescens) using combined thermal infrared and visible light imaging techniques. *Biosystems Engineering, 100*(1), 4–13.

Pertot, I., Caffi, T., Rossi, V., Mugnai, L., Hoffmann, C., Grando, M. S., Gary, C., Lafond, D., Duso, C., Thiery, D., Mazzoni, V., & Anfora, G. (2017). A critical review of plant protection tools for reducing pesticide use on grapevine and new perspectives for the implementation of IPM in viticulture. *Crop Protection, 97*, 70–84.

Popescu, D., Stoican, F., Stamatescu, G., Ichim, L., & Dragana, C. (2020). Advanced UAV–WSN system for intelligent monitoring in precision agriculture. *Sensors, 20*(3), 817.

Radoglou-Grammatikis, P., Sarigiannidis, P., Lagkas, T., & Moscholios, I. (2020). A compilation of UAV applications for precision agriculture. *Computer Networks, 172*, 107148.

Raheman, H., & Snigdharani, E. (2020). Development of a variable-diameter cage wheel for walking tractor and its performance evaluation in soil bin (simulating wetland). *Journal of the Institution of Engineers (India): Series A, 101*(1), 41–48.

Roeber, J. B. W., Pitla, S. K., Kocher, M. F., Luck, J. D., & Hoy, R. M. (2016). Tractor hydraulic power data acquisition system. *Computers and Electronics in Agriculture, 127*, 1–14.

Ruiz-Garcia, L., Lunadei, L., Barreiro, P., & Robla, I. (2009). A review of wireless sensor technologies and applications in agriculture and food industry: State of the art and current trends. *Sensors, 9*(6), 4728–4750.

Rymaniak, Ł., Lijewski, P., Kamińska, M., Fuć, P., Kurc, B., Siedlecki, M., Kalociński, T., & Jagielski, A. (2020). The role of real power output from farm tractor engines in determining their environmental performance in actual operating conditions. *Computers and Electronics in Agriculture, 173*, 105405.

Schönfeld, M. V., Heil, R., & Bittner, L. (2018). Big data on a farm—Smart farming. In *Big data in context* (pp. 109–120).

Shearer, S. A., Pitla, S. K., & Luck, J. D. (2010). Trends in the automation of agricultural field machinery. In *Proceedings of the 21st Annual Meeting of the Club of Bologna, Italy.*

Song, J., Zhang, T. Z., Xu, L. M., & Tang, X. Y. (2006). Research actuality and prospect of picking robot for fruits and vegetables. *Transactions of the Chinese Society of Agricultural Machinery, 37*(05), 158–162.

Sonka, S. (2014). Big data and the Ag sector: More than lots of numbers. *International Food and Agribusiness Management Review, 17*(1030-2016-82967), 1–20.

Sun, H., Wang, S., & Hao, X. (2017). An improved analytic hierarchy process method for the evaluation of agricultural water management in irrigation districts of North China. *Agricultural Water Management, 179*, 324–337.

Sundmaeker, H., Verdouw, C. N., Wolfert, J., & Freire, L. P. (2016). Internet of food and farm 2020. In *Digitising the industry* (Vol. 49, pp. 129–150). River Publishers.

Symeonaki, E., Arvanitis, K., & Piromalis, D. (2020). A context-aware middleware cloud approach for integrating precision farming facilities into the IoT toward agriculture 4.0. *Applied Sciences, 10*(3), 813.

Tian, H., Zheng, W., & Li, H. (2016). Application status and developing trend of open field water-saving internet of things technology. *Transactions of the Chinese Society of Agricultural Engineering, 32*(21), 1–12.

Ting, K. C. (1997). Automation and systems analysis. In E. Goto, K. Kurata, M. Hayashi, & S. Sase (Eds.), *Plant production in closed ecosystems* (pp. 171–187). Kluwer Academic Publishers.

Ting, K. C., Fleisher, D. H., & Rodriguez, L. F. (2003). Concurrent science and engineering for phytomation systems. *Journal of Agricultural Meteorology, 59*(2), 93–101.

Ting, K. C., Lin, T., & Davidson, P. C. (2016). Integrated urban controlled environment agriculture systems. Chapter 2. In T. Kozai, K. Fujiwara, & E. S. Runkle (Eds.), *Plant factory and greenhouse with LED lighting* (pp. 19–36). Springer.

Tsouros, D. C., Bibi, S., & Sarigiannidis, P. G. (2019). A review on UAV-based applications for precision agriculture. *Information, 10*(11), 349.

Tufail, M., Iqbal, J., Tiwana, M. I., Alam, M. S., Khan, Z. A., & Khan, M. T. (2021). Identification of tobacco crop based on machine learning for a precision agricultural sprayer. *IEEE Access, 9*, 23814–23825.

Tzounis, A., Katsoulas, N., Bartzanas, T., & Kittas, C. (2017). Internet of things in agriculture, recent advances and future challenges. *Biosystems Engineering, 164*, 31–48.

Vellidis, G., Tucker, M., Perry, C., Kvien, C., & Bednarz, C. (2008). A real-time wireless smart sensor array for scheduling irrigation. *Computers and Electronics in Agriculture, 61*(1), 44–50.

Wang, L., Lan, Y., Zhang, Y., Zhang, H., Tahir, M. N., Ou, S., Liu, X. T., & Chen, P. (2019). Applications and prospects of agricultural unmanned aerial vehicle obstacle avoidance technology in China. *Sensors, 19*(3), 642.

Wang, R., Liu, L., Xie, C., Yang, P., Li, R., & Zhou, M. (2021). AgriPest: A large-scale domain-specific benchmark dataset for practical agricultural pest detection in the wild. *Sensors, 21*(5), 1601.

Wang, X. Y., Wen, H. J., Li, X. X., Fu, Z. T., Lü, X. J., & Zhang, L. X. (2016). Research progress analysis of mainly agricultural diseases detection and early warning technologies. *Transactions of the Chinese Society of Agricultural Machinery, 47*(9), 266–277.

Wang, Q., Xi, L., Ren, Y., & Ma, X. (2012). Determination of tobacco leaf maturity degree based on computer vision technology. *Transactions of the Chinese Society of Agricultural Engineering, 28*(4), 175–179.

Wikipedia. (2016). *Big data*. Retrieved August 2, 2016, from https://en.wikipedia.org/wiki/Big_data

Wolfert, S., Ge, L., Verdouw, C., & Bogaardt, M. J. (2017). Big data in smart farming—A review. *Agricultural Systems, 153*, 69–80.

Wolfert, S., Goense, D., & Sørensen, C. A. G. (2014, April). A future internet collaboration platform for safe and healthy food from farm to fork. In *2014 Annual SRII Global Conference* (pp. 266–273). IEEE.

Wu, X. H., Du, Y. F., Suh, C. S., Mao, E. R., & Song, Z. H. (2017). On the design and physical validation of a time-frequency controller. *Control Theory & Applications, 34*(03), 329–336.

Wu, Z., Xie, B., Chi, R., & Mao, E. (2018). Active modulation of torque distribution for dual-motor front-and rear-axle drive type electric vehicle based on slip ratio. *Transactions of the Chinese Society of Agricultural Engineering, 34*(15), 66–76.

Xie, B., Wu, Z., & Mao, E. (2018). Development and prospect of key technologies on agricultural tractor. *Transactions of the Chinese Society of Agricultural Machinery, 49*(8), 1–17.

Yan, J. W., Zhao, Y., Zhang, L. W., Su, X. D., Liu, H. Y., Zhang, F. G., Fan, W. G., & He, L. (2019). Recognition of Rosa roxbunghii in natural environment based on improved faster RCNN. *Transactions of the Chinese Society of Agricultural Engineering, 35*(18), 143–150.

Yang, C. (2014). Big data and its potential applications on agricultural production. *Crop, Environment & Bioinformatics, 11*(1), 51–56.

Yang, C., Liu, T. X., & Everitt, J. H. (2008). Estimating cabbage physical parameters using remote sensing technology. *Crop Protection, 27*(1), 25–35.

Yang, Y., Ren, W., Chen, L., Jiang, M., & Yang, Y. (2009). Study on ride comfort of tractor with tandem suspension based on multi-body system dynamics. *Applied Mathematical Modelling, 33*(1), 11–33.

Yin, B., & Liu, F. (2019, November). Construction of e-commerce platform for agricultural tourism products under the background of big data. In *Journal of Physics: Conference Series* (Vol. 1345, No. 6, p. 062014). IOP Publishing.

Zhang, W. Z. (2016). *Study of the plant information acquisition and processing technology based on vision and image*. (Doctoral dissertation, ZheJiang University).

Zhang, Z., Heinemann, P. H., Liu, J., Baugher, T. A., & Schupp, J. R. (2016). The development of mechanical apple harvesting technology: A review. *Transactions of the ASABE, 59*(5), 1165–1180.

Zhang, Z., Liu, H., Meng, Z., & Chen, J. (2019). Deep learning-based automatic recognition network of agricultural machinery images. *Computers and Electronics in Agriculture, 166*, 104978.

Zhang, H. C., Zhou, H. P., Zhen, J. Q., Ge, Y. F., & Li, Y. X. (2020). Research progress and prospect in plant phenotyping platform and image analysis technology. *Transactions of the Chinese Society for Agricultural Machinery, 51*(03), 1–17.

Zhao, Z. H., Che, X. B., & Zhang, J. (2015). Application of internet of things technology in potato late blight control. *China Plant Protection, 35*(007), 37–40.

Zhao, Y., Gong, L., Huang, Y., & Liu, C. (2016). A review of key techniques of vision-based control for harvesting robot. *Computers and Electronics in Agriculture, 127*, 311–323.

Zhao, C. J., Li, M., Yang, X. T., Sun, C. H., Qian, J. P., & Ji, Z. T. (2011). A data-driven model simulating primary infection probabilities of cucumber downy mildew for use in early warning systems in solar greenhouses. *Computers and Electronics in Agriculture, 76*(2), 306–315.

Zhou, Y. C., Xu, T. Y., Zhen, W., Deng, H., & B. (2017). Classification and recognition approaches of tomato main organs based on DCNN. *Transactions of the Chinese Society of Agricultural Engineering, 33*(15), 219–226.

Zilberman, D. (2019). Agricultural economics as a poster child of applied economics: Big data & big issues. *American Journal of Agricultural Economics, 101*(2), 353–364.

Chapter 2
Agricultural Internet of Things

Yao Zhang, Man Zhang, and Minzan Li

2.1 Agricultural Internet of Things Concept

2.1.1 Basic Concepts and Common Technologies of the Internet of Things

The concept of Internet of Things (IoT) was first proposed by the Massachusetts Institute of Technology in 1999. The early IoT, based on radiofrequency identification (RFID) technology and equipment, refers to the network according to the agreed communication protocol, which enable intelligent identification and management of item information to achieve interconnection, exchangeability, and sharing of item information. With the development of technology and application, the connotation of the IoT has been expanded and redefined as follows. IoT is the expanded application and network extension of communication network and Internet, which uses perception technology and intelligent devices to perceive and identify the physical world. Through network transmission interconnection, IoT performs calculation, processing, and knowledge mining to achieve human–object and object–object information interaction and seamless linkage. The goal is to achieve real-time control, precise management, and scientific decision-making over the physical world (ITU Internet Reports, 2005); Internet of Things in 2020, 2008).

The IoT network architecture consists of perception layer, network layer, and application layer, as shown in Fig. 2.1. The perception layer includes perception and control sub-layer and communication extension sub-layer. The perception and control sub-layer can realize the intelligent perception and recognition, information

Y. Zhang · M. Zhang · M. Li (✉)
College of Information and Electrical Engineering, China Agricultural University, Beijing, China
e-mail: limz@cau.edu.cn

S. Ma et al. (eds.), *Sensing, Data Managing, and Control Technologies for Agricultural Systems*, Agriculture Automation and Control,
https://doi.org/10.1007/978-3-031-03834-1_2

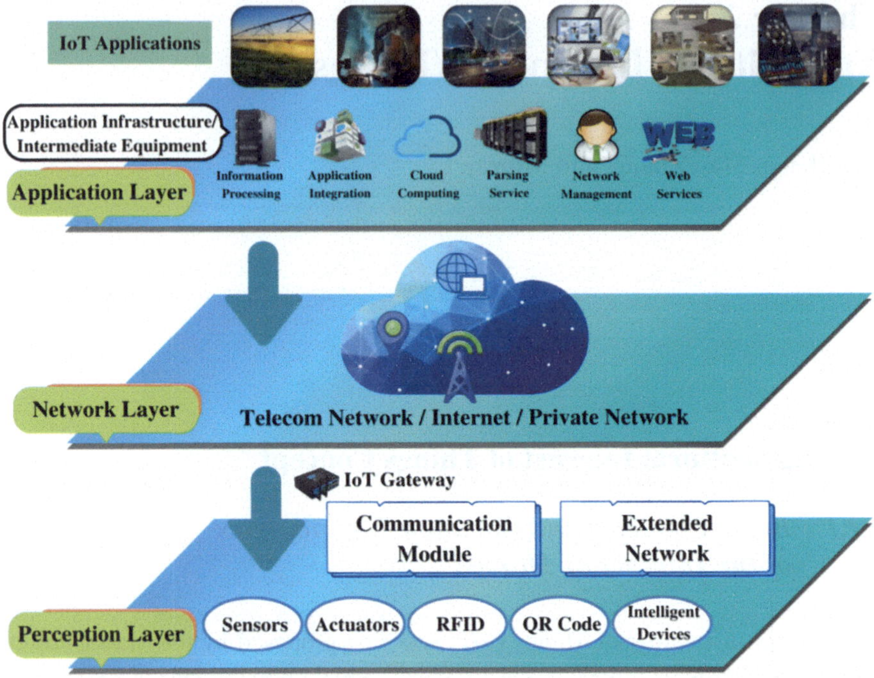

Fig. 2.1 IoT network architecture

collection and processing, and automatic control of real world. The communication extension sub-layer connects physical entities to the network layer and application layer through communication terminal modules directly or after forming an extension network. The network layer includes the access network and the core network, which can realize the transmission, routing, and control of information. The network layer relies on the public telecommunication network and the Internet, as well as the industry-specific communication network. Application layer includes application infrastructure/middleware and various IoT applications. Application infrastructure/middleware provides basic service facilities and resource invocation interfaces for IoT applications, so IOT can be applied in many fields.

The extension of the network to the physical world is a prerequisite for the existence of the IoT.

Therefore, for any extended IoT application, it must rely on the technology of perception level. The common key technologies of IoT mainly include the following areas: perception, control, network communication, microelectronics, computer, software, embedded system, micro-electromechanical, and other technologies. Among them, sensor and RFID technology belong to the first step for the IoT to receive information from physical world. In order to systematically analyze the IoT technology system, the key technologies of IoT can be categorized into the following aspects: perception, network communication, application, network common technologies, and supporting technologies, as shown in Fig. 2.2.

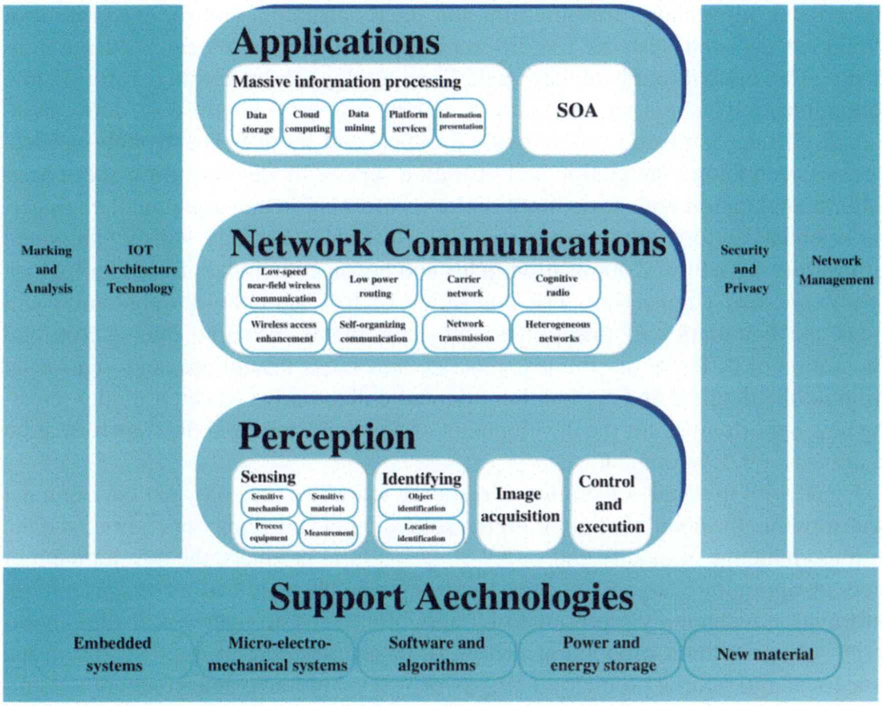

Fig. 2.2 IoT key technologies

2.1.2 Agricultural Internet of Things Concept

The European Smart Systems Integration Technology Platform submitted Internet of Things–Strategic Research Roadmap in 2009, which divided the IoT into 18 major categories. Among them, "Internet of Things in Agriculture and Farming" is one of the most important development directions (Commission of the European Communities, 2009; Internet of Things, 2009). According to the report, agricultural IoT is divided into three layers: information perception, information transmission, and information application. The information perception layer consists of various sensor nodes. Through advanced sensor technology, a variety of parameters to enhance agricultural refinement management can be obtained, such as soil fertility, crop seedling growth, individual animal productivity, health, behavior, and other information. In the information transmission layer, various types of data obtained by sensors are released to the local area network and wide area network through wired or wireless communication protocols. The information application layer fuses and processes the data for making scientific management decisions and controlling the agricultural production process.

The IoT in agricultural application will face a series of scientific and technological problems such as the acquisition of information distributed in a wide area, efficient and reliable information transmission and interconnection, and the integration of intelligent decision systems for different application requirements and environments. The breakthrough of key generic technology in electronic, information, communication technology and industry and the support from low cost, convenient, and easy-to-use hardware and software products and services are strongly needed. Besides, technology integration and operation service mode innovation about agricultural application with agriculture biology, information and equipment engineering scientists' efforts are also required. Information technology will be integrated into various agricultural application fields and become the link between biological, agronomic, and engineering. The innovation of IOT agricultural application will break the boundaries of disciplines and departments, promote intersection and integration of different disciplines, generate new cross-disciplines, and vigorously promote the demand and application oriented collaborative research model, creating new opportunities for the development of new industries and the transformation of agricultural development.

Faced with the dual constraints of resource scarcity and ecological environment deterioration, the contradiction of high resource input and sloppy operation, and the serious challenge of quality and safety of agricultural products, the development of modern agriculture urgently needs to strengthen the application of agricultural information technology represented by agricultural IoT to realize real-time monitoring of agricultural production factors from macroscopic to microscopic in the process of agricultural production, improve the level of fine management of agricultural production and operation, and achieve the purpose of rational use of agricultural resources, reduce production costs, improve the ecological environment, and improve the yield and quality of agricultural products. Based on the principle that information can be perceived at any time, any place, and anything, IoT can support refined process management according to the information and knowledge in all links of agricultural production. In pre-production, IoT can be used to monitor and evaluate agricultural resources such as farmland, climate, water resources, and agricultural materials in real time, providing a basis for scientific utilization and supervision of agricultural resources. In production, IoT can be used to monitor the production process, input use, environmental conditions and implement fine regulation of agricultural production action. In post-production, IoT can be used to connect agricultural products with consumers, so that consumers can transparently understand the production and supply process from the farm to the table. IoT promote the development of e-commerce of agricultural products.

2.2 Basic Technologies for Agricultural IoT

2.2.1 Agricultural Information Sensing Technology

Agricultural information sensing technology refers to the use of agricultural sensors, radiofrequency identification (RFID), bar codes, global navigation satellite systems (GNSS), remote sensing (RS) technology, etc. to collect and acquire information on objects in the agricultural field at any time and any place.

Agricultural Sensors

Agricultural information sensing technology is the core of agricultural IoT. Agricultural sensors are mainly used to collect information in various agricultural applications, including parameters such as light, temperature, water, fertilizer and gas in planting industry; harmful gas content of carbon dioxide, ammonia, and sulfur dioxide, concentration of dust, droplets and aerosols in air, environmental indicators such as temperature and humidity in livestock and poultry breeding industry; dissolved oxygen, pH, ammonia nitrogen, conductivity, and turbidity in aquaculture industry. The first section of this chapter describes several typical agricultural sensors in detail.

Radiofrequency Identification Technology

Radiofrequency identification (RFID), known as electronic tags, refers to the use of radiofrequency signal through space coupling (alternating magnetic field or electromagnetic field) to achieve contactless information transmission and through the information transmitted to achieve the purpose of automatic identification (Finkenzcller, 2003; van Kranenburg, 2008). RFID, emerging from the 1990s, can automatically identify the target through radiofrequency signals and obtain relevant data. The identification process without manual intervention can work in a variety of harsh environments. RFID is mainly used in agriculture in animal tracking and identification, digital breeding, fine crop production, agricultural circulation, etc. (Jeffery et al., 2006; Weizhu et al., 2010).

RFID system is composed of reader, electronic tags, and application software (van Kranenburg, 2008). RFID system working principle diagram as shown in Fig. 2.3. After the electronic tag into the magnetic field, reader transmits a specific frequency of radio wave energy to the electronic tag to drive the electronic tag. By the energy obtained from the induction current, electronic tags send internal data stored in the chip (passive tag or passive tag). Or the tag actively sends a signal at a certain frequency (active tag or active tag). The reader reads and decodes the information and sends it to the central information system for data processing. In terms of communication and energy sensing between RFID reader and electronic tag, it can be roughly divided into two categories including inductive coupling (Inductive Coupling) and backward scattering coupling (Backscatter Coupling). The general low-frequency RFID mostly uses the first type, while the higher frequency mostly uses the second way.

Barcode Technology

Barcode technology is an automatic identification technology integrating barcode theory, photoelectric technology, computer technology, communication technology, and barcode printing technology. Barcode, made up of strips (black, white, and

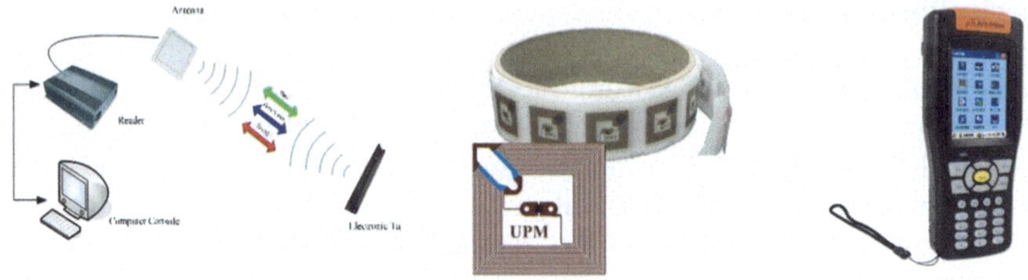

(a) RFID system components **(b)** RFID tags **(c)** RFID card readers

Fig. 2.3 RFID system working principle diagram. (**a**) RFID system components; (**b**) RFID tags; (**c**) RFID card readers

Fig. 2.4 Schematic diagram of 1D barcode and 2D barcode

empty color) with different width and reflectivity according to certain coding rules, is used to express a group of numbers or letters. Barcode technology has been widely used in the quality traceability of agricultural products.

The barcode is divided into one-dimensional (1D) code and two-dimensional (2D) code. Among them, the 1D code is composed of vertical black and white stripes with different thickness. Usually there will be English letters or Arabic numerals under the stripes. 2D code is usually a square structure, not only consists of horizontal and vertical bar codes, and there are also polygonal patterns within the code area. And the texture of the 2D code is also black and white with different thickness. The 2D code is in the form of dot matrix. As shown in Fig. 2.4, the 1D code only contains information in the horizontal direction, the storage capacity is limited, and only numbers could be stored, which could be only used to identify the basic information of goods, such as product name, price, etc. To invoke more information, it is needed to cooperate with the computer database. 2D code, containing information in both horizontal and vertical directions, provides large storage capacity. 2D code can be composed of Chinese characters, letters, numbers, and other information, so it not only has a special identification function but also can display more detailed product content.

2.2.2 *Agricultural Information Transmission Technology*

Agricultural information transmission technology refers to the access of agricultural objects to the transmission network through sensing devices providing the highly reliable information interaction and sharing anytime and anywhere with the help of wired or wireless communication networks. Agricultural information transmission technology can be divided into wireless sensor network technology and mobile communication technology.

Wireless Sensor Network Technology

Wireless sensor network (WSN) is a self-organizing multi hop network system formed by wireless communication. It is composed of a large numbers of sensor nodes deployed in the monitoring area, which are responsible for sensing, collecting, and processing the information of the perceived object in the network coverage area and sending it to the observer (Suman Kumar et al., 2009; Wang et al., 2006).

The typical WSN structure is shown in Fig. 2.5, including sensor node, gateway node, and monitoring software. A large numbers of sensor nodes are distributed in the monitoring area and form a wireless network through self-organization. The data monitored by the sensor node is transmitted hop by hop along other sensor nodes. In the transmission process, the monitoring data may be processed by multiple nodes, routed to the gateway node after multiple hops, and finally transmitted to the monitoring software through Internet, satellite, and other communication methods. Users can configure and manage WSN through monitoring software, publish monitoring tasks, and collect data.

Sensor node is usually a micro embedded system with relatively weak processing, storage, and communication capacity, which is usually powered by battery with limited energy. From the perspective of network function, each sensor node has the dual functions of terminal and router of traditional network nodes. In addition to

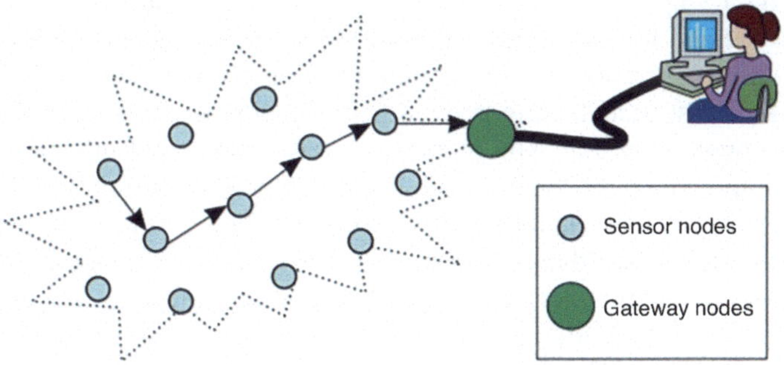

Fig. 2.5 Schematic diagram of wireless sensor network structure

local information collection and data processing, it also needs to store, manage, and integrate the data forwarded by other nodes, and cooperate with other nodes to complete some specific tasks. The gateway node has relatively strong processing, storage, and communication capacity. It is the link connecting WSN, the Internet and other external networks, who realizes the communication protocol conversion between the two protocol stacks, releases the monitoring tasks of the management node, and forwards the collected data to the external network. The gateway node can be either a sensor node with extra functions, with sufficient energy supply, more memory and computing resources, or a gateway device only acting as a wireless communication interface without monitoring capacity. The monitoring software is used to monitor the changes of WSN data information in real time for further analysis and processing according to the collected data information, so as to dig out more valuable information for guiding production practice.

ZigBee technology, based on the IEEE802.15.4 standard, is a technology standard on wireless networking, security, and other applications. ZigBee is widely used in the formation of wireless sensor networks, such as field irrigation, agricultural resource monitoring, aquaculture, agricultural product quality traceability.

Mobile Communication Technology

With the improvement of agricultural informatization level, mobile communication has gradually become an important and key technology for long-distance transmission of agricultural information. Mobile communication has gone through five generations of development. The first-generation mobile communication system (1G) was proposed in the early 1980s, mainly based on cellular structure network, directly using analog voice modulation technology, transmission rate of about 2.4 kbit/s. The second-generation mobile communication system (2G) originated in the early 1990s, using more intensive frequency multiplexing, multiplexing, multiple reuse structure technology, the introduction of smart antenna technology, dual-band, and other technologies. The introduction of GPRs/EDGE technology enabled the organic combination of GSM and computer communication/Internet with data transmission rates up to 115/384 kbps, thus enabling GSM functions to be continuously enhanced and initially equipped with the ability to support multimedia services. The third-generation mobile communication system (3G), also known as IMT 2000, the most basic feature is intelligent signal processing technology, intelligent signal processing unit will become the basic functional module to support voice and multimedia data communications, it can provide a variety of broadband information services that cannot be provided by the first two generations of products, such as high-speed data, slow images and television images. The fourth-generation mobile communication system (4G), which integrates 3G and WLAN and is capable of downloading at 100 Mbps, 2000 times faster than dial-up, and uploading at 20 Mbps. The fifth-generation mobile communication system (5G) adopts a new service-oriented architecture to support flexible deployment and differentiated service scenarios. 5G is capable of downloading at 10 Gbps. The latency of air interface is as low as 1 ms, which can meet real-time applications such as automatic driving and tele-diagnosis.

2.2.3 Agricultural Information Processing Technology

Agricultural information processing, as one of the key technologies of the IoT, is based on agricultural information knowledge and adopts various intelligent computing methods to make objects possess certain intelligence and be able to communicate with users actively or passively. Agricultural information processing technology includes agricultural prediction and early warning, agricultural optimal control, agricultural intelligent decision-making, agricultural diagnostic reasoning, agricultural visual information processing, etc.

Forecasting and Early Warning Technology for Agriculture

Agricultural forecasting is based on actual agricultural information such as soil, environment, meteorological data, crop or animal growth, agricultural production conditions, fertilizers, pesticides, feeds, aerial or satellite images, economic theory and mathematical models, to speculate and estimate the possibility of future development of the research object. Agricultural early warning is to measure the future state of agriculture, forecast the time and space range and harm degree of unusual state, and put forward preventive strategies (Handcock et al., 2009).

Intelligent Control Technology for Agriculture

Agricultural intelligent control is to synthesize and integrate various disciplines such as artificial intelligence, cybernetics, system theory, operational research, and information theory under given constraints in the agricultural field, so that the given performance index of the controlled system can be controlled intelligently.

Intelligent Decision-Making Technology in Agriculture

Intelligent decision-making in agriculture is a specific application of intelligent decision support systems in agriculture, which integrates the knowledge, data, and operations in artificial intelligence (AI), business intelligence (BI), decision support systems (DSS), agricultural knowledge management systems (AKMS), agricultural expert systems (AES), and agricultural management information systems (AMIS).

Agricultural Diagnostic Reasoning Techniques

Agricultural diagnosis refers to the process in which agricultural experts identify the objects based on their characteristic information and use certain diagnostic methods to determine whether the objects are in a healthy state, identify the corresponding causes, and propose ways to change the state or prevent its occurrence, so as to make an objective and realistic conclusion about the object state. Agricultural diagnostic reasoning refers to the construction of a causal network diagnostic reasoning model based on "symptom-disease-cause" by using the knowledge representation method of digital representation and functional description.

Agricultural Visual Processing Technology

Agricultural visual processing refers to the use of image processing technology to process the collected agricultural scene image to realize the recognition and understanding of the target in the agricultural scene. The basic visual information includes brightness, shape, color, texture, etc.

2.3 Agricultural IoT Applications

2.3.1 Application of IoT in Agricultural Information Monitoring

Scientific decision-making and management of the crop growth environment information obtained in real time is an important element of agricultural informatization. Take the "Smart Agricultural Information Platform" supported by China agricultural university as an example, the platform is based on B/S model, consisting of infrastructure layer, data service layer, basic application service layer, service bus layer, business processing layer, and user access layer. It realizes the functions of collecting and storing greenhouse temperature, humidity, light, CO_2, and video information, maintaining basic information, analyzing data and outputting alarms. The application results show that the platform has good stability, perfect functions, and a friendly and convenient human–machine interface, which can realize the effective organization and management of data.

General System Design

The "Smart agriculture information platform" mainly includes "one platform and four systems," namely smart agriculture information platform and precision farming management system for field crop production, fine facility agriculture management system, fine health breeding management system, and traceability management system for agricultural products. The platform is based on B/S mode, and the system consists of an infrastructure layer, data service layer, basic application service layer, service bus layer, business processing layer, and user access layer. Among them, the infrastructure layer mainly includes the hardware and software foundation for information storage and transmission. The data service layer mainly realizes the integration of different types of data originally distributed in the system. The basic application service layer provides many decomposed application services performing a single function, such as permission management, membership management services, etc. The service bus layer uniformly registers the relatively independent basic service objects on the ESB service bus and manages the service life cycle and service interface invocation rules through the ESB. The business processing layer provides a set of business services with related functions established according to the system user roles. The user access layer displays the single application services, composite business services, and integrated data services provided by the bottom layer of SOA architecture to end users through a unified access portal.

1. System Structure Design

   Taking greenhouse as an example, the data collection and remote transmission subsystem are composed of sensor nodes, gateway nodes, and relay routing nodes. The sensor nodes are connected with temperature, humidity, carbon dioxide, and light sensors and deployed in the center of each greenhouse. The nodes transmit the sensed data to the gateway node in a multi-hop manner through ZigBee wireless communication technology, and the gateway node is connected to a local PC through a serial line, on which the stand-alone monitoring software runs, receives, and stores the monitoring data through serial scanning, then processes and analyzes them. The greenhouse administrator can view the monitoring reality.

   The PC is connected to the server depending on the field situation: the PC is connected to the server via the Internet in areas with network; otherwise, it is connected to the server via GPRS. The stand-alone monitoring software uses TCP/IP protocol to transmit greenhouse environmental monitoring data to the information platform in real time. The server consists of a data receiving and storage program, MS SQL Server database, and the information platform. The data receiving and storage program is responsible for listening to the designated port, judging and recognizing the TCP socket connection request from the terminal of stand-alone monitoring software, and storing the received content into MS SQL Server database if it is legal data. MS SQL Server database is responsible for storing the received data and the required basic information for the information platform to access and call. The information platform processes the data in the SQL database and sends the data to the information platform in the form of graphics and charts. The Web-based information platform is a set of web applications, which adopts ASP.NET dynamic web technology and is developed by Visual Studio.net 2008 development tool and C# language.

The video camera is connected to the server platform and PC via Internet network to monitor the growth of greenhouse crops and pests and diseases.

The server platform system is designed in B/S mode. Users can access this Web application simply through a browser to perform management operations such as querying monitoring data, and authorized users can watch the video monitoring images of each greenhouse in real time. The overall structure of the system is shown in Fig. 2.6.

2. System Function Design

   Taking the greenhouse as an example, the software functions of the information platform are divided into five parts: data acquisition, data storage, basic information maintenance, data analysis, and data output. Among them, the data acquisition module adopts TCP socket technology to listen to and receive the data uploaded by the stand-alone monitoring software, and judge whether it is qualified data, and discard it if "no" and store it if "yes." The data storage module can store the received sensor data, historical data of greenhouse, spatial distribution map data, basic information, and user information, etc., which provides the basis for the detailed display and management maintenance of the platform. In the basic information maintenance module, to ensure that users see the latest information, the administrator should update and maintain the basic information at any time, such as setting the standard value with the change of seasons and crops, assigning user rights, updating greenhouse information, etc. The data analysis module can perform statistical analysis on the uploaded data and logically judge whether it exceeds the upper and lower monitoring limits and make a conditional query of data, etc. The data output module can display the uploaded data in real time and output alarm records according to logical judgment.

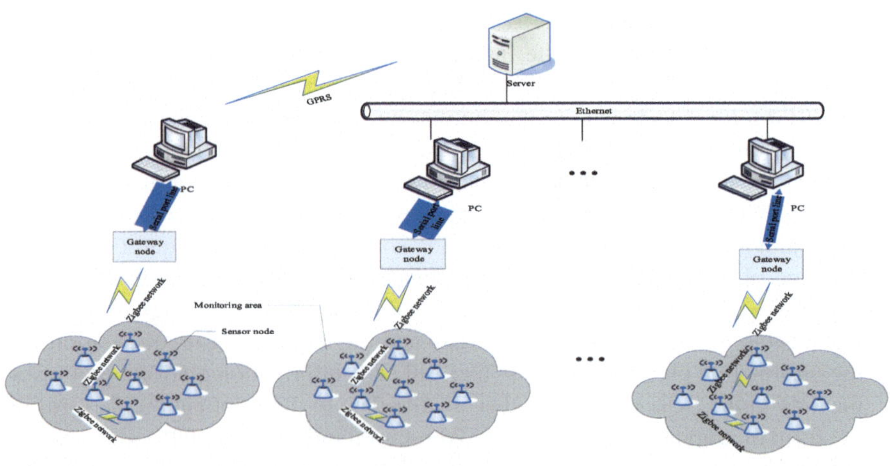

Fig. 2.6 Overall system structure diagram

System Flowchart and Database Design

The smart agriculture information platform mainly includes the design of operation flows, data flows, and database tables.

1. Operation Flow Design
   After authorized users log in, they can view release information, monitoring information, data query, reports, etc. according to different permissions. The general users only have the permission to browse the shared information. The administrator is responsible for assigning user permissions, releasing information, and updating the data.
2. Data Flow Design
   Taking the greenhouse management system as an example, the first data flow divides the functional of system and the data connection between each function. The second layer 2 refine the data flow of each module function.
3. Database Design
   Considering that the data volume will become huge as the collection frequency increases and time lengthens, the medium-sized database—SQL, which has advantages in security, concurrency control capability, data mining, and online operation—is selected for storing data.

Since the information platform function includes four systems with complicated functions, the database tables have complex relationships. Therefore, database tables that conform to the third normal form are selected to eliminate data redundancy, update exceptions, insert exceptions, and delete exceptions.

System Realization

1. Data Storage and Management
   TCP socket technology was used to receive data uploaded by stand-alone monitoring software via Ethernet or GPRS. The database stored the uploaded data and greenhouse environment monitoring standard values, alarm information, basic greenhouse information, user registration information, news information, etc. The system compares the uploaded greenhouse environment monitoring data with the latest monitoring standard values. And if the upper and lower limits are exceeded, alarm information is generated and written in the alarm information table.
2. Data Display and Inquiry
   Curved graphs and tables are used to display real-time monitoring data of each greenhouse node; pie charts and bar charts show greenhouse planting information and greenhouse yield data over the years; tables show greenhouse files, standard values of each greenhouse in different periods, greenhouse planting records, soil information, etc.

2.3.2 Application of IoT in Farmland Moisture Monitoring

In Northern China, droughts occur frequently. It is necessary to monitor soil moisture and implement drought-resistant irrigation in order to prevent the harm of meteorological drought to the growth of winter wheat. Water resources in China are seriously insufficient, but there still exist some problems such as low utilization and serious waste of water resources. Under such circumstances, how to improve the efficiency of water resource utilization is of special practical significance for the development of modern agriculture. By adopting advanced and applicable information communication and sensor technologies, an agricultural water resources information monitoring system for large field agricultural production is established to realize dynamic and accurate monitoring of agricultural resources, thus promoting scientific management and rational utilization of water resources. Therefore, it is necessary to establish a wireless sensor network-based moisture monitoring system for large fields.

Take the example of the precision agriculture demonstration base in Huantai County, Shandong Province, China. The large field moisture monitoring system adopts a combination of fixed monitoring and mobile monitoring schemes. Two farmlands (0.04–0.067 km^2 each) were selected to establish a large field moisture monitoring system. In these two fields, fixed monitoring mode was selected. About 6–10 monitoring nodes were set up to build a field moisture monitoring system through the wireless sensor network. The monitoring nodes converge the collected moisture information to the gateway nodes. Then the gateway nodes use GPRS network or Internet (depending on the site-specific conditions) to send the environmental monitoring information reported by each node to the moisture information management platform to realize the remote collection and monitoring of moisture data. Subsequently, webcams were installed in each of the two farms for monitoring the field sites. The schematic diagram of the system is shown in Fig. 2.7.

Sensor Nodes and Gateway Nodes

The sensor node is shown in Fig. 2.8. This node is solar-powered and is responsible for environmental information collection and short-distance data transmission. This node can input up to four analog or digital signals. In this system, only the FDS soil moisture sensor is connected, with an operating voltage of 5–12 V and an operating current of 35 mA. By calibration, the voltage signal output from the sensor is converted to the volumetric soil moisture content.

In the experimental demonstration base, in addition to the node that can monitor soil water content in real time, a weather station node is also installed to measure the weather information of the monitoring area in real time, and the measurement parameters include air temperature and humidity, rainfall, wind speed, wind direction, light intensity, etc. The weather station node and the soil water content node transmit the real-time measured data to the gateway node using near-range wireless communication respectively. And the gateway node completes the remote transmission of the data (Fig. 2.9).

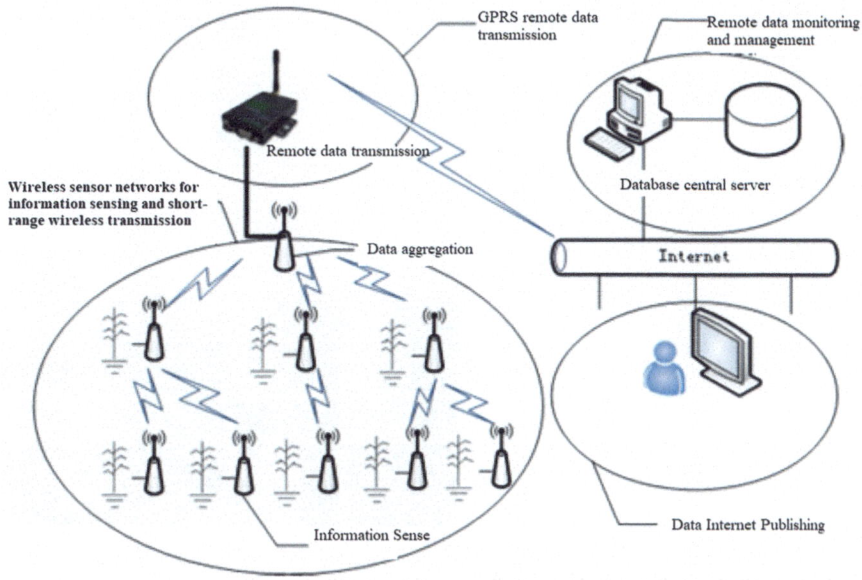

Fig. 2.7 Schematic diagram of the soil moisture monitoring system based on wireless sensor network

Fig. 2.8 Sensor node

The gateway node has the same appearance as the sensor node. While compared with the sensor node, the gateway node has extra GPRS DTU (data transfer unit) modules. The gateway node is responsible for converting the collected sensor node data into serial data and sending it to the GPRS DTU, which then sends serial data to the server.

Fig. 2.9 Weather station node

Video Information Acquisition and Transmission

In the soil moisture monitoring system, a webcam is installed in each farm for real-time monitoring of the field environment. The video information collected by the camera is forwarded to the server through a pair of wireless bridges. As there are many problems in laying high-speed broadband in the field, wireless bridges can effectively replace wired broadband to achieve high-speed transmission of video information from the field to the server.

In this project, the camera was a D-LINK high-speed dome network camera with high resolution and advanced backlight compensation. This camera could capture clear images even under changing illumination conditions and is equipped with a double-layer protective cover and a built-in integrated cloud platform. The cloud platform control program allows remote control for 360° rotation and flexible focal length adjustment. Waveking wireless outdoor bridge, operating in 5 GHz band, built-in 18 dBi patch directional antenna and plastic shell, supports IEEE 802.11A features and IEEE 802.11A/N standard. The wireless bridge integrates the multi-function of wireless AP (Access Point), point-to-point, and point-to-multipoint wireless bridge, with the highest power of 23 dbm (200 mw), which can realize the working modes of single access point connection, multiple access point connection, etc. The installation diagram of the network camera and the wireless bridge is shown in Fig. 2.10. The camera and the bridge of the sending end are installed on the bracket in the farmland. This bridge is responsible for the acquisition and sending of video information. The bridge of the receiving end is installed on the outer wall of the experiment station and is connected to the server by wired method.

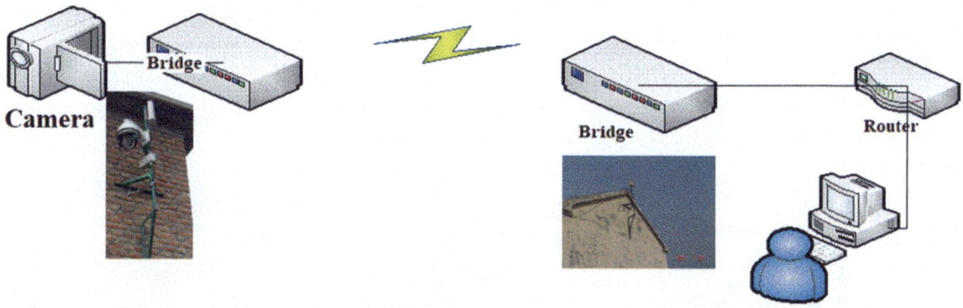

Fig. 2.10 Network camera and wireless bridge

Real-Time Data Monitoring

The information management platform of "Huantai Precision Agriculture Demonstration Project" includes the several parts of field basic information management, soil moisture data collection and management, video monitoring, information announcement, etc. Among them, the basic information management module is responsible for the storage and maintenance of basic information such as soil nutrient distribution map, soil formula fertilization information, area, and facilities. The soil moisture data collection and management module is responsible for managing and analyzing the data collected by the wireless sensor network nodes in real time. And the platform displays the real-time information includes node number, data collection time, and soil moisture content. The user can choose to display all node information or only certain node information. The video monitoring module is responsible for monitoring the field environment, and the user can remotely adjust the focal length and gimbal of the webcam.

2.3.3 Application of IoT in Aquaculture Environmental Monitoring

Aquaculture IoT is an important application area of agricultural IoT. To address the existed problems such as the lack of effective information monitoring technology and methods and the low level of online monitoring and control of water quality, the aquaculture environmental monitoring system (see Fig. 2.11) adopts IoT technology to achieve real-time online monitoring of water quality and environmental information, abnormal alarms, and water quality warnings. Through information transmission channels such as wireless sensor networks, mobile communication networks, and the internet, the abnormal alarms and water quality warnings were notified to the aquaculture managers. Based on water quality monitoring results, aquaculture managers adjust control measures in real time to maintain stable water quality and create a healthy water environment for aquatic products.

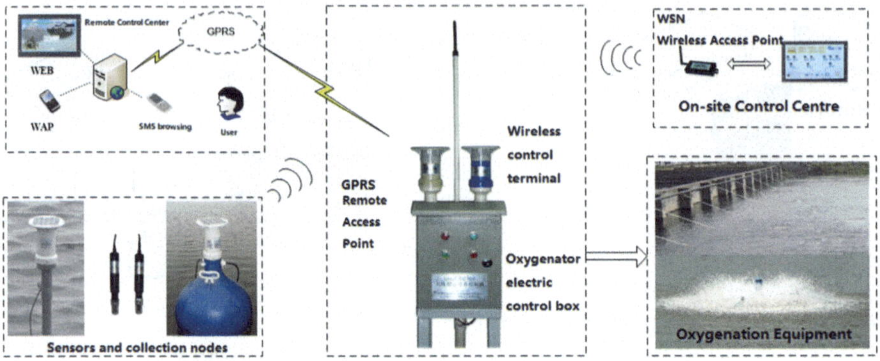

Fig. 2.11 Aquaculture environmental monitoring system

Intelligent Water Quality Sensors

The hardware structure block diagram of the intelligent water quality sensor consists of signal detections and conditioning module, microcontroller, TEDS spreadsheet, bus interface module, power supply, and management module. The microcontroller uses the MSP430F149 from TI, which is a 16-bit RISC structured FLASH-type microcontroller equipped with 12-bit A/D, hardware multipliers, PWM, USART, and other modules. It makes the hardware circuit of the system more integrated and miniaturized. It has a variety of low power consumption modes designed to consume between 0.1 and 400 μA at 1.8–3.6 V and 1 MHz clock conditions, making it ideal for low power consumption product development. The signal conditioning circuit and the bus interface module both use low-voltage, low-power technology, which, together with efficient energy management, enables the entire intelligent sensing system to operate reliably for long periods under battery-powered conditions.

Wireless Oxygen Controller

The wireless dissolved oxygen controller is a key part of the oxygenation control and can drive a variety of oxygenation equipment such as impeller, waterwheel, or micro-hole aeration air compressors. The wireless measurement and control terminal can be configured as a wireless data acquisition node and a wireless control node as required. The wireless control node is the hub that connects the wireless data acquisition node to the site monitoring center. The wireless control node sends the sensed dissolved oxygen information collected by the wireless collection node to the site monitoring center through the wireless network.

The wireless control node can also receive the command requirements sent from the site monitoring center to control electric control box. The output of the electric control box can control all kinds of oxygenators below 10 kw to achieve the

Fig. 2.12 Physical diagram of the wireless oxygenation control system. (**a**) Water quality monitoring point 1, (**b**) Water quality monitoring point 2, (**c**) Water quality control point 1, (**d**) Water quality control point 2, (**e**) Site monitoring center, (**f**) Relay node, (**g**) Video monitoring equipment

automatic control of dissolved oxygen. Figure 2.12 shows the physical diagram of the wireless oxygenation control system.

Wireless Monitoring Network for Aquaculture

The wireless sensor network enables 2.4 GHz short-range communication and GPRS communication, with a 3 km wireless coverage on-site. Intelligent information collection and control technology is used with automatic network routing, self-diagnosis, and intelligent energy management functions. Figure 2.13 shows a diagram of a wireless sensor network.

Intelligent Water Quality Control System

For the intelligent regulation of water quality, the real-time dissolved oxygen volume (RV) and real-time dissolved oxygen variation (RD) are selected as the input to the controller, and the output variable is the oxygenation time (T). Then the corresponding fuzzy control rules are selected to obtain better dynamic characteristics and static quality. It is not difficult to achieve and could meet the requirements of the system. The structure principle of the fuzzy controller is shown in Fig. 2.14.

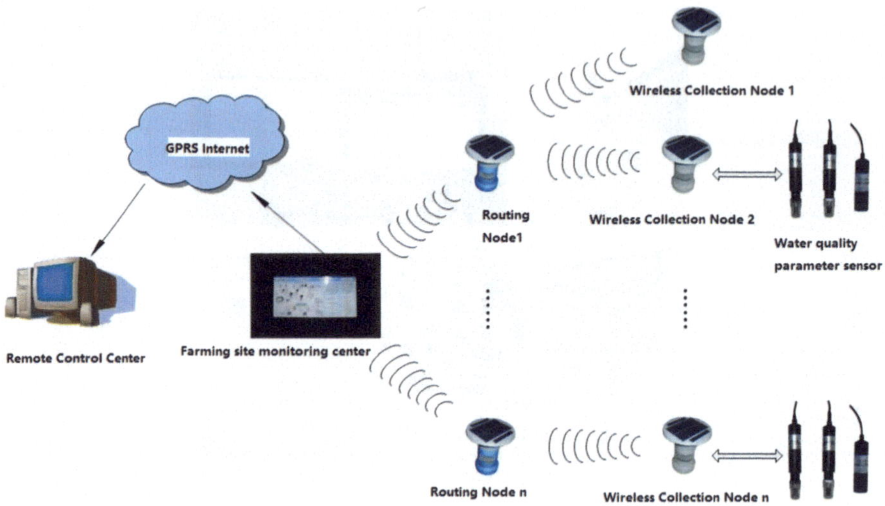

Fig. 2.13 Wireless sensor network

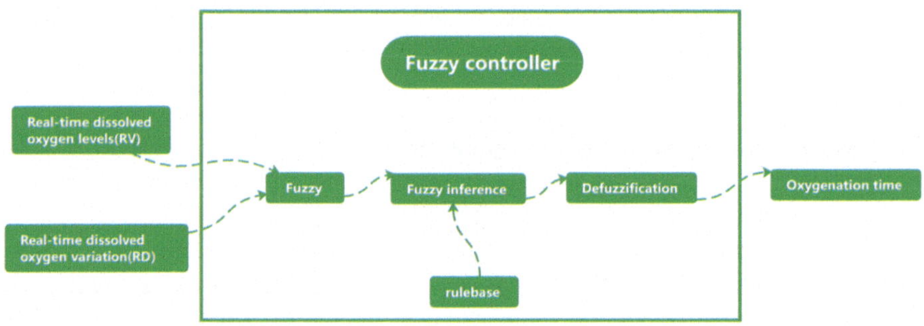

Fig. 2.14 Schematic diagram of the structure of the fuzzy controller

2.3.4 Application of IoT in Agricultural Product Quality Traceability

Aiming at providing traceability basis and means for the whole process supply chain of agricultural products circulation, taking the whole process circulation chain of agricultural products circulation as the foothold, the system comprehensively analyzes the characteristics of various circulating agricultural products and establishes a product quality and safety traceability system from procurement to retail terminal. It realizes the accurate tracking and query of product quality information of the smallest circulation unit. The functional process of agricultural product supervision and traceability system is shown in Fig. 2.15.

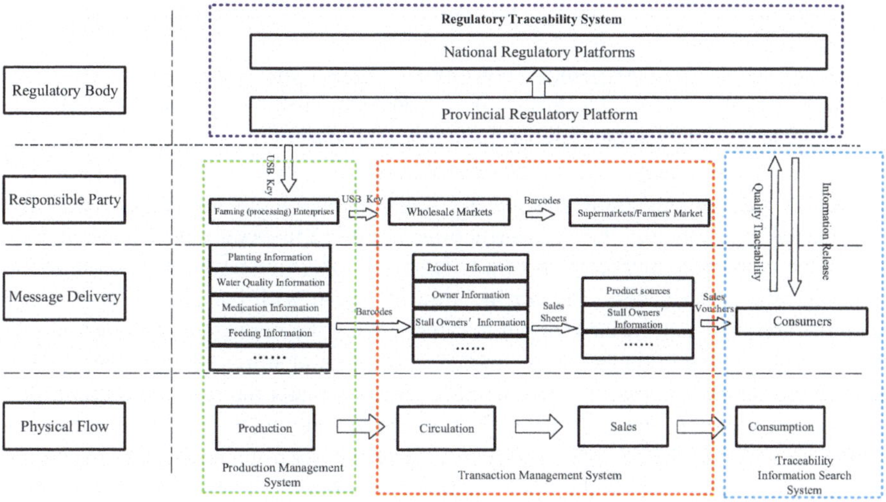

Fig. 2.15 Agricultural products regulatory and traceability system function flow

Production Management Systems

The production management system includes the planting and breeding quality management system and agricultural product processing quality management system developed for planting and breeding managers and processing enterprise users respectively.

Facing the internal management needs of planting and breeding enterprises and aiming at improving the management level of planting and breeding process information and the traceability of planting and breeding process, the planting and breeding quality management system analyzes the production processes of planting and breeding enterprises, such as seedling raising, stocking, feeding, disease prevention, harvesting, transportation, and packaging. The system designs functional modules such as agricultural product planting, breeding production environment, production activities, quality and safety management and sales status to meet the needs of daily management of enterprises. Based on the construction of aquatic product archives information database including basic information, production information, inventory information and sales information, the production management module, inventory management module, and sales module for different users are developed, and each module is integrated to form a safe production management system for agricultural product planting and breeding.

Transaction Management System

Facing the needs of wholesale market management, aiming at realizing product access management and market transaction management, a practical market transaction management system is developed for different modes of wholesale markets, mainly including market access management, market stall management, and transaction management.

Market access management. According to whether the certificate of origin has bar code, the relevant breeder information and product information on the certificate are stored in the central database of the wholesale market in the form of reading or entering, so as to manage the source of products.

Market stall management carries out daily management for each stall in the market, mainly managing basic information, sampling inspection information, etc.

For the wholesale market with a high degree of informatization, according to the principle of market access, the certification of origin with bar code was required from the breeding enterprises (or wholesalers). Then the managers of the wholesale market read the bar code on the certificate of origin and store it in the central database of the wholesale market. The wholesaler downloads the relevant data on that day from the electronic scale through the wireless network, and the wholesalers print one-dimensional bar code product sales orders with manufacturer, wholesale market, wholesaler and product information when trading with customers. Once a product problem occurs, it can be traced back to the wholesaler through the relevant information of the product sales list in the wholesale market.

Regulatory Traceability System

The regulatory traceability platform includes three functional modules: enterprise management, website management, and user management. Enterprise management includes enterprise information upload, enterprise uploading product statistics, SMS platform data statistics, and other functions. Website management includes news system, sampling announcement, enterprise profile, grand view of agricultural products, industry standards, consumer guide, database management, and other functions. At the same time, it meets the different traceability needs of government regulatory departments, enterprise users, and consumers, so as to achieve consumer's satisfaction, improve enterprise management level, and improve the quality and safety of agricultural products. Through modular design and authority division, the regulatory traceability platform can meet the regulatory and traceability needs of regulatory entities at different levels at the ministerial, provincial, municipal, and county levels. It can provide regulatory entities at all levels with detailed responsibility entities of each supply chain of agricultural products, product flow process and agricultural product quality and safety control measures of subordinate regulatory entities. In addition, the number of agricultural product traceability codes and SMS traceability numbers are statistically analyzed through the basic information platform to provide necessary technical support for competent departments at all levels to strengthen management and start risk early warning and emergency responses.

Traceability Information Search System

Through the research of data access general interface, the access protocol of computer network, wireless communication network, and telephone network to the same database was studied. The general API supporting SMS gateway, PSTN gateway, and IP gateway are developed. And the multi-mode query based on the central traceability information database is realized. The traceability information query system on the basis of the traceability information of each link system module takes the product label bar code and product traceability code as the query means and carries out traceability information query through a variety of traceability information query methods such as website, POS machine, SMS, and voice phone.

2.4 Summary

Agricultural IoT technology brings good opportunities for the development of smart agriculture. The application of IoT in smart agriculture can expand the development potential of agriculture, which is conducive to promoting the sustainable development of agriculture. IOT, together with other emerging technologies, provides reliable application technologies for the development of smart agriculture, which not only improve agricultural operation mode, but also ensure the safety and efficiency of agricultural production.

References

Commission of the European Communities. *Internet of things—An action plan for Europe [R], Brussels*, 18.6.2009.

Finkenzcller, K. (2003). *RFID handbook* (2nd ed.). John Wiley & Sons Ltd. England.

Handcock, R. N., Swain, D. L., & Bishop-Hurley, G. J. (2009). Monitoring animal behaviour and environmental interactions using wireless sensor networks, GPS collars and satellite remote sensing [J]. *Sensors, 9*, 3586–3603.

Internet of things—Strategic research roadmap. EoPSS, 15 September 2009.

Internet of things in 2020—Roadmap for the future, Version 1.1. EoPSS, 27 May 2008.

ITU. (2005). *ITU Internet Reports 2005: The internet of things—Executive summary*.

Jeffery, S. R., Garofalakis, M., & Franklin, M. J. (2006). Adaptive cleaning for RFID data streams. In *Proceedings of 32nd International Conference on Very Large Data Bases* (pp. 163–174).

Suman Kumar, S. S., Iyengar, R. L., Wiggins, U., Sekhon, K., Chakraborty, P., & Dora, R. (2009). *Application of sensor networks for monitoring of rice plants: A case Study[M]*, IRA-DSN 2009

van Kranenburg, R. (2008). *The internet of things—A critique of ambient technology and the all-seeing network of RFID[R]*,. Insitute of Network Cultures, Amsterdam. September 2008

Wang, N., Zhang, N., & Wang, M. (2006). Wireless sensors in agriculture and food industry—Recent development and future perspective. *Computers and Electronics in Agriculture, 50*, 1–14.

Weizhu, C., Limin, Z., Hong, Z., et al. (2010). Studies of EPC encoding and privacy of RFID tag in traceability systems. In *World Automation Congress* (pp. 383–387).

Chapter 3
Applied Machine Vision Technologies in Specialty Crop Production

Manoj Karkee and Uddhav Bhattarai

3.1 Introduction

Vision is one of the most important sensory inputs, animals use to make sense of the environment and be able to traverse around and manipulate objects of interest (e.g., finding and collecting food). A wide range of electromagnetic spectrums is used by the animal vision to perform these tasks. Human vision, as an example, operates using a section of the electromagnetic spectrum called visible light (wavelength range ~380 nm to ~750 nm). The light coming from the environment and objects is allowed to enter our eyes through cornea and lens, which hits the back of the eyes called retina exciting photo-receptor cells called rods and cones. The signals thus generated are transmitted to the brain by our optic nerves and processed by a dedicated section of the brain to characterize the environment and objects of interest and estimate their locations using stereo-vision techniques that relies on the disparity between object locations recorded by two eyes.

Similar to human vision, machine vision is one of the most widely used sensor systems for collecting and processing data that can be used to develop automated systems in a wide range of industries such as manufacturing, defense, medicine and healthcare, construction, and agriculture. When it comes to agriculture, machine vision systems have and will continue to be one of the most impactful technologies in monitoring plant growth and health, detecting pest stresses, and detecting and localizing various types of objects in plants (also called plant canopies) and crop fields for automated/robotic operations such as precision chemical application, plant training, pruning and thinning, and crop harvesting. A wide range of vision

M. Karkee (✉) · U. Bhattarai
Department of Biological Systems Engineering, Washington State University, IAREC, Prosser, WA, USA
e-mail: manoj.karkee@wsu.edu

S. Ma et al. (eds.), *Sensing, Data Managing, and Control Technologies for Agricultural Systems*, Agriculture Automation and Control,
https://doi.org/10.1007/978-3-031-03834-1_3

sensors, sensing systems, and data/image processing techniques have been investigated and used in the past several decades, covering various applications in production agriculture, including specialty crops such as fruit and vegetables (more discussion in Gongal et al. (2015)). Specifically, color cameras, hyper- and multispectral cameras, thermal sensors, ultrasonic sensors, and various types of 3D measurement techniques have been used in the past to collect noncontact information that can be used to characterize (e.g., color, shape, size, texture) and locate plants using various types of data and image processing techniques including machine learning such as Deep Neural Networks.

A machine vision system includes sensors that collect/guide different spectrums of light using lenses, record the intensity of light at certain wavelength using analog (historically) or digital media (current systems), and analyze and interpret the data/images to convert into specific information (Fig. 3.1). Machine vision systems provide the most fundamental information on the target objects and environment that could be used for monitoring crops over space and time, analyzing the impact of crop and pest management practices, and providing capabilities for robotic systems to operate in the field to achieve specific tasks such as following corn rows and herding animals (navigational robotics), and picking apples (manipulative robotics).

As depicted by a simplified representation of the vision system pipeline in Fig. 3.1, a machine vision system (to be referred to as a "vision system" in the text

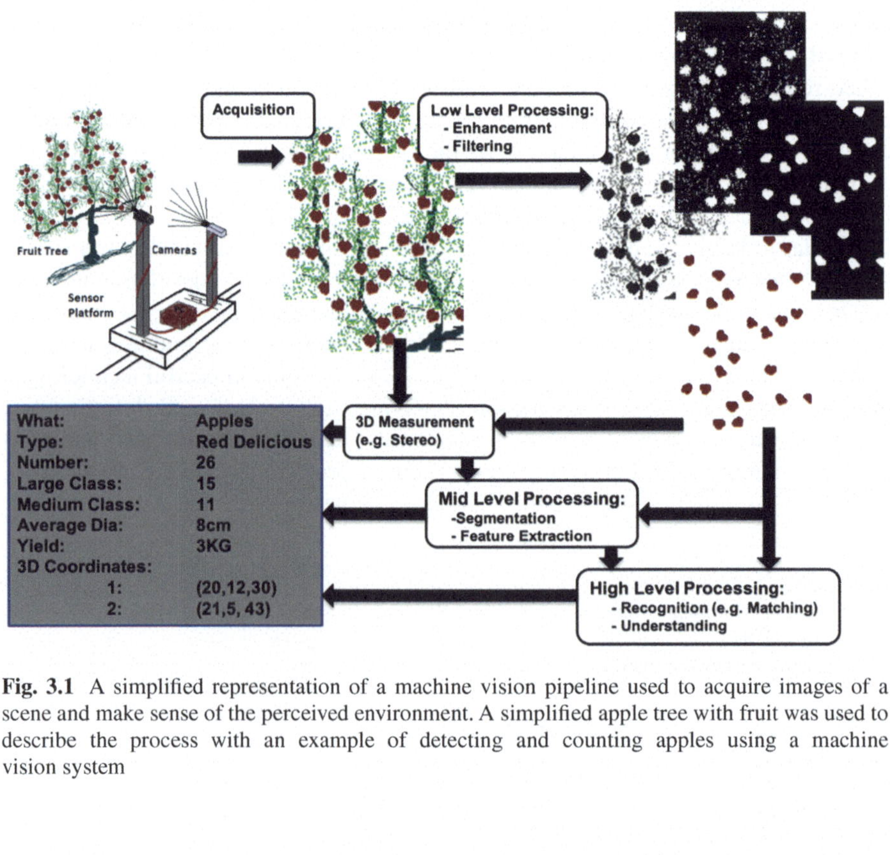

Fig. 3.1 A simplified representation of a machine vision pipeline used to acquire images of a scene and make sense of the perceived environment. A simplified apple tree with fruit was used to describe the process with an example of detecting and counting apples using a machine vision system

to follow) works by first acquiring images using different types of sensors or cameras (top left corner in Fig. 3.1). Raw images then go through a range of image processing techniques, which can roughly be categorized into pre-processing or low-level processing, mid-level processing, and high-level processing steps (Fig. 3.1). However, it is challenging to define a distinct line separating those categories, and the lines have been further blurred by the introduction of end-to-end deep-learning networks. The first set of processing steps includes image enhancement and filtering techniques to reduce the noises and improve sharpness as necessary. These steps, collectively, are generally referred to be the low-level processing or pre-processing.

The next level of processing, also referred to as mid-level processing, includes segmentation and feature extraction techniques. In this stage, we estimate various characteristics/features of the objects of interest including color, shape, size, and texture. The object attributes or parameters estimated at the mid-level processing can then be used as inputs to what is called high-level processing steps where the focus will be in making sense of (or recognizing) the environment and objects. Some of the processing techniques used in this stage include pattern recognition, template matching, and classification. Location/3D information could be used in different ways to help improve the outcome of each of these processing steps and can be combined with the information from high-level processing to assist with the downstream applications of the vision system, such as robotic operation in the agricultural fields (e.g., picking apples and weeding corn). However, there are other kinds of applications of machine vision systems where 3D information may not be as critical, such as detecting and counting the number of flowers in a given tree and tracking birds as they enter and leave an agricultural field.

The objective of this chapter is to enable readers to understand the basics of imaging and image processing techniques and to be able to use such techniques to solve research and engineering challenges. We will specifically discuss basic techniques of binary and color (RGB) image processing including morphological operations, image enhancement techniques, and unsupervised and supervised classification techniques. The chapter will also introduce specific examples of how these concepts and techniques have been used in the latest research and development efforts around the world, specifically in specialty crops such as tree fruits, crops, and vegetables. Spectral sensing and 3D measurement techniques are outside of the scope of this chapter. Readers interested in those concepts and their applications in agriculture can refer to Chap. 4 of this book as well as Karkee and Zhang (2021).

3.2 Image Acquisition

Image acquisition is a process of projecting a real-word scene onto an image space where two types of information is used to describe the scene: (1) geometric information or location of all the object points in the environment and (2) radiometric information of all the object points represented by the light intensity in the image space. Sensors or cameras used in acquiring images/data operate using the electromagnetic spectrum of varying wavelengths. The sensors generally have a mechanism, as mentioned before, to provide a small opening from which electromagnetic waves enter into the camera housing. The wave then travels through a mechanism (e.g., a simple lens) that bends the rays such that they hit and excite an analog or digital photo-receptive surface (sometimes also called a sensor), which can then be recorded to represent the intensity of the light received. Some sensors record electromagnetic wave originating from only one point in the space/scene at a time over a range of wavelengths, which are called spectroscopic sensors. Point sensors can then be used to scan lines and space using mechanical systems to swing the sensor in two directions. These types of sensors are called scanners. Some examples of scanning sensors including Laser or LiDAR, RADAR, and CT Scanners. Often, the light intensities are recorded in spatial grids (Fig. 3.2), and multiple layers of such grids are used to record light intensity at a specific wavelength of the spectrum. These sensors are called imagers or imaging sensors. For example, in a color (RGB) camera, light intensities in red, green, and blue regions are recorded in individual layers creating a three-channel image. Other imaging sensors include X-ray machines, infrared cameras, thermal cameras, and multi/hyperspectral imaging systems.

In general, vision sensors can be categorized into two encompassing groups: active and passive sensors. Active sensors are those that have their own light source and can operate irrespective of the availability of external light sources. These sensors emit the light and receive and record the intensity of the same once it gets reflected from the target surfaces. Examples of such sensors include Laser/LiDAR,

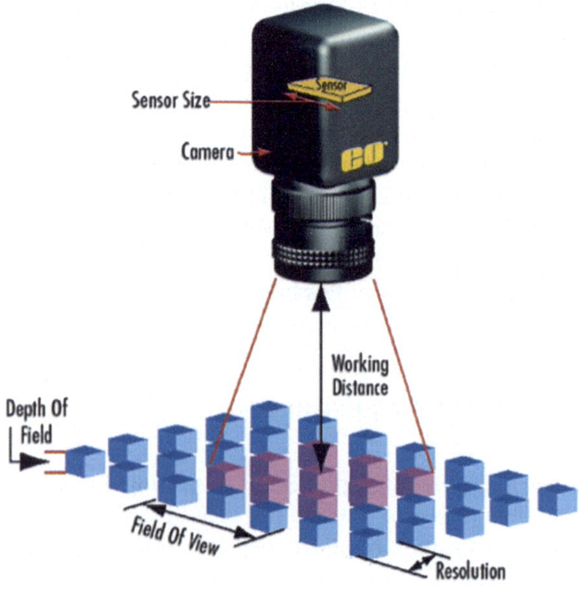

Fig. 3.2 Working principle of an imaging sensor. Field of view represents the space in the environment being imaged by the sensing system (or camera), and "depth of field" represents the range of distances from the camera that will be in focus. Gap between individual pixels is just for demonstrating the discrete nature of the sensing system; no such gaps exist in the actual images acquired. (Image from http://www.evaluationengineering.com/)

RADAR, and 3D cameras with structured lights. Contrary to the active sensors, passive sensors do not have a light source and are designed to record the intensity of the light reflected by the objects based on external light sources such as the sun or other artificial lights. Various color/RGB cameras and spectral sensors such as thermal cameras fall in this category. These vision sensors or cameras can be installed in various platforms for acquiring images including satellites, manned and unmanned aerial systems, balloons, ground vehicles, fixed elevated platforms, and tripods. Quite commonly, particularly when mobile device-based imaging is used, human hands are also used as the sensor platform.

3.2.1 Pinhole and Lens Camera Models

The basic principle of an imaging sensor or camera can be described by a pinhole camera (Fig. 3.3a), a concept used by the earliest cameras developed. In such a camera, a tiny hole (called pinhole) is created on one face of a completed bounded cube. All six surfaces of the cube are opaque so that no light can enter into the box other than from the pinhole. The inside surface of the opposite side of the pinhole would be painted with a photo-receptive material. When the pinhole is opened for a brief duration, light rays from the environment in front of the pinhole enter into the camera housing through the pinhole and excite the film coated with the photo-receptive media. As the hole is really tiny, in principle, only one ray of light from a given location in the environment can enter the housing as shown in Fig. 3.3, thus creating a reasonably crisp image of the environment. However, in practice, lights from multiple sources close to each other can enter and be projected onto the same location. In addition, diffraction of the light as it passes through the sharp hole leads to image blurring. To address this challenge, an improved camera model has been developed using one or more lenses. A simple camera model using one convex lens is depicted in Fig. 3.3b in which light originated from a given point in the space hits the sensor surface at the same location irrespective of the path it takes from the

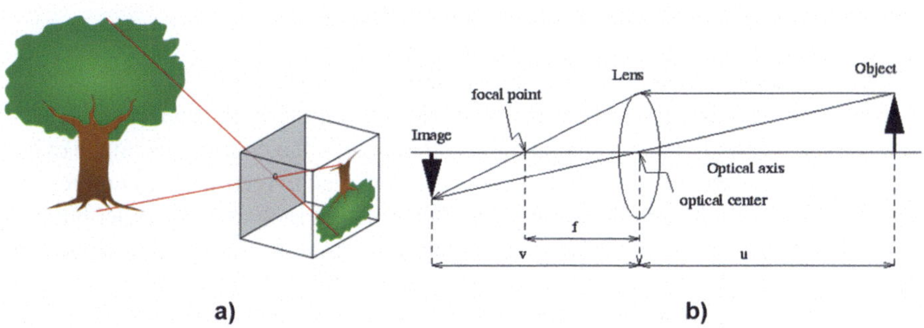

Fig. 3.3 (**a**) A pinhole camera model. (Image from TMMY PHTOG, "How Pinhole Camera Works.." licensed under CC BY 2.0); and (**b**) a lens camera model

environment to the sensor through different locations in the lens surface. This technique, therefore, minimizes the impact of multi-source projection and diffraction issues faced with the pinhole model. The object distance, imaging sensor distance, and focal length of the lens are related, in this model, by Eq. 3.1. This camera design, however, faces the challenge of chromatic aberration (a phenomenon that describes a varying level of bending of light as it passes through prisms depending on the wavelength), leading to image blurring. This issue can be minimized using a long focal length (narrow) lens, designing more complex lens systems using both convex and concave lenses, and using image processing techniques.

$$\frac{1}{u}+\frac{1}{v}=\frac{1}{f} \tag{3.1}$$

3.2.2 Image Representation

An image, often also called an intensity or gray level image, can be defined as $I(x,y)$, where x and y are the spatial coordinates, and I is the intensity of the light recorded at the specific locations, representing the projection of the real world. This representation of the images is in the continuous domain, assuming x and y are the continuous variables (Fig. 3.4). As discussed before, such analog images in the continuous domain could be created and developed by allowing light waves to hit the traditional photo-receptive films. Images can be represented in the discrete domain by digitizing them from the continuous domain or acquiring directly in the discrete form using digital sensors. Digital sensors used to acquire images often consist of charge-coupled devices (CCDs), which are integrated circuits with arrays (discrete cells) of photoreceptive surfaces (Fig. 3.5). As the photons in the incident light hit those surfaces, they free a proportional number of electrons. The free electrons are integrated together over the shutter opening duration, and amplified and quantized to represent the intensity level at each cell. In the digital form, images are represented by $I(r,c)$, where I is the quantized intensity level, and r and c represent the discretized location in the space at rth row and cth column.

Discretization and quantization, as mentioned before, are two critical steps to create digital images used in all modern vision systems (Fig. 3.6). Discretization is the process of representing a continuous world (within the field of view) by a two-dimensional array of discrete spatial units called pixels. If a continuous image exists and needs to be digitized, a digitizing system is designed to record the light intensity at regularly spaced discrete locations (e.g., 300 dots per inch). On the other hand, if a digital imaging system or camera is used, image formation starts with a physically discretized photo-receptive surface as discussed before. A lot of modern digital imaging systems such as color cameras have more than 5000 × 4000 discrete sampling locations (or pixels) covering the space being imaged. Following the discretization of the spatial or geometric information represented by an image into

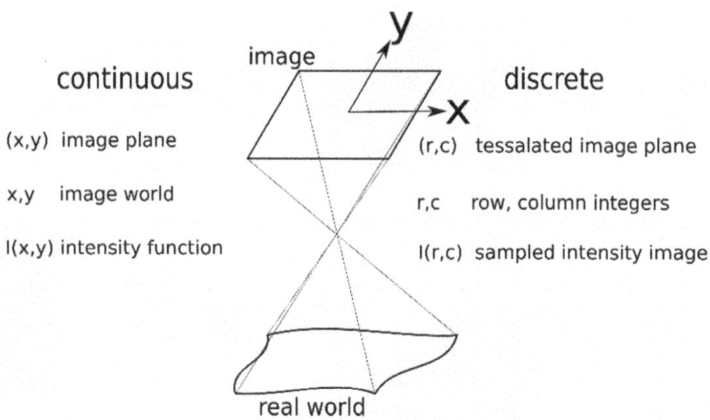

Fig. 3.4 Schematic representation of the real world with continuous and discrete images

Fig. 3.5 A grided representation of sensors used by various imaging systems such as color cameras

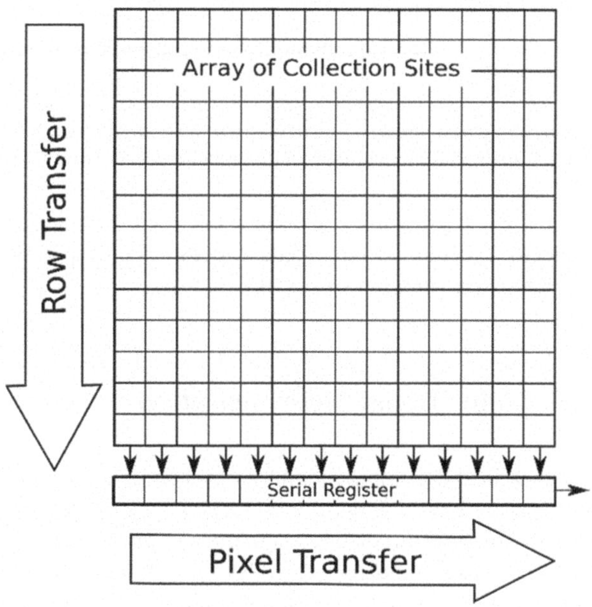

individual pixels, radiometric information, or the intensity value at each pixel needs to be stored in a digital register of a certain size (a specific length of bits), leading to a finite number of discrete intensity levels to be preserved. This process is called quantization (Fig. 3.6, right). For example, if the pixel intensity is represented by a 4-bit register, only 16 discrete intensity levels (0–15) would be possible. In modern computers with powerful processing units and large storage capacities, quantization is no more a limiting issue as millions of intensity levels can be represented without much impact on the practical implementation. However, the fineness of the discretization process (pixel resolution) remains to be an important parameter optimized

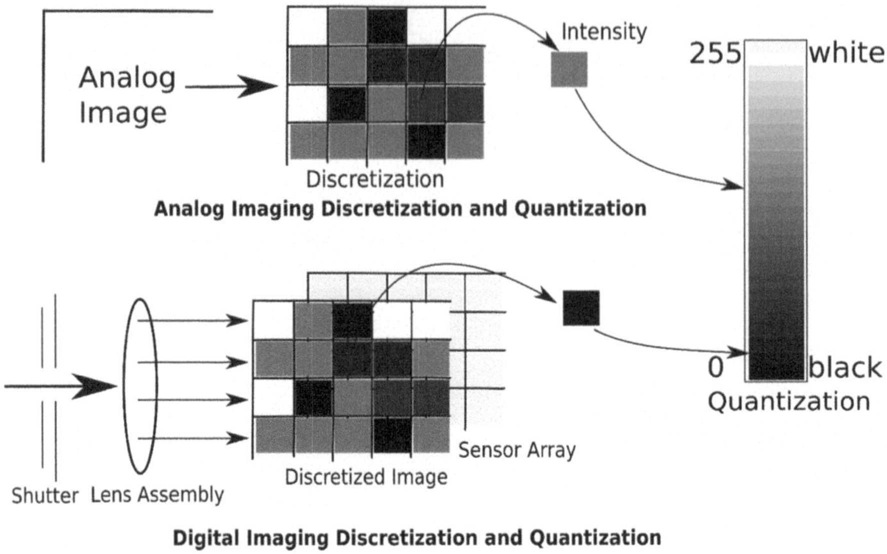

Fig. 3.6 Discretization and quantization in analog (top) and digital (bottom) imaging system. Analog imaging requires a separate discretization pipeline. Since imaging sensors are placed in discrete array, digital imaging enables embedded discretization within the imaging sensor

for specific applications to balance between the accuracy and computational performance of the vision system as computational time is often exponentially dependent on image size (number of pixels).

3.3 Color Image Representation

In the earliest development phase of the imaging sensors (e.g., pinhole cameras), the information would generally be recorded as the presence or absence of light in the visible range of the electromagnetic spectrum as it hit a photo-receptive film surface. This technique provided the most fundamental representation of an image, called a black-and-white or binary image. Such a binary representation of images is used widely in modern machine vision systems as well (discussed in Sect. 3.4). The binary images used these days are generally created using color or multi-spectral images using specific thresholding, filtering, segmentation, and/or classification techniques. As the sensing technology improved, the intensity of the light received by the photo-receptive film or the sensor was represented in continuous scale from black to white, thus creating grayscale images. When the entire range of electromagnetic spectrum received is allowed to hit a given location of the sensor, it creates a single channel grayscale image. In modern imaging systems (e.g., color cameras), only a selected range of the electromagnetic spectrum is allowed to excite the sensor at a time, thus creating a number of image channels, each representing light intensity levels at a certain frequency (or wavelength) called a monochromatic image. One of the most common imaging systems used in modern machine vision systems is RGB or color cameras, which include three specific image channels recorded in red, green, and blue frequencies of the visible light. These color images are represented in various ways facilitating different types of image processing techniques as discussed below.

3.3.1 Red Green Blue (RGB) and CMYK Color Models

In the RGB color model/space, images are represented as the combination of three monochromatic images, each for one primary color: red, green, and blue (see Fig. 3.7a). The intensity values of primary colors can be varied to represent arbitrary colors in the visible spectrum. The RGB model is the most common way of representing color images. In a digital form, RGB color representation generally contains 24 bits/pixel information, 8 bits for each red, green, and blue channel. The RGB model is spatially represented in a Cartesian coordinate system in the form of a unity cube where each corner along the axis is defined as one of the primary color components, as shown in Fig. 3.8a. The RGB model is an additive color model where red, green, and blue lights are added at a certain proportion to produce the visible spectrum of color starting from black going all the way to white. This model is widely used in image representation on electronic devices such as cameras and computers.

Complementary to the RGB model, CMYK is a subtractive color model/space used primarily for printing purposes (see Fig. 3.7b). For example, cyan is the absence of red color in white light. Similarly, magenta and yellow are the absence of green and blue colors from white light. In other words, the RGB model interprets color in terms of the reflected light spectrum while CMYK interprets color in terms of the absorbed light spectrum. Hence, painting cyan, magenta, and yellow color results in black color that absorbs the visible color spectrum of red, green, and blue.

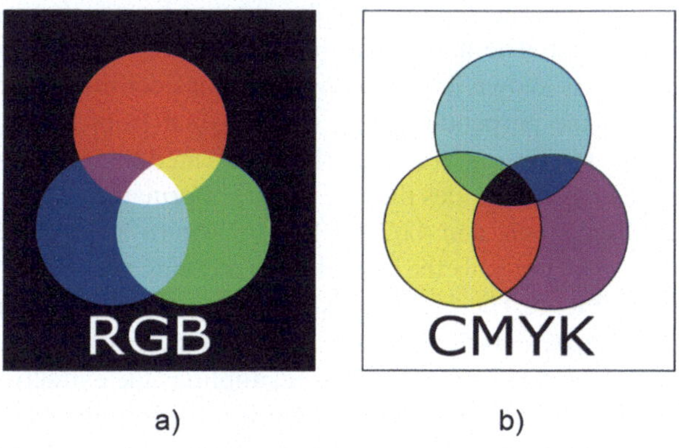

Fig. 3.7 (**a**) RGB and (**b**) CMYK color space representation. Notice that RGB color space starts from the dark background while CMYK starts from the white background. The combination of R, G, B results in white color while the same with C, M, Y results in black color

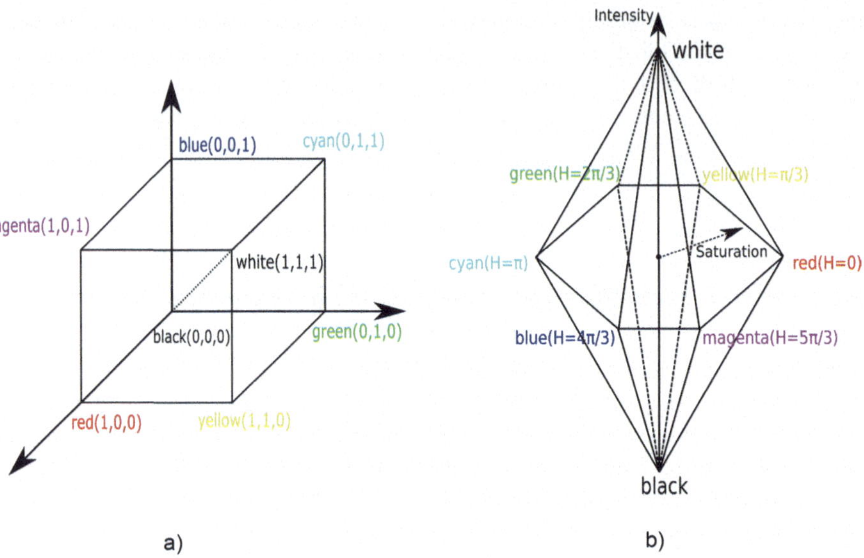

Fig. 3.8 (**a**) RGB color space representation using a unity cube. (**b**) Double hexagonal pyramid representation of HSI color space. HSI color space is generated by projecting RGB cube into a plane perpendicular to the line connecting black (0,0,0) and white (1,1,1) color coordinates

While the RGB model adds red, green, and blue colors, the CMYK model masks the red, green, and blue colors (adds cyan, magenta, yellow, and black) to form the desired image.

3.3.2 HSI Color Model/Space

The HSI color space represents colors in terms of Hue (H), Saturation (S), and Intensity (I) in the cylindrical coordinate system. Hue represents the color defined by the prominent frequency of the spectrum, whereas saturation describes the purity of the color, accounting for the amount by which a given color is diluted by the white light. The intensity value defines the extent of brightness or darkness of the image. As shown in Fig. 3.8b, let us consider the projection of the RGB color cube in a plane perpendicular to the diagonal from (0,0,0) to (1,1,1). The projection results in a hexagon (generally represented as a color wheel), and the periphery of the hexagon provides the hue (color) information that ranges from 0 to 2π (0 for red, $2\pi/3$ for green, and $4\pi/3$ for blue) (Shapiro & Stockman, 2000). The line running from the center to the boundary represents the saturation of the color that ranges from 0 to 1. The center axis (the line connecting from (0,0,0) to (1,1,1) in RGB cube) represents the intensity, and it ranges from 0 to 1. RGB images are often converted into HSI color space as it simplifies the extraction of individual colors based on hue. As intensity is decoupled from color information, sometimes HSI space is preferred to RGB images as it is robust against varying lighting and illumination conditions. Furthermore, color representation is relatable to human interpretation since hue and saturation are similar to how the human eye interacts with colors.

Different color models have been used by vision systems investigated for agricultural applications. Zhou et al. (2012) used RGB images to detect green apples while they leveraged HSI color space to deal with varying illumination to detect red apples. The intermediate image channels were computed as the ratio of the difference between red and green, and green and blue channels, and later converted to HSI color space followed by thresholding in saturation channel ($S > 0.4$) to segment red apples from the background.

HSV (hue-saturation-value) and HSL (hue-saturation-lightness) are other color spaces used in agricultural vision systems whose color space representations are similar but not completely identical to HSI color space. Wang et al. (2013) used HSV color space to detect red and green apples in images acquired from commercial orchards. The red apple pixels were segmented by thresholding hue channel ($0^\circ \leq \text{hue} \leq 349^\circ$ or $349^\circ \leq \text{hue} \leq 360^\circ$) followed by thresholding in saturation or value channel (saturation ≤ 0.1 or value ≤ 0.1) to remove background pixels, and a post-processing step was used to segment red apples. Furthermore, to detect green apples, the hue channel was first thresholded ($49^\circ \leq \text{hue} \leq 75^\circ$) to remove dark background followed by thresholding in the saturation channel (saturation ≥ 0.8) to segment green apple pixels.

3.3.3 *CIELAB Color Model/Space*

The CIELAB is one of the most popular color spaces defined by Commission Internationale de l'éclairage (CIE) in 1976 to create a standard for color communication regardless of the devices being used. The letter/variables L^*, a^*, and b^* are used to define the CIELAB color space, where L^* represents the intensity from black to white, and a^* and b^* represent the color directions or chromaticity coordinates (see Fig. 3.9). The a^* axis ranges from $+a^*$ representing red to $-a^*$ representing green colors. Similarly, b^* axis ranges from $+b^*$ representing yellow and $-b^*$ representing blue colors (Distante et al., 2020). Variable L^* ranges from zero to 100 while a^* and b^* are unbounded with no limits but usually range from −60 to +60 for practical purposes. CIELAB space parameters (L^*, a^*, and b^*) are defined based on the nonlinear transformation of the RGB model to achieve uniformity in color representation. Wachs et al. (2010) converted RGB image to CIELAB color space and used K-means clustering in a^* and b^* channels to detect green apples in commercial orchards. CIELAB color space can also be used to derive another color space known as CIE LCh color space, where L^* is the lightness component same as that of CIELAB, and C^* and h are chroma and hue angle derived from a^* and b^* (Distante et al., 2020).

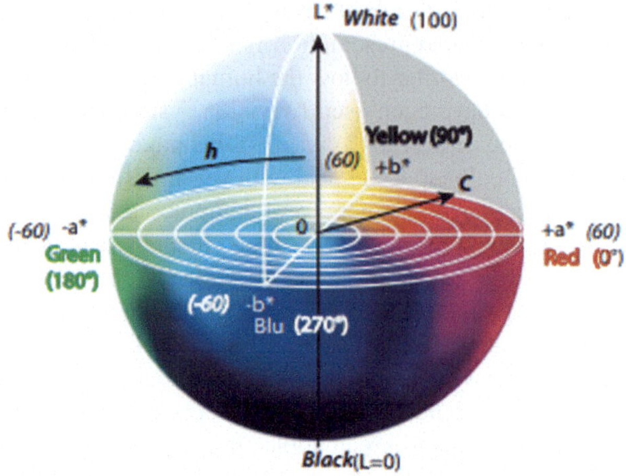

Fig. 3.9 A graphical representation of CIELAB and CIE LCh color space. (Image from Green Mamba :)--<, "cielab" licensed under CC BY-ND 2.0)

3.4 Binary Image Processing

As discussed before, with the advent of digital imaging technologies, images are captured by exciting semiconductor-based photosensors (e.g., CCDs), and a range of intensity values (often represented by 8 bits, 256 levels of intensity) are represented for each pixel of an image. Monochromatic (single channel) or multi-spectral (e.g., RGB, three channels) images recorded that way can then be converted to binary images using specific techniques such as classification discussed later (Sects. 3.6 and 3.7), where foreground (objects of interest) is generally represented by 1s and the background by 0s resulting in 1 bit/pixel information. In this section, various techniques used in analyzing binary images for delineating, improving, and characterizing objects of interest will be described.

3.4.1 Morphological Operations

Morphological operations are a collection of nonlinear image processing techniques often used to analyze binary images in which images are manipulated based on object shape and size. These operations are performed over a certain neighborhood of a target pixel location and is repeated over the entire image moving from top-left corner to bottom-right corner. More specifically, when performing morphological operations, a binary image is operated (certain binary operation such as AND or OR) by another small-sized binary image known as a structuring element. The pixel in the binary image corresponding to the center pixel of the structuring element is manipulated in each iteration, which is repeated to cover the entire image. A structuring element can be of arbitrary size and structure. Commonly used structuring elements include a rectangular box or circular disk. Morphological operations are widely used in machine vision systems for removing noisy areas, filling small holes, and bridging unwanted gaps. In addition, morphological algorithms/operations can be used for shape matching or pattern recognition, boundary extraction, region filling, connected component extraction, convex hull, thinning, thickening, skeletonization, and pruning of objects or regions. Major morphological operations include dilation, erosion, closing, opening, and Hit or Miss transform, which will be discussed below. In MATLAB (Mathworks Inc., Natick, MA), *bwmorph()* can be used to perform various morphological operations.

Dilation Dilation extends the boundary of objects in selected parts by one or more pixels depending on the shape and size of the structuring element used, which helps fill small holes and broken areas in binary images while also increasing the object size slightly (Fig. 3.10a, b). While moving structuring element from one place to the next, if any of the pixels in the image overlapping with the structuring element has a binary value 1, the center pixel (or the pixel over the origin of the structuring element) is converted to binary 1. Mathematically, dilation is represented by a logical OR operation between the binary image and structuring element as given by Eq. 3.2.

$$A \oplus B = \bigcup_{b_i \in B} A_{b_i} \tag{3.2}$$

Erosion As the name suggests, erosion removes some of the object pixels from the edges of those objects, which can be used to remove small anomalies or noisy areas in the binary images (Fig. 3.10a, c). The operation, however, reduces the object size slightly. In this operation, the image pixel at the center of the structuring element remains 1 only when all the image pixels overlapping with the structuring element

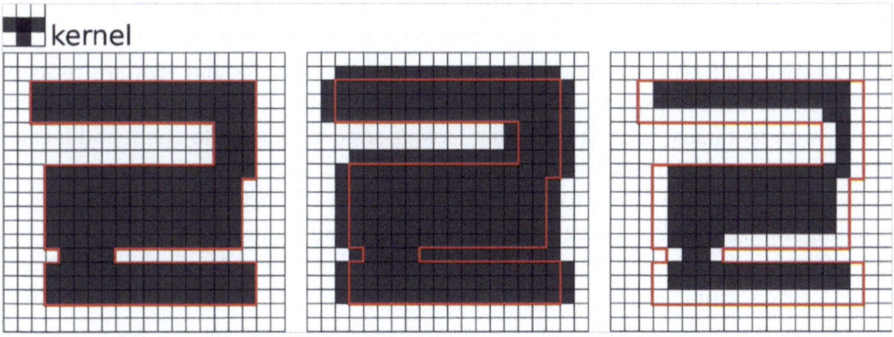

Fig. 3.10 Effect of morphological operations on a sample binary image; (**a**) Original image and a structuring element (dark cell in the structuring element indicates the origin); (**b**) Dilated image; and (**c**) Eroded image

are 1s; otherwise the pixel is turned to 0. Mathematically, this operation is represented by Eq. 3.3.

$$A \ominus B = \{p \mid B_p \subseteq A\} \tag{3.3}$$

Closing and Opening Closing and opening are the derived operations using dilation and erosion. Morphological closing is a combination of a dilation followed by erosion, while the opening is erosion followed by dilation. Opening removes the gap connected by thin bridges, small anomalies, or islands while keeping the remaining parts in their initial size. Closing, on the other hand, is often used to fill small unwanted holes in the images while keeping the overall object shape and size intact.

Hit or Miss Transform Morphological hit or miss transform is used for detecting objects with specific shapes or to find specific patterns in images. While operations like erosion or dilation do not consider the image background, the hit or miss transform uses foreground and background information to detect object shape. It is performed by performing erosion using nonoverlapping structuring elements for foreground and background. Let us consider $I(x, y)$ is the input image and $S_f(m, n)$ and $S_b(m, n)$ are two non-overlapping structuring elements for the foreground and the background. Mathematically, Hit or Miss Transform is represented by Eq. 3.4 (Fig. 3.11).

$$I(x,y) \odot S(m,n) = \left(I(x,y) \ominus S_f(m,n)\right) \& \left(\tilde{I}(x,y) \ominus S_b(m,n)\right) \tag{3.4}$$

where $\tilde{I}(x,y)$ is the complement of input binary image.

3.4.2 *Object Counting and Geometric Feature Extraction*

Once foreground objects are segmented out and their shape and size are refined using various processing techniques including morphological operations described above, various binary image processing techniques can be used to count and locate objects and extract their geometric features.

In a situation when only a total count of objects in a binary image is necessary, a counting algorithm like "corner detection" or "driving around the block" could be used. The corner detection algorithm works by assessing 4-pixel regions in an image using a row scanning technique and counting the number of convex and concave corners present in the entire image. A convex or external corner is defined by a cluster of four neighboring pixels with one object pixel (binary 1s) and three background pixels (0s), whereas a concave or internal corner is defined by three object

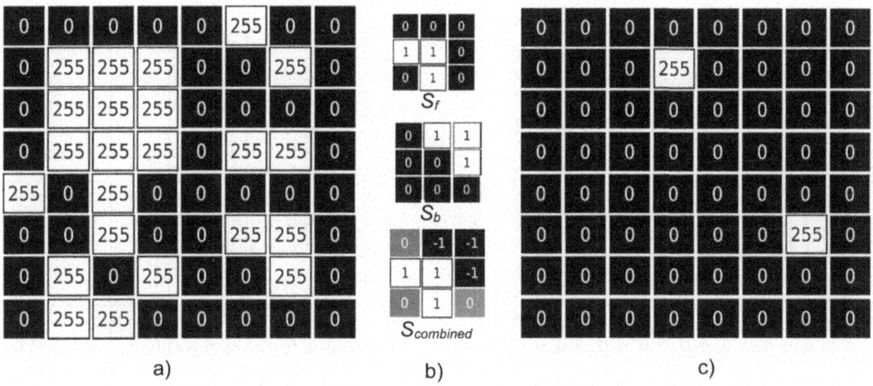

Fig. 3.11 Hit or Miss transform to detect top right corner in a binary image. (**a**) Input binary image. (**b**) S_f and S_b represent 3 × 3 kernel for foreground and background image. S_f and S_b were merged resulting in combined kernel (S_{combined}). Zeros in S_{combined} represent don't care pixels. (**c**) Resulting binary image with top right corners

pixels (1s) and one background pixel. The number of objects would then be given by Eq. 3.5.

$$\text{Number of objects} = \frac{\text{total external corners} - \text{total internal corners}}{4} \tag{3.5}$$

Similar to corner detection and counting, "driving around the block" algorithm works by first detecting the top-left corner of a given object and following the object boundary in the clockwise direction. Right and left turns made along the way until the starting point is reached are counted. The total number of objects would then be counted the same way as defined by Eq. 3.5.

A wider goal of object detection, localization in the image, and counting can be achieved using a technique called connected component labeling. The concept can be described using the following pseudo-code.

When B is a Binary Image; $B(r,c) = B(r',c') = v$; and $v = 0$ or 1
$B(r,c)$ is connected to $B(r',c')$ with respect to v if
$B(r,c) = B(r_0,c_0) = B(r_1,c_1) = B(r2,c2) \ldots = B(rn,cn) = B(r',c') = v$
$B(r_{i-1},c_{i-1})$ neighbors to $B(r_{i-1}, c_{i-1})$ for all $i=1$

A labeled image would have integer values uniquely defining each object (or connected component) in a binary image. Both four and eight connectivity can be used to define connected components or delineate individual objects. One example binary image and the result of a labeling technique is depicted in Fig. 3.12. A few different algorithms can be used to label the connected components in binary images. One of those algorithms is called recursive labeling, which is presented in Box 3.1. Alternatively, a more commonly used *Row-by-Row Labeling* algorithm can be used, which relies on two passes over the specific images. The first pass is used

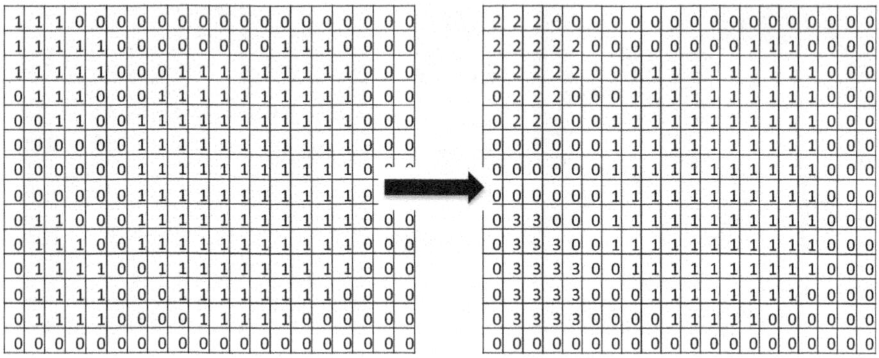

Fig. 3.12 Example of binary images showing the input (left) and the output (right) of a connected component labeling technique

to record equivalences and assign temporary labels, and the second pass is used to replace each temporary label with the label of its equivalence class. More details on these algorithms can be found in Shapiro and Stockman (2000). In MATLAB, binary image labeling and geometric feature extraction can be achieved using a library function *bwlabel()*.

Box 3.1 Recursive Labeling of Connected Objects

1. Negate the binary image
2. Start from top-left
3. If a foreground pixel
    (a) Put label on the pixel
    (b) Find foreground neighbor(s) in a row scanning pattern
    (c) If there is a neighbor, label it and find its neighbors
    (d) Continue this recursive search until all the connected pixels are found (similar to depth first search technique)
4. Repeat for all pixels

Once the objects are labeled, various geometric features of each object can be extracted to describe those objects, help identify specific kinds of objects, and/or classify objects into different groups. Some of the major geometric features include area, perimeter, length, width, aspect ratio, elongation, roundness, and eccentricity. Features such as area (A), perimeter (P), length (L), and width (W) can be calculated using the number of pixels occupied by the objects and the distance between specific points along the object boundary. Compactness is another important geometric feature used to define object shape, which is given by Eq. 3.6.

$$\text{Compactness}(C) = \frac{P^2}{A} \quad (3.6)$$

3.5 Image Enhancement and Spatial Filtering

Image enhancement is a pixel-level processing technique (considering one pixel at a time) that can alter how the overall intensities are distributed and can enhance various features or regions of the objects for better viewing as well as improved performance of down-stream image processing techniques. Spatial filtering techniques, on the other hand, often operate using intensities of a number of neighboring pixels to manipulate the intensity of a particular pixel (often the center pixel in the neighborhood used). The neighborhood used in such filtering techniques is defined by a square matrix of varying size (e.g., 3 × 3 pixels, 5 × 5 pixels, or 7 × 7 pixels), also known as a filter, mask, or kernel. The intensity of the pixels corresponding to the filter's center is manipulated by sliding the filter iteratively throughout the image in a row scanning fashion, also called masking. Spatial filtering can be linear where the computed pixel intensity is the sum of products (convolution) of filter coefficient and image intensity or nonlinear operation such as finding the median of the pixel intensities in the target neighborhood. Based on the type of information/signals, these filters allow signals to pass through. Spatial filters can be categorized as smoothing or low-pass filters, sharpening or high-pass filters, and band-pass filters. In the following several subsections, some of the most widely used image enhancement and spatial filtering techniques are described.

3.5.1 Gamma Correction

Gamma correction controls the brightness and contrast of an image by manipulating the pixels independently. Let *I*out and I_{in} be the input and output intensity values, respectively. Gamma correction is performed as $I_{out} = (I_{in})^{\gamma}$. When the gamma, γ is smaller than 1, darker regions in an image become lighter, thus better exposing the objects in those regions (Fig. 3.13a, b). When an appropriate γ is selected in the range of 0–1 for a specific image, specific details in the background can be better visualized. However, the picture starts to wash out if the gamma value is decreased further. When γ is greater than 1, the shadows become darker, increasing the contrast of the image (Fig. 3.13a, c), whereas no impact on the image would be noticed when γ is equal to 1.

Fig. 3.13 Image enhancement with gamma correction; (**a**) original image; (**b**) improved object exposure in darker area with $\gamma = 0.5$; and (**c**) contrast improvement with $\gamma = 1.5$

3.5.2 Histogram Equalization

Histogram equalization is one of the most widely used image enhancement techniques used to improve the global contrast of an image (Fig. 3.14). Histogram equalization is accomplished by evenly spreading out the most frequent pixel intensities across the entire dynamic range of an image (all possible intensity levels of the given image). This operation improves the contrast and dynamic range of the image by both spreading the pixel intensities and introducing intensity levels that were not present in the original image.

3.5.3 Smoothing Filters

As discussed before, smoothing filters remove high-frequency signals/information (or abrupt changes in the intensities) in an image which would be essential to suppress noisy pixel intensities and improve the smoothness of an image. However, smoothing filters blur the images and over-application of the filter may suppress desired image features such as sharp edges. The three most widely used smoothing filters in machine vision systems include a mean filter, a median filter, and a Gaussian filter.

Averaging/Mean Filter Output of an average filter is the average intensity over the intensities of pixels within the neighborhood defined by the kernel. A standard averaging or a weighted average could be used in manipulating target pixel intensities with these filters (Fig. 3.15). Filters with weighted averaging perform more natural smoothing by providing larger weight to the pixels that are closer (spatially) to the center pixel while decreasing the weight as the distance from the center pixel increases.

Fig. 3.14 Top—Original image and its histogram showing a narrow dynamic range; Bottom—Equalized histogram and enhanced image with improved contrast

Fig. 3.15 (**a**) Original image. (**b**) Mean filtered image with kernel size 7 × 7 using standard averaging

Median Filter The output of the median filter is the median intensity of the pixels in the neighborhood defined by the filter (Fig. 3.16). Median filters are easy to implement and are effective in removing impulse noise (e.g., salt and pepper noise).

Gaussian Filter Gaussian filter is a bell-shaped filter that provides the highest weight to the pixels near the center of the kernel while transitioning to the reduced weights to the pixels farther from the center pixel using a Gaussian function characteristics. Since the kernel is bell-shaped, the Gaussian filter provides a smooth blur-

Fig. 3.16 (**a**) Original image suffering from salt and pepper noise; (**b**) median filtered image with kernel size 7 × 7

Fig. 3.17 (**a**) Original image; (**b**) Gaussian filtered images with kernel size 7 x 7 and $\sigma = 2$; and (**c**) corresponding image with kernel size 7 x 7 and $\sigma = 4$

ring result. For the given Gaussian kernel, the extent of blurring can be controlled by varying the standard deviation (σ) of the Gaussian function used. A larger value of σ represents smaller peak weight with gentler changes to weights always from the center/peak leading to greater blurring and vice versa (Fig. 3.17).

3.5.4 *Sharpening Filters*

Sharpening filters allow only the high-frequency signals/information to pass and therefore highlight the edges and fine details in an image while suppressing regions with minimal changes in the intensities. Mathematically, image sharpening corresponds to a differentiation operation. Some of the simplest sharpening filters are Sobel Operator, Prewitt operator, and Robert's operator. Sobel and Prewitt operators use 3 × 3 kernels to detect horizontal and vertical edges by calculating the vertical or horizontal derivative over 2D images (Fig. 3.18). However, unlike the Sobel operator, the Prewitt operator does not emphasize the pixel closer to the kernel center. Furthermore, Robert's operators are designed to respond maximally to the edges running at 45°.

-1	0	1
-2	0	2
-1	0	1

a)

-1	0	1
-1	0	1
-1	0	1

b)

0	1
-1	0

1	0
0	-1

c)

Fig. 3.18 Edge detection operators; (**a**) Sobel operator; (**b**) Prewitt operator; and (**c**) Roberts operator. Both Sobel and Prewitt operators detect vertical edges. Horizontal edges can be detected by rotating the kernel by 90^0

Another commonly used sharpening filter is the Laplacian filter, which uses the second derivative operation in the discrete domain. Let us consider a simple 1D array defined as *f*(*x*), the first- and second-order derivatives are given as

$$\frac{\delta f(x)}{\delta x} = f(x+1) - f(x)$$

$$\frac{\delta^2 f(x)}{\delta x^2} = f(x+1) - 2f(x) + f(x-1)$$

For a 2D image, the Laplacian filter is defined by the second-order partial derivative

$$\begin{aligned} \nabla^2 f(x,y) &= \frac{\delta^2 f(x,y)}{\delta x^2} + \frac{\delta^2 f(x,y)}{\delta y^2} \\ &= f(x+1,y) - 2f(x,y) + f(x-1,y) + f(x,y+1) - 2f(x,y) + 2f(x,y-1) \\ &= f(x+1,y) + f(x-1,y) + f(x,y+1) + 2f(x,y-1) - 4f(x,y) \end{aligned} \tag{3.7}$$

In the discrete form with a 3 × 3 kernel, the partial second-order derivative can be represented by Eq. 3.7 and graphically by Fig. 3.19. As can be seen from the figure, the sum of coefficients is zero, and therefore when there exist very low changes in gray/intensity levels in the input image, the intensity value in the output image corresponding to the filter's center will be close to zero, thus suppressing smooth or low-frequency regions in the image. In addition to the above kernel, there are variants of the Laplacian filters with varying weights in the center and neighboring pixels. See Shapiro and Stockman (2000) for more details on all types of spatial filtering techniques.

0	1	0
1	-4	1
0	1	0

Fig. 3.19 Kernel representation of a Laplacian high pass filter. Notice that the sum of coefficients is zero. If there exist very low changes in gray/intensity levels in the input image, the intensity value in the output image corresponding to the filter's center will be close to zero, thus suppressing smooth or low frequency regions in the image

3.5.5 Edge Detection

Edges are the regions in an image with abrupt variation in intensities (high-frequency components). Edges can represent the boundary between two objects or an object of interest and the background. Different techniques are used for highlighting sharp variation in pixel intensities and detecting object edges in 2D images. These techniques include different operators such as Sobel operator, Prewitt operator, Robert operator, and Laplacian of Gaussian. Laplacian of Gaussian (LoG) uses a Gaussian filter to smooth images and remove unwanted noise before applying a second derivative-based Laplacian filter that strongly emphasizes high-frequency changes in pixel intensities.

In addition to the above operators, another popular and robust edge detection algorithm known as the Canny edge detector is widely being used in different applications. Developed in 1986 by John F. Canny, the Canny edge detector uses multi-stage algorithms to detect and highlight edges in image. Because of its efficient edge detection capability, Canny detectors are still used in machine vision systems with a broad spectrum of applications. The algorithm's multi-stage detection process is listed in Box 3.2. The canny edge detector is a powerful tool for edge detection and is used widely as a processing algorithm for different vision applications in agriculture, such as object detection, segmentation, manipulation. Septiarini et al. (2020) used a Canny edge detector along with different pre- and post-processing operations to detect oil palm fruit. In another approach, Luo et al. (2016) applied a Canny edge detector in grape cluster images to extract highly probable berries objects followed by berry detection (by fitting circles) and refinement to delineate individual berries (Fig. 3.20).

Fig. 3.20 Detecting grapes using a Canny edge detector; (**a**) A grape cluster image; (**b**) Edge detection using a Canny edge detector; and (**c**) Approximate berry detection using a post-processing technique. (Images from Luo et al. (2016))

Box 3.2 Steps Used in Canny Edge Detector Algorithm

1. Apply image smoothing and noise removal with Gaussian Filter
2. Sharpen the image to enhance most substantial intensity gradients (high strength edges) (can use Sobel, Prewitt, or Roberts operator)
3. Compute magnitude and direction of gradients
4. Suppress non-maxima to thin the edges and remove unwanted artifacts
5. Perform hysteresis thresholding to remove and accept lower and higher intensity gradients and perform edge linking to find and locate the final edges

3.6 Unsupervised Learning or Clustering

Various approaches have been developed to further process information/images to identify/recognize pixels or objects and group those with similar characteristics together. These approaches constitute a part of machine learning and are being used in vision systems applied to various fields, including agriculture. Based on the requirement of input data and the way data are analyzed, these approaches can be broadly categorized into two branches: unsupervised learning and supervised learning.

Unsupervised learning approaches make inferences from the input data without using information from the labels or output. These algorithms do not take feedback on whether the estimations are accurate. The main goal of unsupervised learning is to estimate the structure or the patterns of the input dataset. On the other hand, supervised learning requires examples with labeled input and output so that the learning method can make necessary changes to model parameters/features to look into in the input dataset to come up with correct predictions. The main goal of supervised learning is to train the model based on ground truth or example data such that the same model can be used to predict output in the situation when a new dataset without known output is introduced. Since supervised learning algorithms/models are trained based on correct outputs to given inputs, they often result in more reliable predictions/classification outcomes compared to the same with unsupervised methods. However, the supervised algorithms require a longer training time and run the risk of overfitting to the training dataset leading to unexpected or erroneous results when applied to datasets outside the training samples. On the other hand, the unsupervised learning approaches do not require prior knowledge about the dataset and are generally simpler and faster to apply. However, these approaches often lack the level of reliability since they do not have a feedback system to test the accuracy/usefulness of the obtained results. Unsupervised approaches are commonly used in dataset clustering, dimensionality reduction, similarity detection, and anomaly detection. In this section, we discuss different unsupervised (e.g., K-means Clustering, Iterative Self-Organizing Data Analysis Technique (ISODATA)) and supervised learning (Support Vector Machine (SVM), Bayesian Classifier, Artificial Neural Network).

3.6.1 K-Means Clustering

K-means clustering is a simple unsupervised learning approach to cluster data points with similar characteristics into a number of groups, K (MacQueen, 1967). K-means clustering works iteratively to assign each data point to one of the K groups based on the feature similarity of the candidate data point with the mean characteristics of the points already belonging to specific groups. As there are K clusters, there will be K number of centroids of the data points, each representing one data cluster. The clustering algorithm attempts to minimize the within-cluster variance of the participating data points. Initially, K random data clusters (if not specified by the user) and corresponding centroids are specified in the feature space. Each data point is assigned to a cluster based on the Euclidean distance between the data point and individual cluster centroids. Once all the data points are explicitly assigned to individual clusters, a new centroid for each cluster is computed, and the process is repeated until there are no/minimum cluster re-assignments or no/minimum change of centroids. Alternatively, the process is stopped when a minimum decrease in the sum of squared error between the cluster centroids and the corresponding data points is achieved (Ullman & Rajaraman, 2012). K-means clustering is simple to implement, flexible to adapt to the changes in dataset/new dataset, suitable for large datasets, and guarantees to provide convergence. K-means clustering has commonly been used in machine vision systems proposed for agricultural applications. For example, Wachs et al. (2010) used K-means clustering in a and b channels of LAB color space in thermal and color images to detect green apples. Bulanon et al. (2004) also implemented K-means clustering to detect apples in color images transformed to chromaticity space. There also are some limitations of using K-means clustering. The algorithm is sensitive to outliers and applies only to the datasets where the mean can be defined. Furthermore, the clustering result could vary based on the initial seed, which might/might not reflect the real-world clustering.

3.6.2 Iterative Self-Organizing Data Analysis Technique (ISODATA)

ISODATA clustering is another unsupervised learning algorithm that works in a similar way to K-means clustering. However, ISODATA clustering allows dynamic change in the number of clusters, while in K-means clustering the number of clusters is defined beforehand and remains constant. Furthermore, the ISODATA clustering allows cluster modification by splitting and merging the clusters (Ball & Hall, 1967). Clusters are split if the within-cluster variance is greater than a given threshold, whereas the clusters are merged if the distance between the centroids of clusters is less than a given threshold.

3.6.3 Supervised Learning or Classification Approaches

As discussed before, supervised learning approaches require labeled pairs of input and output data to train learning models such that the models can be later used for accurately predicting the output when an unseen input is presented. To implement supervised learning approaches, the available input-output dataset is generally divided into three categories: training, validation, and test datasets. The training dataset is used for training the models to learn relevant features, whereas the validation dataset is used to make sure the model is not overfitted to the training samples. Model performance can further be assessed using the test dataset that would include samples not previously presented to the model during the training process. Supervised approaches have been widely used in agriculture applications such as object (flower, fruit, weed, branches) detection, segmentation and classification, yield estimation, soil and water management, crop quality assessment, livestock management, and disease detection, among others.

3.6.4 Support Vector Machine (SVM)

Support vector machine is a supervised statistical learning algorithm used for classification, regression, and outlier detection tasks among other. SVMs are commonly used to solve classification problems by computing decision boundaries, also known as a hyperplane. A hyperplane is an $n - 1$ dimensional subspace for n-dimensional feature space representing the objects of interest. For example, for a set of points in 2D space, a hyperplane is a 1D line with a particular slope and intercept that separates the data points most optimally (see Fig. 3.21). Support vectors, which are the data points nearest to the hyperplane, are the most critical components of an SVM. Support vectors are located in the feature space such that they are the most difficult data points to classify into one group or another and therefore provide the basis for defining the optimum location of the hyperplane. The distance between the hyperplane and support vectors is known as margin.

Let us consider a set of data points that belong to classes C_1 and C_2 as shown in Fig. 3.21. As discussed earlier, for the given data points in 2D space, the objective is to determine a 1D hyperplane (straight line for this case) to classify these points. The equation of the straight line is given as $X_2 = mX_1 + c$, where m and c are slope and intercept of the line. The equation of 1D line (hyperplane) can be represented in a more generalized form as $w_1X_1 + w_2X_2 + b = 0$ (Gonzalez & Woods, 2018). Any point in X_1X_2 plane with $w_1X_1 + w_2X_2 + b > 0$ belongs to class C_1, and any point with $w_1X_1 + w_2X_2 + b < 0$ belong to class C_2. In terms of the SVM, the coefficients of X_1 i. e. (w_1) and X_2 i. e. (w_2) are considered weights. The weights define the coordinate of the vector perpendicular to the hyperplane. The intercept value b is considered a bias value. Similar to the intercept value b , the bias value displaces the strict requirement that the hyperplane should be passing through the origin. Similar to the example of 2D space, the decision boundary for an n-dimensional space can be determined by $n - 1$ dimensional hyperplane which can be represented as

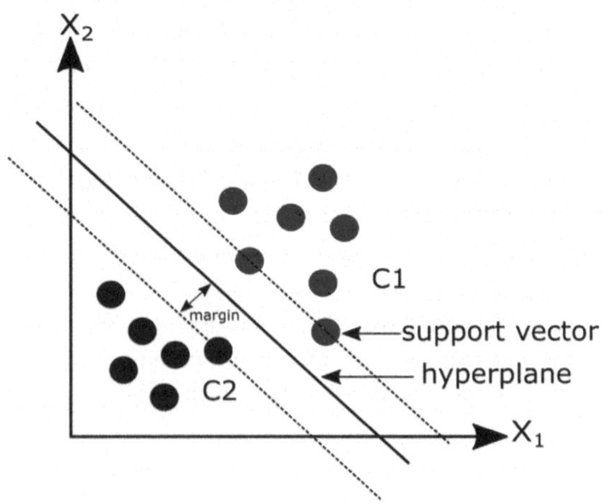

Fig. 3.21 Implementation of SVM in 2D space. Hyperplane in 2D space is specified by a line

$$
\begin{aligned}
&w_1X_1 + w_2X_2 + w_3X_3 \cdots + w_nX_n + b = 0 \\
&\Rightarrow \sum_{i=1}^{n} w_iX_i + b = 0
\end{aligned}
\tag{3.8}
$$

The objective of training an SVM is to determine a hyperplane such that the distance between the support vectors and the hyperplane is maximized. In simple cases, it is easy to have a linear hyperplane between the classes. However, datasets might be nonlinearly separable with some features corresponding to the wrong class in the real world. To address the nonlinearly separable cases, SVM uses a technique called "kernel trick". Kernel trick transforms the data points in low dimensional space to higher-dimensional space such that the data points are separable in a more simplified manner. Some of the commonly used kernel tricks involve the use of polynomial and Radial-basis function (RBF).

SVM techniques, in general, are robust and generalizable, and are one of the most popular classification approaches used in agriculture. Qiang et al. (2014) used multiclass SVM to segment fruit, branch, and leaves in citrus field images and achieved 92.4% accuracy in counting citrus. The proposed SVM approach leveraged the RBF kernel to handle nonlinearity followed by morphological operation for refining the segmented results. Ji et al. (2012), Kong et al. (2010), and Wang et al. (2009) used different types of SVMs to detect and classify apple fruit. Color and shape features were extracted by segmenting apples using median filter and seeded region growing. An SVM model was then trained using three types of object features: color, shape, or combination of color and shape in conjunction with three different kernel functions (Poly, RBF, Sigmoid). Combining color and shape features with RBF kernel function achieved the highest accuracy (Ji et al., 2012; Kong et al., 2010; Wang et al., 2009). Rakun et al. (2011) also proposed a three-stage apple detection using segmentation in HSI color space for possible fruit regions, followed by refining segmentation results using SVM for texture feature analysis and 3D reconstruction for shape analysis.

3.6.5 Nearest Neighbor Classifier

The nearest neighbor is a supervised classification approach in which every data point in the training set are compared against each other in the feature space to find the closest (nearest) neighbors. Commonly used techniques to estimate the distance or closeness in the feature space include Euclidean distance, Manhattan distance, Hamming distance, and Cosine similarity. Since the nearest neighbor classifier uses distance as the parameter to classify objects, it is also known as a minimum-distance classifier. The simplest form of the nearest neighbor classifier uses 1-Nearest Neighbor (NN), where the closest neighbor is used as a reference to classify unknown data. A generalized form of this algorithm is called K-NN in which K-nearest neighbors are found based on one of the similarity measures mentioned before and the majority voting rule is used to classify new observations to a class. When a K-NN technique is used to classify individual pixels in image (semantic segmentation), the training and test images are compared pixel-by-pixel. Although K-NN does not require training like SVM, for a large number of datasets, it takes a significant amount of time to find the neighbor/s. This algorithm has also been used commonly in developing machine vision applications for agriculture. For example, Linker et al. (2012) leveraged a K-NN classifier to classify pixels based on color and texture features to find green apple seed regions followed by post-processing to segment green apples. The algorithm was able to correctly detect 85% of visible apples on images acquired in varying lighting conditions. Seng and Mirisaee (2009) used K-NN to classify apple, banana, strawberry, lemon, durian, and watermelon with classification accuracy up to 90%. Various color and geometric features such as area, shape, and perimeter were used to define objects of interest (fruit) and classify them to specific classes.

3.6.6 Decision Trees

As the name suggests, decision trees are tree-like structures commonly used for classification and regression tasks. A decision tree starts from a root node and iteratively tests object attributes in the internal node/decision node to create branches until leaf nodes are attained. Each leaf node is assigned a class label for classification tasks, while root node and decision nodes handle attribute tests to separate objects based on different characteristics (Tan et al., 2006). In simple terms, a series of questions are asked about specific attributes of objects, and a decision is made to move to a specific branch based on the feature value. The process is repeated with follow-up questions until the conclusion about the object class is reached. Because of the simplicity, decision trees are one of the widely used supervised learning algorithms. However, the technique is prone to overfitting since it has a low bias but high variance leading to a change in output for a small change in the input variable. Furthermore, when the tree starts to grow deep, it focuses on minor details to make decisions instead of looking into overall generalized patterns/features.

3.6.7 Random Forest (RF)

Random forest is a supervised statistical learning algorithm that creates an uncorrelated forest of decision trees. RF is an ensemble learning technique that combines a set of models (e.g., decision trees) and aggregates the output to find the result. The technique leverages bagging, also known as the Bootstrap Aggregation ensemble technique where the random sample in the training dataset is selected with replacement. Individual data points can be chosen more than once and act as input to multiple decision tree models. Each decision tree is trained separately, and the final result of RF is obtained using the majority voting method, also known as aggregation. RF overcomes the overfitting problem in decision trees by combining the results based on majority voting. Furthermore, each decision tree can be trained independently, allowing parallelization during computation. However, random forest is considered complex and takes longer compared to decision trees. Some example studies using this technique in agriculture include Zawbaa et al. (2014) which implemented a RF model to classify three kinds of fruit: apples, strawberries, and oranges. Geometric and color features were extracted, and Scale Invariant Feature Transform (SIFT) was used to improve the decision-making. It was found that RF performed better compared to K-NN and SVM learning algorithms. In another work, Ishikawa et al. (2018) implemented RF to classify strawberry shapes to nine different classes. It was found that the geometric features (length of strawberry contour line, fruit area, fruit length, fruit width, fruit width/length ratio), Elliptic Fourier descriptor, and Chain code subtraction descriptor were instrumental in fruit shape classification.

3.6.8 Bayesian Classifier

The Bayesian classifier uses probability distribution to classify objects to the class it is most likely to belong. To explain this concept, let us take the Iris dataset as an example. The Iris dataset consists of three flower varieties (Setosa, Versicolor, Virginica) which provides ground truth (actual) measurements of flower sepal length, petal length, and width. The objective of the Bayesian classifier is to predict the flower class for specific flowers with given petal length, septal length, and width using the highest conditional probability that the flower belongs to a specific class. Mathematically, let us consider $\boldsymbol{X} = (X_1, X_2, X_3 \ldots X_n)$ represent n features, and Y is the class label. The objective of the classifier is to determine $P(Y = y_i | \boldsymbol{X})$; the probability that the object with a specific feature belongs to class $Y = y_i$. The Bayes theorem is used to compute the conditional probability, which is given as:

$$P(Y = y_i \mid \boldsymbol{X}) = \frac{P(\boldsymbol{X} \mid Y = y_i) P(Y = y_i)}{P(\boldsymbol{X})} \tag{3.9}$$

Here,

$P(Y = y_i | \boldsymbol{X})$: Probability that the feature $\boldsymbol{X}$ belongs to the class y_i.
$P(\boldsymbol{X} | Y = y_i)$: Probability that class y_i will exhibit feature $\boldsymbol{X}$. Provides information on the distribution of feature $\boldsymbol{X}$ in class y_i; also known as class conditional probability.
$P(Y = y_i)$: Probability of occurrence of class y_i out of all classes. Provides information on the frequency of occurrence of class; also known as a priori probability.
$P(\boldsymbol{X})$: Probability of occurrence of feature $\boldsymbol{X} = (X_1, X_2, X_3 \ldots X_n)$. Since we are looking into a particular feature, the probability of occurrence of the particular feature

is the same for all classes. This value just acts as a scaling factor and can be ignored.

It is generally assumed that the class conditional probability can be safely represented by the Gaussian probability density function (Gonzalez & Woods, 2018). For the n-dimensional feature vector, the n-dimensional multivariate class conditional probability can be given by Eq. 3.10.

$$P\left(\boldsymbol{X} \mid Y = y_i\right) = \frac{1}{\left(2\pi\right)^{\frac{n}{2}} \left|\boldsymbol{C}\right|^{\frac{1}{2}}} e^{-\frac{1}{2}\left(X-\mu_i\right)^T C^{-1}\left(X-\mu_i\right)} \tag{3.10}$$

where $\boldsymbol{\mu} = E(\boldsymbol{X}) = \{\mu_1, \mu_2, \ldots \mu_n\}$ is the mean vector, and $\boldsymbol{C}$ is the covariance matrix.

The Bayesian classifier above is known as Naive Bayes because it assumes that the input features are independent. However, this assumption is quite unrealistic because it is virtually impossible to find such independence in real-life data.

A large number of machine vision studies based on Bayesian Classifiers can be found with applications in agriculture. Amatya et al. (2016) used a Bayesian classifier in RGB images to segment cherry tree branches. Image pixels were classified into four classes: branch, cherry, leaf, and background. The feature vector created based on red, green, and blue intensity values of each object class in a manually selected sample region was used to compute the class conditional probability density function. Class priori probability was calculated as a percentage of feature vectors corresponding to a specific class with respect to the total number of feature vectors. Morphological operations and a curve-fitting method were used to refine and connect the segmented branches, which achieved a branch detection accuracy of 89.2%.

3.6.9 Artificial Neural Network (ANN) and Deep Learning

The performance of the traditional machine learning algorithms was primarily based on feature engineering, where the best features representing the objects must be carefully selected and passed to the learning algorithms. Classical machine learning achieved some success by training the models based on hand-crafted features. However, since the features are hand-coded, the generalizability would be challenging if the objects are presented from varying environments or viewpoints. Especially in agriculture, the problem becomes challenging because of the variability due to canopy architectures (e.g., canopy density, canopy size, varietal differences), occlusions (due to branches, leaves, trellis wires, other fruit/flowers), and environmental factors (varying lighting conditions, wind). Unlike traditional machine learning approaches, deep learning approaches can extract useful feature information from raw data. Deep learning, which is an extended and improved version of ANN, leverages hierarchical feature learning where higher-level complex features can be formed by combining simple lower-level features (Bengio, 2009). Learning the features from raw data and developing complex features from simple features make deep learning approaches powerful and has shown improved performance in computer vision tasks. In recent years, ANN and Deep Learning have been some of the most widely used supervised learning techniques in a wide variety of agricultural applications as well because of the increased accuracy and the robustness to deal with variation in environmental conditions. However, ANN-based techniques are often criticized for unexplainability and uncertainty as these are considered as back-box models that do not clearly explain how the models achieved what they achieved. More discussion on ANN and deep learning are included in Chap. 7.

3.7 Major Challenges and Opportunities

Accurate, reliable, and robust machine vision system is a key for the success of all types of automated or robotic operations in agriculture including crop monitoring, phenotyping, crop management, and harvesting. However, unstable, uncertain, and variable outdoor environment present challenges for machine vision systems to achieve the desired level of accuracy, robustness, and reliability. Unstructured and uncertain plant canopy structures and variable shape, size, color, and location of the objects of interest such as fruit and branches are other limiting factors in field conditions. To address these challenges, machine vision systems developed for agricultural applications need to use novel approaches to optimize sensing and lighting conditions including the use of cameras that have improved performance in unstable lighting, minimizing lighting variability by introducing mechanical structures to block outside lighting, and use of artificial lighting. Other aspects to consider for improving vision system performance in field conditions include calibration of outputs with sun incident angle and experimentation with time-of-the-day. In addition, novel algorithmic approaches such as dynamic exposure adjustment and exposure fusion could be helpful. Newer, low-cost consumer cameras capable of collecting both color and 3D information (RGB-D cameras; e.g., Zed 2, Stereolabs Inc.) have shown, in recent years, to be practically adopted in orchards environment. These low-cost sensors and AI techniques such as deep learning show huge potential for continual development and widespread application of machine vision technologies in farming.

Furthermore, the application of machine vision systems in the agricultural environment is limited by the fact that a large proportion of the objects of interest are partially or fully occluded by the same and other types of objects such as fruit, branches, and foliage, which then makes it challenging for detecting and locating those objects. One way to address such challenges could be to use spectral sensors such as hyperspectral or thermal sensing systems that may be able to penetrate slightly past the surface to detect hidden objects behind foliage. But more research and development would be essential to validate this concept. Another important way the agricultural environment could be improved for machine vision application is to design planting and canopy structures that can present most of the objects of interest on the outer, visible, and accessible canopy surfaces. For example, there have been continuous improvement and adoption of tree fruit canopies in recent years to create structured, narrow canopies where all the canopy parts (e.g. fruit and branches) are visible. Continuous work in genetics, breeding, and horticultural studies, in close collaboration with engineering studies, is crucial to further improve crop varieties, cropping systems, and canopy architectures that can facilitate the wider application of machine vision systems in farming.

It is also noted that ubiquitous and widespread use of cell phone and other mobile devices has created a huge potential for developing machine vision applications that could be accessible and affordable to farmers with all scales, types, geographic locations, and weather conditions from around the world. Mobile devices these days come with high-resolution cameras, 3D measurement capabilities, and powerful computational units allowing developers to offer highly impactful applications for farming such as crop-load estimation, crop stress monitoring, disease and insect identification, and irrigation control.

Successful development and adoption of machine vision tools have a great potential for improved accuracy and efficiency in all aspects of farming including crop condition and cropping environment monitoring, making timely farming decisions, automating various farming operations, and improving worker productivity and health and safety, thus leading to reduced use of scarce farming resources such as water, nutrients and labor, increased yield and quality, and better environmental stewardship. Consequently, farmers would realize higher net returns and farming industries would realize more sustainable (economically, socially, and environmentally) production systems to feed the growing global population.

References

Amatya, S., Karkee, M., Gongal, A., Zhang, Q., & Whiting, M. D. (2016). Detection of cherry tree branches with full foliage in planar architecture for automated sweet-cherry harvesting. *Biosystems Engineering, 146*, 3–15.

Ball, G. H., & Hall, D. J. (1967). A clustering technique for summarizing multivariate data. *Behavioral Science, 12*(2), 153–155.

Bengio, Y. (2009). *Learning deep architectures for AI*. Now Publishers Inc..

Bulanon, D. M., Kataoka, T., Okamoto, H., & Hata, S. (2004). Development of a real-time machine vision system for the apple harvesting robot. In *SICE 2004 Annual Conference* (Vol. 1, pp. 595–598). IEEE.

Distante, A., Distante, C., & Distante, W. (2020). *Handbook of image processing and computer vision*. Springer.

Gongal, A., Amatya, S., Karkee, M., Zhang, Q., & Lewis, K. (2015). Sensors and systems for fruit detection and localization: A review. *Computers and Electronics in Agriculture, 116*, 8–19.

Gonzalez, R. C., & Woods, R. E. (2018). *Digital image processing* (4th ed., global edition).

Ishikawa, T., Hayashi, A., Nagamatsu, S., Kyutoku, Y., Dan, I., Wada, T., Oku, K., Saeki, Y., Uto, T., & Tanabata, T. (2018). Classification of strawberry fruit shape by machine learning. *International Archives of the Photogrammetry, Remote Sensing & Spatial Information Sciences, 42*(2).

Jain, R., Kasturi, R., & Schunck, B. G. (1995). *Machine vision* (Vol. 5). McGraw-Hill.
Ji, W., Zhao, D., Cheng, F., Xu, B., Zhang, Y., & Wang, J. (2012). Automatic recognition vision system guided for apple harvesting robot. *Computers & Electrical Engineering, 38*(5), 1186–1195.
Karkee, M., & Zhang, Q. (2021). *Fundamentals of agricultural and field robotics*. Springer Nature.
Kong, D., Zhao, D., Zhang, Y., Wang, J., & Zhang, H. (2010). Research of apple harvesting robot based on least square support vector machine. In *2010 International Conference on Electrical and Control Engineering* (pp. 1590–1593). IEEE.
Linker, R., Cohen, O., & Naor, A. (2012). Determination of the number of green apples in RGB images recorded in orchards. *Computers and Electronics in Agriculture, 81*, 45–57.
Luo, L., Tang, Y., Zou, X., Ye, M., Feng, W., & Li, G. (2016). Vision-based extraction of spatial information in grape clusters for harvesting robots. *Biosystems Engineering, 151*, 90–104.
MacQueen, J. (1967). Some methods for classification and analysis of multivariate observations. In *Proceedings of the Fifth Berkeley Symposium on Mathematical Statistics and Probability* (Vol. 1, pp. 281–297). University of California Press.
Qiang, L., Jianrong, C., Bin, L., Lie, D., & Yajing, Z. (2014). Identification of fruit and branch in natural scenes for citrus harvesting robot using machine vision and support vector machine. *International Journal of Agricultural and Biological Engineering, 7*(2), 115–121.
Rakun, J., Stajnko, D., & Zazula, D. (2011). Detecting fruits in natural scenes by using spatial-frequency based texture analysis and multiview geometry. *Computers and Electronics in Agriculture, 76*(1), 80–88.
Seng, W. C., & Mirisaee, S. H. (2009). A new method for fruits recognition system. In *2009 International Conference on Electrical Engineering and Informatics* (Vol. 1, pp. 130–134). IEEE.
Septiarini, A., Hamdani, H., Hatta, H. R., & Anwar, K. (2020). Automatic image segmentation of oil palm fruits by applying the contour-based approach. *Scientia Horticulturae, 261*, 108939.
Shapiro, L., & Stockman, G. (2000). *Computer vision*. Prentice Hall.
Tan, P.-N., Steinbach, M., & Kumar, V. (2006). Classification: Basic concepts, decision trees, and model evaluation. *Introduction to Data Mining, 1*, 145–205.
Ullman, J. D., & Rajaraman, A. (2012). Clustering. In *Mining of massive datasets* (pp. 241–280).
Wachs, J. P., Stern, H. I., Burks, T., & Alchanatis, V. (2010). Low and high-level visual feature-based apple detection from multi-modal images. *Precision Agriculture, 11*(6), 717–735.
Wang, Q., Nuske, S., Bergerman, M., & Singh, S. (2013). Automated crop yield estimation for apple orchards. In *Experimental robotics* (pp. 745–758). Springer.
Wang, J., Zhao, D., Ji, W., Tu, J., & Zhang, Y. (2009). Application of support vector machine to apple recognition using in apple harvesting robot. In *2009 International Conference on Information and Automation* (pp. 1110–1115).
Zawbaa, H. M., Hazman, M., Abbass, M., & Hassanien, A. E. (2014). Automatic fruit classification using random forest algorithm. In *2014 14th International Conference on Hybrid Intelligent Systems* (pp. 164–168). IEEE.
Zhou, R., Damerow, L., Sun, Y., & Blanke, M. M. (2012). Using colour features of cv. 'Gala' apple fruits in an orchard in image processing to predict yield. *Precision Agriculture, 13*(5), 568–580.

Chapter 4
Imaging Technology for High-Throughput Plant Phenotyping

Jing Zhou, Chin Nee Vong, and Jianfeng Zhou

4.1 Introduction

The development of new crop varieties with improved traits through crop breeding is one of the most important solutions to produce sufficient food, feed, and fiber for the estimated population of more than nine billion in 2050. It is expected that agricultural production needs to double its current growth rate to meet the world's demand (Hincks, 2018). Plant phenotyping is essential to plant breeding programs aiming to select elite cultivars from candidate cultivars (Fasoula et al., 2020). As a counterpart to *genotype*, the term *phenotype* is the functional plant body formed during plant growth and development from the dynamic interactions between plant genotype (G) and their growing environments (E), i.e., G × E. The term *phenotyping* is referred to as the set of methodologies and protocols used to measure physiological and morphological characteristics of plants, such as chlorophyll content, three-dimensional architecture, and composition at different scales (Fiorani & Schurr, 2013).

Plant breeding aims to develop new crop varieties with improved traits, including high yield potential, high food quality, and resilience to biotic and abiotic stresses due to adverse environments (Staton, 2017). Breeding strategies require accurate quantification of plant phenotypes to evaluate plant performance, characterize germplasm and experimental populations to identify valuable genes/QTLs (Mir et al., 2019), or train prediction models in both conventional or molecular breeding (Jarquin et al., 2020). Conventional plant phenotyping methods measure crop traits

J. Zhou
Biological Systems Engineering, University of Wisconsin-Madison, Madison, WI, USA

C. N. Vong · J. Zhou (✉)
Division of Plant Science and Technology, University of Missouri, Columbia, MO, USA
e-mail: zhoujianf@missouri.edu

S. Ma et al. (eds.), *Sensing, Data Managing, and Control Technologies for Agricultural Systems*, Agriculture Automation and Control,
https://doi.org/10.1007/978-3-031-03834-1_4

based on visual observations and manual tools, which are time-consuming, labor-intensive, and sometimes destructive. Conventional breeding programs need immense human resources to sample a large population of crop plants, which has become one of the major bottlenecks hindering crop breeding and functional genomics studies (Fasoula et al., 2020). Thanks to the advances and reduced costs in sensors, computer vision, and machine learning technologies, plant phenotyping methods have been elevated to a high-throughput manner during the last decade. Over the years, a diversified range of imaging sensors/systems, data processing and analyzing methods, and their successful applications are continuously emerging and developing. This chapter will briefly introduce the high-throughput plant phenotyping (HTPP)-related technologies and their applications. We will also discuss the opportunities and challenges of using imaging technology in high-throughput plant phenotyping to transform conventional crop breeding to next-generation breeding programs.

4.2 High-Throughput Plant Phenotyping Technology

In the field of plant breeding and genetics, plant phenomics refers to the multidisciplinary approaches of high-throughput acquisition and analysis of multidimensional phenotypes of plants on an organism-wide scale in their development (Yang et al., 2020). Phenomics has been advanced for studying plant response to biotic and abiotic stresses (Ganthaler et al., 2018; Hasan et al., 2020; Zhou et al., 2020), dissecting dynamic changes in plant structure and functions (Jahnke et al., 2009), and developing cultivars with better adaptation to stress-prone environments for plant breeding (Dwivedi et al., 2013). Phenomics aims to bridge the gap between high-density genomic data acquired by emerging sequencing technology and the low-density of phenomic data from the traditional phenotypic approaches.

A high-throughput plant phenotyping (HTPP) system is an essential tool to plant phenomics and is generally considered as an integrated system that can collect massive amounts of phenotypic data from hundreds of thousands of plants regularly (days to months) with a high degree of automation (Li et al., 2021; Yang et al., 2017b). With this broad definition, a wide range of HTPP systems has been developed under various scenarios, including controlled environments and field conditions (Fig. 4.1) to characterize plant features at different scales (roots, leaves, and canopies, Fig. 4.2). The general architecture of a HTPP system includes supportive hardware (platforms), sensing systems (cameras and controllers), data acquisition and management systems, and computational software (data processing and analysis).

High-throughput plant phenotyping systems under controlled environments are often accompanied by precise control of environmental factors (such as temperature, humidity, wind speed, light intensity, and nutrient content) and experimental treatments. Some examples of such systems include growth chamber systems (Jansen et al., 2009) and robotic mobile systems in a greenhouse (Pereyra-Irujo

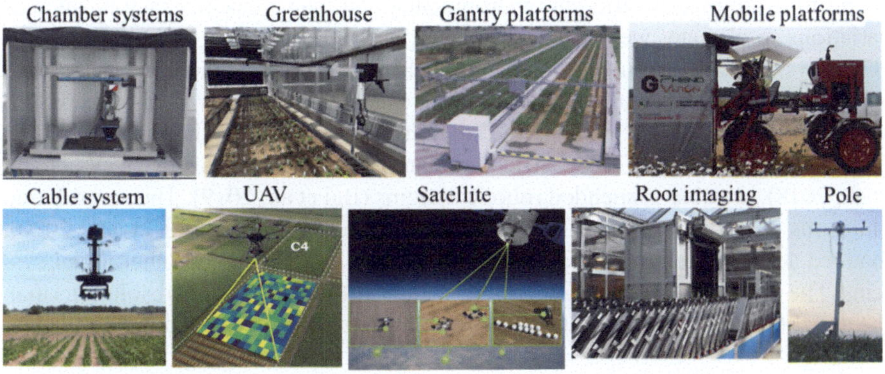

Fig. 4.1 Examples of high-throughput plant phenotyping (HTPP) systems used in controlled environments and field conditions

Fig. 4.2 Plant traits derived from image features acquired at different scales. (**a**) Soybean roots in a rhizobox system (Martins et al., 2020). (**b**) Top and side views of a 3D point cloud model of a soybean plant under a greenhouse environment (Zhou et al., 2019). (**c**) An aerial image of three two-row soybean plots taken at 15 m in a breeding field (Zhou et al., 2020). (**d**) An orthomosaic image of a 4-ha soybean field taken at 30 m in a breeding field

et al., 2012; Zhou et al., 2018). The HTPP systems in controlled environments can be used to capture plant responses to specific variables in a non-destructive, highly repeatable, and high-resolution manner, including structural, physiological, and biochemical traits of plant roots and shoots. On the other hand, the field conditions are an intricate mix of different environmental and managemental factors, such as variations in climate, weather, and soil properties, resulting in a large variety of HTPP systems for various purposes. The field HTPP systems include ground- and aerial-based platforms equipped with various imaging sensors. Ground-based HTPP systems include pole or tower-based imaging stations (Shafiekhani et al., 2017), gantry platforms (Vadez et al., 2015), unmanned ground vehicles or UGVs (Jiang et al., 2018), or cable-suspended robotic systems (Bai et al., 2019). The aerial-based platforms include unmanned aerial vehicles or UAVs (Moghimi et al., 2020), manned aerial vehicles (Yang & Hoffmann, 2015), satellites, and other aerial systems. The ground HTPP generally provides data with higher spatial resolutions than the aerial HTPP but has limitations in field coverage and data collection efficiency.

4.3 Imaging Sensors and Systems

When incident radiation (light source or natural light) hits the surface of an object, the radiation will either be absorbed, transmitted, or reflected (Lillesand et al., 2004). The reflected radiation can be expressed as the electromagnetic spectrum (EMS) with a range of wavelengths (Fig. 4.3). Different ranges of wavelength (waveband) are denoted to different names such as visible region (400–700 nm), near-infrared (NIR, 700–1000 nm), short-wave infrared (SWIR, 900–2500 nm), and long-wave infrared (LWIR, 7.5–14 μm) (Silván-Cárdenas et al., 2015). Imaging sensors in HTPP systems are the devices that can capture electromagnetic spectrum and convert them into image pixels of different values, i.e., large pixel values in an image indicate strong EMS.

Plant leaves and canopy can generally reflect less visible light (about 400–700 nm in wavelength) but reflect more light in the near-infrared (NIR, about 800–1400 nm) range. As shown in Fig. 4.4a, leaves of healthy plants usually absorb more blue and red light to conduct photosynthesis, create chlorophyll, and reflect more green light, making them look greener than those of unhealthy plants. On the other hand, healthy plants reflect more NIR light than unhealthy plants. The different reflective energy to light at different wavelengths (Fig. 4.4b) makes spectral reflectance an effective,

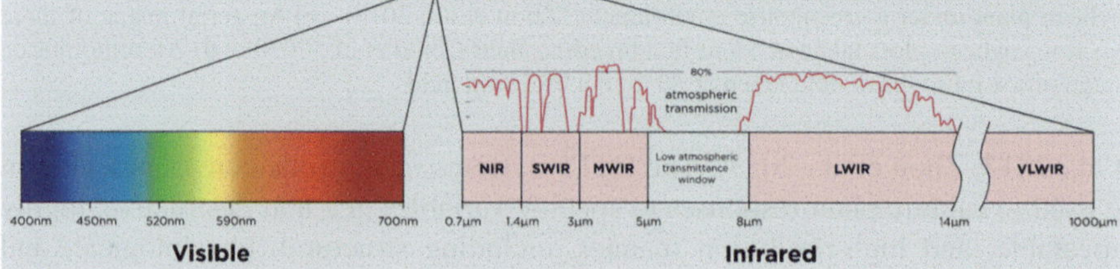

Fig. 4.3 Electromagnetic spectrum (EMS) scheme (nm)

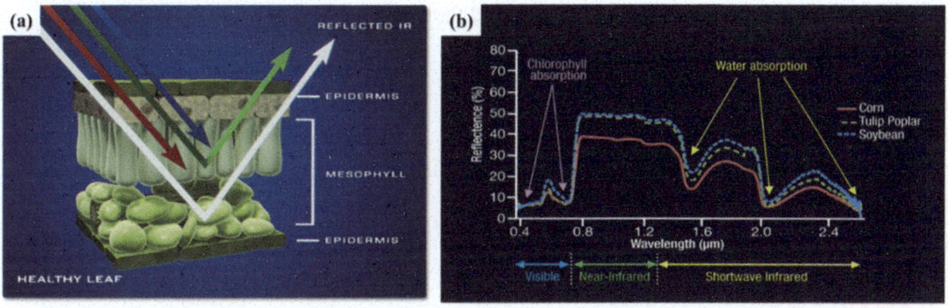

Fig. 4.4 Spectrum reflectance of plant leaves. Image courtesy to NASA Science. (**a**) Healthy vegetation absorbs blue- and red-light energy for conducting photosynthesis and creating chlorophyll. (**b**) An example of spectral signatures of vegetation in reflectance

non-contact and non-destructive tool to detect different chemical and physical characteristics in plants. Spectral images have been used widely to quantify plant physical, chemical, and biological characteristics (Jin et al., 2020). The widely used imaging sensors used in HTPP systems include visible cameras, multispectral cameras, hyperspectral cameras, infrared thermal imagers, fluorescence imaging sensors, depth cameras (RGB-D), and tomography imaging sensors. The imaging sensors vary in cost, working principles, processing methods, and functionalities, leading to various applications in HTPP system (Table 4.1).

4.3.1 Visible Cameras

Visible cameras (also called red-green-blue, or RGB cameras) consist of three sensor arrays to detect EMS energy in red (typical 550–650 nm), green (typical 470–600 nm), and blue (typical 420–530 nm) spectral bands to produce digital images (Kolláth et al., 2020). Visible cameras use either charge-coupled devices (CCD) or complementary metal-oxide-semiconductor (CMOS) sensors to capture the reflectance or transmittance of light. Image sensors turn spectral energies into electrical signals proportional to their energy level, which can be further converted to digital numbers, i.e., pixels (e.g., 0–255 for an 8-bit camera). Therefore, image pixels can capture plant responses and quantify plant geometric information (e.g., leaf length and area) and color information (naturally due to their physiological and biochemical characteristics). The visible cameras are the most widely used imaging sensors in HTPP systems due to the features of low cost, high resolution, user-friendly operation, lightweight, and adaptability to various working conditions (Yang et al., 2017b).

High-resolution images from visible cameras can be used to build three-dimensional (3D) point cloud data of plants. Sequential and highly overlapping images collected from UAVs or robotic systems can be processed to develop digital elevation models (DEMs). Commonly used image processing methods for DEMs

Table 4.1 Imaging sensors in HTPP systems and their applications

Sensors	Visible RGB camera	Spectral camera	Thermal camera	Fluorescence camera	Depth camera (RGB-D)	Tomographic imaging
EMS region (nm)	400–700	400–2500	7.5–13 × 10^3	400–700	400–700	< 400
Plant traits						
Plant height	Hassan et al. (2019)	Borges et al. (2021)	–	–	Wang et al. (2020)	–
Leaf area/LAI	Raj et al. (2021)	Zhang et al. (2021)	–	–	Martinez-Guanter et al. (2019)	–
Plant lodging	Wang et al. (2021)	Chauhan et al. (2019)	Cao et al. (2021)	–	–	Wu et al. (2021)
Photosynthesis pigment, biochemical contents	Brambilla et al. (2021)	Suarez et al. (2021)	Maimaitijiang et al. (2017)	Zhao et al. (2021)	–	–
Water content	–	Mwinuka et al. (2021)	Jin et al. (2021)	–	–	–
Biotic stress	Naik et al. (2017)	Bebronne et al. (2020)	Mastrodimos et al. (2019)	Konanz et al. (2014)	–	–
Abiotic stress	Zhou et al. (2021b)	Zhou et al. (2020)	Zhou et al. (2021a)	Gomes et al. (2012)	–	–
Root architecture system	Eberle et al. (2020)	Bodner et al. (2021)	–	–	–	Phalempin et al. (2021)
Seed quality, composition and size	–	Choi et al. (2021)	Rojas-Lima et al. (2021)	Li et al. (2019)	–	Hughes et al. (2017)

include feature registration (e.g., Scale-Invariant Feature Transform, or SIFT) and stereovision (e.g., Structure from Motion, or SfM). The DEMs include a 2D matrix of depth information in each pixel used to build a 4D matrix containing both color and position information (Zhou et al., 2018). Three-dimensional geometric traits, such as plant height and leaf angles, can be acquired from the 3D models of plants (Cao et al., 2019).

Nevertheless, visible cameras have some limitations. First, image pixels are dependent on passive light radiations in conjunction with cameras parameters (e.g., sensor size, bit depth, shooting speed, aperture, and focal length). Inconsistent or ununiform light conditions during data collection will cause challenges in quantifying variations of plant traits using imaging features. Another limitation is geometric distortions (e.g., vignetting and perspective distortions) by the camera lens, which can cause inaccurate measurements of plant dimensions (Kim, 2008). Therefore, specific procedures are required before and after image collection, such as setting ground control points (Han et al., 2019), using color standard boards (Del Pozo et al., 2014), performing distortion calibration (Feng et al., 2020), and using calibration models (Deng et al., 2018). Moreover, the visible cameras only sense the spectral reflectance wavelength in the range of 400–700 nm, limiting the ability to discover the more complex biochemical and physiological plant traits.

4.3.2 Spectral Cameras

Spectral features of plant vegetation can be quantified using spectral cameras that consist of sensor arrays sensitive to light in the spectral bands beyond the visible range. Currently, spectral cameras can sense light of reflectance or transmittance in the spectra range of 400–2500 nm, usually called visible-NIR or VNIR. It can be seen from Fig. 4.4b that spectral reflectance to vegetation at the edge of the NIR range (700–750 nm, i.e., red-edge) has the steepest responses in the VNIR spectra, and the differences among plants have been signified in this range. Multiple spectral bands are combined to develop different vegetation indices to signify the differences in spectral reflectance features (Humplík et al., 2015). For example, the widely used vegetation index normalized difference vegetation index (NDVI) uses spectral bands in red and NIR spectral range. Vegetation indices are also used to build predictive models using machine learning techniques to predict complex crop traits, such as yield (Zhou et al., 2018). Spectral cameras are categorized to multispectral cameras and hyperspectral cameras according to the number of discrete wavebands or channels.

Multispectral Cameras

Multispectral cameras consist of sensors that are sensitive to a relatively small number (less than 10) of discrete wavebands in the VNIR range (Humplík et al., 2015). The spectral wavebands are usually selected based on research results that indicate their effectiveness in representing important crop traits. Some widely used wavebands include blue (450–520 nm), green (520–600 nm), red (630–690 nm), red edge (700–730 nm), and NIR (760–900 nm) (Hunt Jr et al., 2013; Thenkabail & Lyon, 2016). Depending on the waveband width (number of wavelengths) of every single channel, spectral cameras can also be divided into narrow waveband (e.g., <50 nm) or broad waveband (>50 nm) cameras (Hunt et al., 2005). For example, the multispectral camera Micasense RedEdge-M+ (Micasense, Seattle, WA, USA) consists of five narrow spectral bands of blue (475 ± 20 nm), green (560 ± 20 nm), red (668 ± 10 nm), red edge (717 ± 10 nm bandwidth), and NIR (842 ± 40 nm bandwidth). The narrow-band spectral cameras are usually more accurate to pick up the differences of spectral signatures for different plants; however, they are more expensive than broad-band spectral cameras.

Multispectral cameras have disadvantages of relatively higher costs and lower image resolution than visible cameras to acquire high-resolution information of plants (Li et al., 2014; Xie & Yang, 2020). Moreover, multispectral cameras have fewer wavebands than hyperspectral cameras, restricting their potential to discover novel plant traits that are important to breeders. Similar to the visible cameras, data acquired from multispectral cameras can be affected by illumination conditions and should be calibrated using reference standards, such as a calibration board, in field studies under changing lighting conditions due to cloud and shadows (Jin et al., 2020).

Hyperspectral Camera

Hyperspectral cameras can be considered as more powerful "multispectral cameras" that include as many as more than 300 narrow wavebands in the range of 400–2500 nm (Li et al., 2014). There are two different types of hyperspectral cameras based on their data collection modes, i.e., pushbroom (or line scanning) and snapshot (snapshotting) hyperspectral cameras. Pushbroom hyperspectral cameras consist of a line of spectroscopic sensors that acquire images using a line-by-line scanning method when the cameras are moving above a scene. On the other hand, a snapshot hyperspectral camera consists of a matrix of spectroscopic sensors that are able to acquire images of a scene without moving the camera. Pushbroom hyperspectral cameras usually have more narrow spectral wavebands or higher spectral resolution comparing to snapshot cameras. However, pushbroom cameras require stabilized mounts and smooth movements to "reconstruct" the image, which becomes a limitation for some scenarios. For example, the pushbroom cameras are unsuitable for ground-based mobile platforms with large vibration in field conditions. In addition, a consistent artificial light source is always needed to provide extra lights for hyperspectral cameras if used in indoor environments or other low-light conditions.

Hyperspectral cameras can capture a large number of narrow-band spectral information of crops, which are closely associated with the chemical and physiological information of crops. Hyperspectral imagery data can be processed using big data processing and analytic technologies, such as machine learning and deep learning, to quantify and predict plant photochemical and physiological features (Pandey et al., 2017; Yang et al., 2017a), health status (Knauer et al., 2017; López-Maestresalas et al., 2016), and biomass/yield (Liang et al., 2018). Current UAV-based hyperspectral imagers are primarily using 50–270 narrow spectral bands of VNIR (400–1100 nm) due to the weight and cost. Adding a short-wave infrared spectral imager (SWIR, 900–2500 nm) may greatly improve spectral range and capacity to detect additional responses of crops to environments. However, SWIR imaging sensors are made of indium gallium arsenide (inGaAs) and usually are heavy and expensive. The SWIR bands have a minimum amount of atmospheric disturbance or noise and the ability to separate different ground materials, thereby helping in feature extraction accurately (Swathandran & Aslam, 2019). Studies have shown great potential in quantifying plant response to different stresses using VNIR cameras (Rascher et al., 2011; Thomas et al., 2017; Yuan et al., 2014). The SWIR channel shows specific reflectance for vegetation water content (Hunt et al., 2011) and soil moisture (Olsen et al., 2013). The most important applications of the SWIR include agricultural management by assessing the crop stress by the reflectance of different pigments in the leaves along with crop moisture estimation and mapping as well as quantifying the crop residue and predicting the quality of the soil (Galloza et al., 2013; Hively et al., 2018; Serbin et al., 2009). With regard to the spectral reflectance differences of moisture absorption properties, various drought indices using the backscatter energy from near-infrared (NIR) and shortwave-infrared (SWIR) channels have been formulated to estimate vegetation water content using satellite remote sensing (Ji et al., 2011; Vescovo et al., 2012; Wang & Qu, 2007), which may serve as reliable indicators for crop drought stresses, and potentially to be used in selecting drought-resistant varieties.

There are some challenges in using hyperspectral imaging systems in plant breeding. Compared to visible and multispectral cameras, hyperspectral imaging systems are expensive, heavy to integrate with UAVs, and complicated to operate (to acquire high-quality data). In addition, hyperspectral imagery data are usually in a large volume, complicated to process and need more computing resources (Jin et al., 2020; Li et al., 2014; Thenkabail & Lyon, 2016). Some widely used data processing methods include principal component analysis, derivative analysis, partial least squares, etc. (Thenkabail & Lyon, 2016). Similarly, when collecting data using hyperspectral cameras, frequent light calibration using a reference board is needed for outdoor field-scale studies (Jin et al., 2020). Thus, hyperspectral imaging is challenging to scale up for usage in large-scale field phenotyping.

4.3.3 Infrared Thermal Imager

Thermal imaging allows visualization of the energy of infrared (IR) radiation of any objects with a temperature above absolute zero (−273 °C). The principle of thermal infrared cameras is to capture long-wave IR radiation (7.5–14 μm) emitted by crops and convert such radiation to electrical signals (Jones, 2004). Plant temperature measurement has been primarily used to study plant water relations, and specifically stomatal conductance, because a major determinant of leaf temperature is the rate of evaporation or transpiration from the leaf (Jones, 2004). During evaporation, a substantial amount of energy is required to convert liquid water in leaves to water vapor, and this energy is then taken away from the leaves resulting in reduced temperature. When a plant experiences abiotic or biotic stresses, the transpiration rate may reduce, leading to higher temperature than those without stresses. Therefore, thermal cameras are useful in detecting water-related stresses (Balota & Oakes, 2017). For example, canopy temperature depression, defined as the temperature difference between crop canopy and air, is highly correlated with the canopy water mass and can be used to quantify crop response to abiotic and biotic stresses (Ludovisi et al., 2017).

Thermal images can provide spatiotemporal temperature information of plants and help understand the interaction between plant, water, soil, and environments. Thermal cameras have become a widely used tool in research programs and commercial applications of plant breeding and precision agriculture. For example, Jin et al. (2021) found that UAV-based thermal imagery is a useful tool to assess the variations in rice growth due to soil water availability and water use efficiency. Zhou et al. (2020, 2021a) used thermal cameras to evaluate soybean responses under water stresses (i.e., flooding and drought) to identify stress-tolerant genotypes by comparing canopy temperature among all varieties. In practice, thermal images can be used simultaneously with visible and spectral images that help segment crops from images with complex backgrounds (e.g., weeds, crop residues) (Leinonen & Jones, 2004; Möller et al., 2007).

The limitations of thermal cameras include that thermal measurement is heavily affected by ambient conditions, such as air temperature, humidity, and wind speed (Jin et al., 2020; Li et al., 2014). It is required to conduct pre-calibrations and post-corrections based on solar position and environmental factors, including transient wind and cloud cover for temporal data (i.e., data collected at different days) (Li et al., 2014; Zhang & Zhang, 2018). In addition, the orientation of plant leaves towards the incident radiation and camera angle needs to consider during data processing and analysis (Jones et al., 2009). Furthermore, additional procedures are required to ensure measurement accuracy, for example, preheating cameras for about 20 min before using them to reduce temperature drift (Jin et al., 2020).

4.3.4 Fluorescence Imaging

Chlorophyll fluorescence of plants is the light re-emitted by the chlorophyll molecules during the returning from excited to nonexcited states (Maxwell & Johnson, 2000). The yield of chlorophyll fluorescence depends on the efficiency of converting absorbed light to fluorescence. For healthy plants, the majority of absorbed light by chlorophyll molecules is used for photosynthetic quantum conversion, and only a small portion is de-excited via emission as heat or as red and far-red chlorophyll fluorescence. The ability of photosynthetic quantum conversion declines for plants under stress, with a concomitant increase in red and far-red chlorophyll fluorescence (Lichtenthaler & Miehé, 1997). Therefore, the analysis of chlorophyll fluorescence re-emitted from plant leaves can obtain information about plant health status and has been used as an important tool in the research of plant breeding and physiology (Halbritter et al., 2020).

Fluorescence imaging sensors (cameras) are used to capture the re-emitted proportion of irradiation in a short wavelength such as ultraviolet (UV) light (340–360 nm) by plants (Li et al., 2014). Recently, a high-resolution ultraviolet

(UV) laser-induced fluorescence (LIF) imaging system was developed to image all four fluorescence bands: blue, green, red, and far-red (Ortiz-Bustos et al., 2016). The inverse relationship between photosynthetic performance and chlorophyll fluorescence analysis has made a large contribution to the understanding of photosynthesis and electron transport reactions.

A fluorescence sensing system usually consists of one or more CCD cameras with filters to capture fluorescence signals (Wang et al., 2018). Active light sources, such as pulsed lasers, pulsed flashlight lamps, or light-emitting diodes (LEDs), are used to provide irradiation (excitation) in specific wavelengths (Baker, 2008). The Pulse Amplitude Modulated (PAM) fluorometry method by Schreiber et al. (1986) has been widely adopted in practical applications. The PAM uses a short (e.g., 1 μs) pulse of light (also called a dark adaptation) to excite a target object and measure the minimum fluorescence value (Fo) of the object, which is then exposed to a saturating pulse of light to measure the maximum amount of fluorescence (Fm). The difference between Fo and Fm is defined as variable fluorescence (Fv), and the ratio of Fv/Fm provides a measure of photochemical efficiency (Schreiber et al., 1986). The Fv/Fm ratio depends on the metabolic capacity of plants and is highly sensitive for plant photosynthetic activity yielding parameters closely related to photosynthetic functions (Serôdio et al., 2018). Therefore, fluorescence imaging is commonly used to detect stress symptoms induced by pathogen attack (Chaerle et al., 2007), monitor stress responses (Baker, 2008), measure physiological phenomena relating to photosynthesis, metabolism, and growth-related traits (Baker & Rosenqvist, 2004; Lenk et al., 2006).

The applications of fluorescence imaging systems are used primarily in indoor environments and for single leaves or seedling levels of crops due to the requirement on active light sources (Li et al., 2014). In addition, the fluorescence imagery data are sensitive to ambient conditions and generally not available for ground and aerial phenotyping platforms for field-scale studies and applications (Jin et al., 2020).

4.3.5 Depth Imaging Camera

Depth cameras are used to measure the distance of the object from the camera, or "depth," using the principle of "Time-of-Flight (ToF)." Depth cameras consist of signal emitters, receivers, and control modules. The distance is calculated based on the travel time of emitted signals from the emitters and returned signed captured by the receiver (Vázquez-Arellano et al., 2016). The widely used types of signals include laser light and ultrasonic waves, corresponding to two major types of ToF sensors, i.e., Light Detection and Ranging (LiDAR) and ultrasonic sensors. Some depth cameras have both visible imaging sensors and depth sensors (i.e., RGB-D camera or LiDAR camera) to simultaneously acquire both types of information (color and depth).

Depth cameras are widely used to build 3D models of plants to quantify different traits, such as the structure of plant organs (Paulus et al., 2014; Wang et al., 2017), leaf area index (LAI) (Kjaer & Ottosen, 2015), and plant height (Holman et al., 2016; Hu et al., 2018; Jimenez-Berni et al., 2018). For example, plant structures, such as leaf area and stem length, extracted from 3D models were used to detect water stress in barley. The LAI of different rapeseed genotypes was calculated based on their leaf area extracted from a LiDAR sensor to study the interaction of genotypes and environments (Kjaer & Ottosen, 2015). Furthermore, plant height, ground cover, and above-ground biomass of wheat in fields were obtained using LiDAR sensors mounted on UAV to study their growth rate temporally (Holman et al., 2016; Jimenez-Berni et al., 2018). In addition, plant height from UAV-based LiDAR sensors was used to estimate yield in a sorghum breeding program (Hu et al., 2018). Some limitations of depth cameras are their sensitivity to vibration and outdoor illumination, low resolution, and complex data processing (Jin et al., 2020; Li et al., 2014).

4.3.6 Tomographic Imaging

Tomography is imaging objects by sections or sectioning through the use of penetrating waves (e.g., X-ray). For instance, nuclear magnetic resonance imaging (MRI) detects ***nuclear magnetic resonance signals*** to form images (Li et al., 2014). The usage of MRI includes obtaining non-destructive 3D structures of plant organs, seeds (Melkus et al., 2011), root system architecture in or near natural soil (Moradi et al., 2010), and entire plants (Van As & Van Duynhoven, 2013). Furthermore, they are utilized to depict 3D representations of water distribution to noninvasively quantify the water content, water diffusion, and water transport of plants or plant organs (Windt et al., 2006). Another type of tomographic imaging is positron emission tomography (PET), which is a nuclear imaging technique producing 3D images of a functional process by detecting pairs of ***gamma rays*** emitted indirectly from a positron-emitting radionuclide. For instance, the transport of ^{11}C-labeled photoassimilates during the CO_2 consumption in photosynthesis can be imaged regularly in 3D by PET. This imaging mode can dissect transport domains in plant organs to deliver quantitative parameters, including transport velocities and lateral loss rate along transport paths (Bühler et al., 2011). PET can be used in conjunction with MRI to obtain structural and functional traits and analyze the transport of water and labeled compounds independently (Crosson et al., 2010).

X-ray computed tomography (X-ray CT) utilizes computer-processed ***X-rays*** to generate tomographic images of particular parts of the scanned object. It can also produce a 3D image of the inner part of an object from a series of 2D radiographic images captured around a single axis of rotation. X-ray CT enables such measurements in a scalable fashion due to its non-direct contact, relatively quick implementation, and inherently non-destructive. It has been widely used to detect and quantify the inner structures of shoots, organs, and roots from the cellular to the whole-organ scale of plants. The examples comprise assessing the structure of xylem vessels and the frequency of embolism in maize leaves rapidly (Ryu et al., 2016) and providing detailed internal three-dimensional (3D) phenotypic information of flowers (Tracy et al., 2017), grains (Hughes et al., 2017), spikes (Strange et al., 2014; Zhang et al., 2018), and stalks (Zhang et al., 2018). Moreover, it is used in root lodging studies to determine the structure of roots in soil (Flavel et al., 2012), quantify compaction (Tracy et al., 2015), and investigate the effect of the rhizosphere on soil hydraulic properties (Daly et al., 2015), which are important to develop better crops. However, the tomographic imaging technologies are usually expensive and time-consuming (e.g., ≥60 min for PET and 40–60 min for MRI), which highly restrict their usage for large-scale studies and adoption in field applications (Yang et al., 2020).

4.4 Image Processing and Analytics

Processing and analyzing image data acquired by HTPP systems is essential to translate sensor data (imagery) to crop traits that breeders can use directly in their breeding programs. Many different methods have been developed using advanced machine learning methods for image processing and analytics.

4.4.1 Image Processing

Although the processing methods are different for individual studies, some essential steps are in common, including geometric and radiometric corrections, segmentation of individual plants, and removal of unnecessary information (image background). The image processing pipelines that are commonly used in existing studies include two categories: (1) image processing based on orthomosaic images (including 3D models) of a whole field and (2) processing individual image frames. Figure 4.5 provides an example of image processing pipelines for field phenotyping studies in our research group (Feng et al., 2020; Zhou et al., 2020).

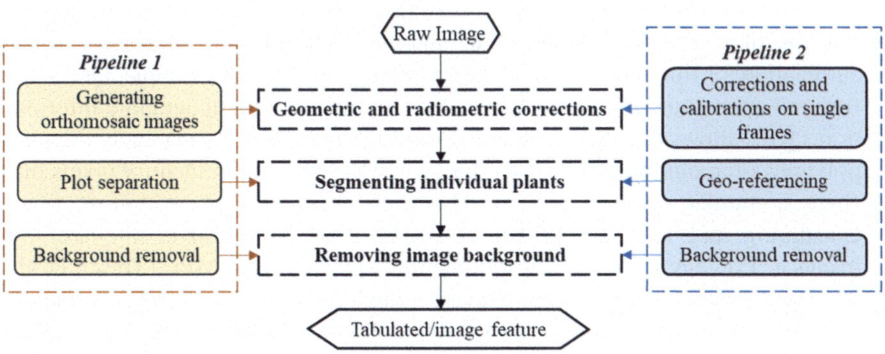

Fig. 4.5 Pipelines for processing images collected using HTTP systems in field conditions

1. Image processing based on 2D orthomosaic images

   Sequential images collected by UAV systems or robotic systems usually cover a portion of a field or testbed. One of the widely used methods to process the image is building 2D orthomosaic images of the test field using commercial software packages, such as Pix4Dmapper (Pix4D, Lausanne, Switzerland) and Agisoft MetaShape Professional (Agisoft, St. Petersburg, Russia). These packages are able to retrieve image points with depth information from sequential images with sufficient overlaps (over 70%) and reconstruct the image points into a point cloud by matching the points with similarities. The point cloud data can be further processed to build digital elevation models (DEMs) containing depth and (3D) structural information. Meanwhile, geometric deformation and radiometric variations are also corrected during image processing.

   Image background removal is an essential step of differentiating the pixels into two classes: plant and background (soil and residues). The common methods of background removal include threshold-based and machine learning-based methods. The threshold-based method divides image pixels into different categories (plants and non-plant materials) using proper thresholds developed in original (i.e., RGB) or converted (e.g., HSV or any vegetation index) color spaces. Thresholds determine the accuracy of background removal. Generally, a large threshold causes the losses of pixels of plants, while a small threshold causes the inclusion of background information in the image. Various techniques have been applied to optimize the values of threshold, including dynamic thresholding (Reid & Searcy, 1987), hysteresis thresholding (Marchant et al., 1998), Otsu's thresholding (Otsu, 1979), and entropy of a histogram (Tellaeche et al., 2008). Although the threshold-based methods are effective when there is a clear difference between plants and background, they cannot perform well in the conditions where plants and background are similar or images have severe light variations at data collection. The machine learning-based methods have been investigated to improve the binary classification under various illumination conditions, such as fuzzy clustering (Meyer et al., 2004), support vector machines (Zhou et al., 2019), and decision trees (Guo et al., 2013), to improve the accuracy of background removal.

   The orthomosaic images of a whole need to be segmented to separate each row or plot of different genotypes. Various methods and tools have been used towards plot separation at different automation levels, i.e., manually, semi-automation, or fully automation (Chen & Zhang, 2020). For example, QGIS is a free and open-source software for processing images with geographic information. QGIS allows users to specify an area of interest (AOI) by manually drawing polygons or assigning pixels as a reference, which is used to identify pixels that share a similar spectral pattern. However, manual segmentation is time-consuming and laborious. Other commercial or open-source software, for instance, Progeny (https://www.plotphenix.com/) and EasyMPE (Tresch et al., 2019) can grid the scene images with the knowledge of essential field parameters (e.g., size of the plot, number of rows and columns). However, they still require manual operations to identify field boundaries and adjust for misalignment.

Machine learning-based methods have emerged to achieve fully automated segmentation. One example is the Trainable Weka Segmentation (TWS) (Arganda-Carreras et al., 2017), a segmentation tool that consists of supervised learning algorithms and can learn pixel-based features from a provided training dataset. However, the training process requires manual labeling of a large number of images, which is labor-intensive.

2. Processing individual images

   Another pipeline of image processing is to process individual image frames directly without building orthomosaic images. The direct processing pipeline may provide a timely and cost-effective alternative for extracting plant features. In this pipeline, pre-processing has to be performed on each original image frame, including radiometric calibration, geometric correction, and image enhancement. Geometric deformation of images includes optical distortion due to the camera lens and perspective distortion due to the orientation of camera lenses relative to the scene. Both distortions will affect the ground sampling distance (GSD) in each image frame and, consequently, their geometric measurement precision. The optical distortion is often corrected through the calibration procedure by characterizing the intrinsic (e.g., the shape of the camera lens) and extrinsic (i.e., pose) parameters using a standard checkerboard. The perspective distortion can be corrected using inverse perspective transformation methods using detectable standard objects (e.g., the plot row spacing or plot row length) in an image (An et al., 2016; Feng et al., 2020). Corrected images are enhanced using image enhancement technologies, such as contrast adjustment and image filtering, to reduce the influence of luminance, such as sunlight and shadow (Jeon, 2014). Background removal and plot separation are conducted after correction. All segmented images are geo-referenced based on either GPS or relative coordinates. The internal coordinate system of an image (pixel positions) can be associated with locations in physical space by knowing the center position of the image and its GSD.

4.4.2 Imagery Data Analytics and Modeling

Imagery data acquired by HTPP systems can provide multidimensional information of plants or plant organs and characterize crop phenotypic traits in a high spatiotemporal resolution. Data for different research purposes require reliable methods to remove irrelevant or redundant information and tune model arguments to reach acceptable results eventually. Traditional regression-based models analyze image features to correlate and predict crop traits, which have the advantage of good interpretability. However, simple models are difficult to explain the multivariate relationships between predictors and responses (Parmley et al., 2019). In recent years, with the advance in high-performing computers and Application Programming Interfaces (APIs, e.g., Keras in Python), machine learning (ML), and deep learning (DL), models have been extensively implemented for processing imagery data from HTTP systems in plant breeding applications.

Classic machine learning models, such as support vector machine (SVM), tree-based models, and Linear/Quadratic Discriminant Analysis (L/QDA), have been used to estimate crop traits (e.g., yield and plant height) and classify categorical traits (e.g., stress injury scores and maturity groups). The performance of classic ML models depends on the quality of input features (Chandrashekar & Sahin, 2014), which are usually manually selected. Therefore, an unsupervised ML method, principal component analysis (PCA), is often used to reduce feature dimensions by creating principal components (linear combinations of relevant features) that explain the most variations in the dataset. Other supervised ML models, such as Lasso (least absolute shrinkage and selection operator) and ridge regression, can also reduce the dimension of datasets by assigning extremely small coefficients (even 0) to the variables that had minor contributions to the model performance (James et al., 2013).

The advance in high-performance computing has resulted in the fast development and implementation of DL models to analyze imagery data. Deep learning is the most important branch of ML allowing hierarchical data learning. One of the advantages of DL is that it does not require manual feature selection, but the algorithms will determine the important features during the training process by adjusting a set of parameters associated with the input features. There are a few classic DL models widely applied in HTPP, including artificial neural networks (ANNs) for tabulated input similar to ML models, convolutional neural networks (CNNs) for image analysis, recurrent neural networks (RNNs) for sequence or time data processing, and auto-encoder networks (AENs) (Goodfellow et al., 2016). However, DL models require a large dataset for training, testing, and validation. They are also computationally expensive and have poor interpretability.

4.5 Opportunities and Challenges of Imaging Technology for Plant Breeding

The development and innovation of HTPP technology need collaborative efforts from the multidisciplinary cooperation of experts in plant breeding, genetics, engineering, and computer science (Li et al., 2021). The emerging technologies in big data analytics and artificial intelligence (AI) provide great opportunities for implementing imaging technology in plant breeding using next-generation HTPP systems. Imaging technology will be more successful in plant breeding with the advances in UAV, robotics, high-performance computing, deep learning, and artificial intelligence. The importance and feasibility of imaging systems in the quantification of crop traits have been demonstrated in numerous studies, as summarized in the review papers (Lu et al., 2020; Yang et al., 2020; Zhao et al., 2019). In modern plant breeding programs, image-based HTPP systems have become standard tools to quantify crop traits and assist decision-making on selecting elite genotypes. It is believed that imaging technologies, especially hyperspectral imaging, have great potential to discover the insides of plant response to environments. By integrating with plant genomic information acquired by high-throughput plant genotyping technology, image-based high-density plant traits, or plant phenomics, will greatly enhance the capacity of developing precision breeding programs. Imaging technology will be a great tool for transferring conventional plant breeding to next-generation plant breeding programs.

Although image-based HTPP can efficiently acquire big data set of plants, the large amount of data start overwhelming researchers and breeders. Therefore, there is a critical need to develop efficient and effective data management, processing, and analytic systems (pipelines) for HTPP systems, which can translate imagery data to useful crop traits important to breeders. Different breeding programs or research teams have been developing various software, tools, and online database; however, it is still challenging for breeders to apply and customize these tools for their specific research purposes. Thus, more user-friendly or "breeders preferred" platforms are still needed to bridge the gap between breeders and the research community of "phenotypers." Our vision in addressing the challenges includes developing AI-based edge computing to process data during collection, as well as automated and cloud-based data processing and analytic pipelines for real-time data processing and analyzing. We do believe imaging technology will be one of the most important tools in the next-generation plant breeding.

4.6 Summary

In this chapter, we briefly introduced imaging technology in high-throughput plant phenotyping. Various phenotyping hardware platforms have been developed and applied for plant breeding in controlled environments and field conditions. Commonly used imaging sensors, such as visible, spectral, infrared thermal, fluorescence, depth camera (RGB-D), and tomographic imaging, and their applications in plant breeding were discussed. We summarized the image processing and analytic methods that have been developed and applied in current studies and their pros and cons. At last, we discussed the opportunities and challenges in using imaging technology for high-throughput plant phenotyping.

References

An, N., Palmer, C. M., Baker, R. L., Markelz, R. C., Ta, J., Covington, M. F., Maloof, J. N., Welch, S. M., & Weinig, C. (2016). Plant high-throughput phenotyping using photogrammetry and imaging techniques to measure leaf length and rosette area. *Computers and Electronics in Agriculture, 127*, 376–394.

Arganda-Carreras, I., Kaynig, V., Rueden, C., Eliceiri, K. W., Schindelin, J., Cardona, A., & Sebastian Seung, H. (2017). Trainable Weka segmentation: A machine learning tool for microscopy pixel classification. *Bioinformatics, 33*(15), 2424–2426.

Bai, G., Ge, Y., Scoby, D., Leavitt, B., Stoerger, V., Kirchgessner, N., Irmak, S., Graef, G., Schnable, J., & Awada, T. (2019). NU-Spidercam: A large-scale, cable-driven, integrated sensing and robotic system for advanced phenotyping, remote sensing, and agronomic research. *Computers and Electronics in Agriculture, 160*, 71–81.

Baker, N. (2008). Chlorophyll fluorescence: A probe of photosynthesis in vivo. *Annual Review of Plant Biology, 59*, 89–113.

Baker, N., & Rosenqvist, E. (2004). Applications of chlorophyll fluorescence can improve crop production strategies: An examination of future possibilities. *Journal of Experimental Botany, 55*(403), 1607–1621.

Balota, M., & Oakes, J. (2017). UAV remote sensing for phenotyping drought tolerance in peanuts. In *Proceedings of SPIE—The International Society for Optical Engineering*.

Bebronne, R., Carlier, A., Meurs, R., Leemans, V., Vermeulen, P., Dumont, B., & Mercatoris, B. (2020). In-field proximal sensing of septoria tritici blotch, stripe rust and brown rust in winter wheat by means of reflectance and textural features from multispectral imagery. *Biosystems Engineering, 197*, 257–269.

Bodner, G., Alsalem, M., & Nakhforoosh, A. (2021). Root system phenotying of soil-grown plants via RGB and hyperspectral imaging. In *Methods in molecular biology* (pp. 245–268).

Borges, M. V. V., de Oliveira Garcia, J., Batista, T. S., Silva, A. N. M., Baio, F. H. R., da Silva Junior, C. A., de Azevedo, G. B., de Oliveira Sousa Azevedo, G. T., Teodoro, L. P. R., & Teodoro, P. E. (2021). High-throughput phenotyping of two plant-size traits of Eucalyptus species using neural networks. *Journal of Forestry Research, 33*, 591–599.

Brambilla, M., Romano, E., Buccheri, M., Cutini, M., Toscano, P., Cacini, S., Massa, D., Ferri, S., Monarca, D., Fedrizzi, M., Burchi, G., & Bisaglia, C. (2021). Application of a low-cost RGB sensor to detect basil (Ocimum basilicum L.) nutritional status at pilot scale level. *Precision Agriculture, 22*(3), 734–753.

Bühler, J., Huber, G., Schmid, F., & Blümler, P. (2011). Analytical model for long-distance tracer-transport in plants. *Journal of Theoretical Biology, 270*(1), 70–79.

Cao, W., Qiao, Z., Gao, Z., Lu, S., & Tian, F. (2021). Use of unmanned aerial vehicle imagery and a hybrid algorithm combining a watershed algorithm and adaptive threshold segmentation to extract wheat lodging. *Physics and Chemistry of the Earth, Parts A/B/C, 123*, 103016.

Cao, W., Zhou, J., Yuan, Y., Ye, H., Nguyen, H. T., Chen, J., & Zhou, J. (2019). Quantifying variation in soybean due to flood using a low-cost 3D imaging system. *Sensors, 19*(12), 2682.

Chaerle, L., Hagenbeek, D., De Bruyne, E., & Van Der Straeten, D. (2007). Chlorophyll fluorescence imaging for disease-resistance screening of sugar beet. *Plant Cell, Tissue and Organ Culture, 91*(2), 97–106.

Chandrashekar, G. & F. Sahin (2014). A survey on feature selection methods. Computers & Electrical Engineering 40(1), 16–28.

Chauhan, S., Darvishzadeh, R., Lu, Y., Stroppiana, D., Boschetti, M., Pepe, M., & Nelson, A. (2019). Wheat lodging assessment using multispectral uav data. In *International archives of the photogrammetry, remote sensing and spatial information sciences—ISPRS archives*.

Chen, C. J., & Zhang, Z. (2020). GRID: A python package for field plot phenotyping using aerial images. *Remote Sensing, 12*(11), 1697.

Choi, J. Y., Kim, H. C., & Moon, K. D. (2021). Geographical origin discriminant analysis of Chia seeds (Salvia hispanica L.) using hyperspectral imaging. *Journal of Food Composition and Analysis, 101*.

Crosson, B., Ford, A., McGregor, K. M., Meinzer, M., Cheshkov, S., Li, X., Walker-Batson, D., & Briggs, R. W. (2010). Functional imaging and related techniques: An introduction for rehabilitation researchers. *Journal of Rehabilitation Research and Development, 47*(2), vii–xxxiv.

Daly, K. R., Mooney, S. J., Bennett, M. J., Crout, N. M., Roose, T., & Tracy, S. R. (2015). Assessing the influence of the rhizosphere on soil hydraulic properties using X-ray computed tomography and numerical modelling. *Journal of Experimental Botany, 66*(8), 2305–2314.

Del Pozo, S., Rodríguez-Gonzálvez, P., Hernández-López, D., & Felipe-García, B. (2014). Vicarious radiometric calibration of a multispectral camera on board an unmanned aerial system. *Remote Sensing, 6*(3), 1918–1937.

Deng, L., Hao, X., Mao, Z., Yan, Y., Sun, J., & Zhang, A. (2018). A subband radiometric calibration method for UAV-based multispectral remote sensing. *IEEE Journal of Selected Topics in Applied Earth Observations and Remote Sensing, 11*(8), 2869–2880.

Dwivedi, S., Sahrawat, K., Upadhyaya, H., & Ortiz, R. (2013). Chapter one—Food, nutrition and agrobiodiversity under global climate change. In D. L. Sparks (Ed.), *Advances in agronomy* (pp. 1–128). Academic Press.

Eberle, S., Gilli, C., Fleury, Y., & Camps, C. (2020). Hairy root disease: Digitized images based method to monitor the hairy root development on eggplants growing on soilless substrate in greenhouse. In *Acta Horticulturae* (pp. 313–321).

Fasoula, D. A., Ioannides, I. M., & Omirou, M. (2020). Phenotyping and plant breeding: Overcoming the barriers. *Frontiers in Plant Science, 10*, 1713.

Feng, A., Zhou, J., Vories, E., & Sudduth, K. A. (2020). Evaluation of cotton emergence using UAV-based imagery and deep learning. *Computers and Electronics in Agriculture, 177*, 105711.

Fiorani, F., & Schurr, U. (2013). Future scenarios for plant phenotyping. *Annual Review of Plant Biology, 64*(1), 267–291.

Flavel, R. J., Guppy, C. N., Tighe, M., Watt, M., McNeill, A., & Young, I. M. (2012). Non-destructive quantification of cereal roots in soil using high-resolution X-ray tomography. *Journal of Experimental Botany, 63*(7), 2503–2511.

Galloza, M. S., Crawford, M. M., & Heathman, G. C. (2013). Crop residue modeling and mapping using Landsat, ALI, Hyperion and airborne remote sensing data. *IEEE Journal of Selected Topics in Applied Earth Observations and Remote Sensing, 6*(2), 446–456.

Ganthaler, A., Losso, A., & Mayr, S. (2018). Using image analysis for quantitative assessment of needle bladder rust disease of Norway spruce. *Plant Pathology, 67*(5), 1122–1130.

Gomes, M. T. G., da Luz, A. C., dos Santos, M. R., Batitucci, M. C. P., Silva, D. M., & Falqueto, A. R. (2012). Drought tolerance of passion fruit plants assessed by the OJIP chlorophyll a fluorescence transient. *Scientia Horticulturae, 142*, 49–56.

Goodfellow, I., et al. (2016). Deep learning, MIT Press.

Guo, W., Rage, U. K., & Ninomiya, S. (2013). Illumination invariant segmentation of vegetation for time series wheat images based on decision tree model. *Computers and Electronics in Agriculture, 96*, 58–66.

Halbritter, A. H., De Boeck, H. J., Eycott, A. E., Reinsch, S., Robinson, D. A., Vicca, S., Berauer, B., Christiansen, C. T., Estiarte, M., Grünzweig, J. M., Gya, R., Hansen, K., Jentsch, A., Lee, H., Linder, S., Marshall, J., Peñuelas, J., Kappel Schmidt, I., Stuart-Haëntjens, E., … Vandvik, V. (2020). The handbook for standardized field and laboratory measurements in terrestrial climate change experiments and observational studies (ClimEx). *Methods in Ecology and Evolution, 11*(1), 22–37.

Han, X., Thomasson, J., Xiang, Y., Gharakhani, H., Yadav, P., & Rooney, W. (2019). Multifunctional ground control points with a wireless network for communication with a UAV. *Sensors, 19*(13), 2852.

Hasan, R. I., Yusuf, S., & Alzubaidi, L. (2020). Review of the state of the art of deep learning for plant diseases: A broad analysis and discussion. *Plants, 9*.

Hassan, M. A., Yang, M., Fu, L., Rasheed, A., Zheng, B., Xia, X., Xiao, Y., & He, Z. (2019). Accuracy assessment of plant height using an unmanned aerial vehicle for quantitative genomic analysis in bread wheat. *Plant Methods, 15*(1), 37.

Hincks, J. (2018). The world is headed for a food security crisis. Here's how we can avert it. *Times*. Retrieved from https://www.un.org/development/desa/en/news/population/world-population-prospects-2017.html

Hively, W., Lamb, B., Daughtry, C., Shermeyer, J., McCarty, G., & Quemada, M. (2018). Mapping crop residue and tillage intensity using WorldView-3 satellite shortwave infrared residue indices. *Remote Sensing, 10*(10), 1657.

Holman, F., Riche, A., Michalski, A., Castle, M., Wooster, M., & Hawkesford, M. (2016). High throughput field phenotyping of wheat plant height and growth rate in field plot trials using UAV based remote sensing. *Remote Sensing, 8*(12), 1031.

Hu, P., Chapman, S. C., Wang, X., Potgieter, A., Duan, T., Jordan, D., Guo, Y., & Zheng, B. (2018). Estimation of plant height using a high throughput phenotyping platform based on unmanned aerial vehicle and self-calibration: Example for sorghum breeding. *European Journal of Agronomy, 95*, 24–32.

Hughes, N., Askew, K., Scotson, C. P., Williams, K., Sauze, C., Corke, F., Doonan, J. H., & Nibau, C. (2017). Non-destructive, high-content analysis of wheat grain traits using X-ray micro computed tomography. *Plant Methods, 13*(1), 76.

Humplík, J. F., Lazár, D., Husičková, A., & Spíchal, L. (2015). Automated phenotyping of plant shoots using imaging methods for analysis of plant stress responses–a review. *Plant Methods, 11*(1), 29.

Hunt, E. R., Cavigelli, M., Daughtry, C. S., Mcmurtrey, J. E., & Walthall, C. L. (2005). Evaluation of digital photography from model aircraft for remote sensing of crop biomass and nitrogen status. *Precision Agriculture, 6*(4), 359–378.

Hunt, E. R., Jr., Doraiswamy, P. C., McMurtrey, J. E., Daughtry, C. S., Perry, E. M., & Akhmedov, B. (2013). A visible band index for remote sensing leaf chlorophyll content at the canopy scale. *International Journal of Applied Earth Observation and Geoinformation, 21*, 103–112.

Hunt, E., Li, L., Yilmaz, M., & Jackson, T. (2011). Comparison of vegetation water contents derived from shortwave-infrared and passive-microwave sensors over central Iowa. *Remote Sensing of Environment, 115*, 2376–2383.

Jahnke, S., Menzel, M. I., Van Dusschoten, D., Roeb, G. W., Bühler, J., Minwuyelet, S., Blümler, P., Temperton, V. M., Hombach, T., & Streun, M. (2009). Combined MRI–PET dissects dynamic changes in plant structures and functions. *The Plant Journal, 59*(4), 634–644.

James, G., et al. (2013). An introduction to statistical learning, Springer.

Jansen, M., Gilmer, F., Biskup, B., Nagel, K. A., Rascher, U., Fischbach, A., Briem, S., Dreissen, G., Tittmann, S., & Braun, S. (2009). Simultaneous phenotyping of leaf growth and chlorophyll fluorescence via GROWSCREEN FLUORO allows detection of stress tolerance in Arabidopsis thaliana and other rosette plants. *Functional Plant Biology, 36*(11), 902–914.

Jarquin, D., Howard, R., Xavier, A., & Choudhury, S. D. (2020). Predicting yield by modeling interactions between canopy coverage image data, genotypic and environmental information for soybeans. In *Intelligent image analysis for plant phenotyping* (pp. 267–286). CRC Press.

Jeon, G. (2014). Color image enhancement by histogram equalization in heterogeneous color space. Int. J. *Multimedia Ubiquitous Eng, 9*(7), 309–318.

Ji, L., Zhang, L., Wylie, B. K., & Rover, J. (2011). On the terminology of the spectral vegetation index (NIR− SWIR)/(NIR+ SWIR). *International Journal of Remote Sensing, 32*(21), 6901–6909.

Jiang, Y., Li, C., Robertson, J. S., Sun, S., Xu, R., & Paterson, A. H. (2018). GPhenoVision: A ground mobile system with multi-modal imaging for field-based high throughput phenotyping of cotton. *Scientific Reports, 8*(1), 1213.

Jimenez-Berni, J. A., Deery, D. M., Rozas-Larraondo, P., Condon, A. G., Rebetzke, G. J., James, R. A., Bovill, W. D., Furbank, R. T., & Sirault, X. R. R. (2018). High throughput determination of plant height, ground cover, and above-ground biomass in wheat with LiDAR. *Frontiers in Plant Science, 9*(1), 237–255.

Jin, H., Köppl, C. J., Fischer, B. M. C., Rojas-Conejo, J., Johnson, M. S., Morillas, L., Lyon, S. W., Durán-Quesada, A. M., Suárez-Serrano, A., Manzoni, S., & Garcia, M. (2021). Drone-based hyperspectral and thermal imagery for quantifying upland rice productivity and water use efficiency after biochar application. *Remote Sensing, 13*(10), 1866.

Jin, X., Zarco-Tejada, P. J., Schmidhalter, U., Reynolds, M. P., Hawkesford, M. J., Varshney, R. K., Yang, T., Nie, C., Li, Z., & Ming, B. (2020). High-throughput estimation of crop traits: A review of ground and aerial phenotyping platforms. *IEEE Geoscience and Remote Sensing Magazine, 9*(1), 200–231.

Jones, H. G. (2004). Application of thermal imaging and infrared sensing in plant physiology and ecophysiology. In *Advances in botanical research* (pp. 107–163). Academic Press.

Jones, H. G., Serraj, R., Loveys, B. R., Xiong, L., Wheaton, A., & Price, A. H. (2009). Thermal infrared imaging of crop canopies for the remote diagnosis and quantification of plant responses to water stress in the field. *Functional Plant Biology, 36*(11), 978–989.

Kim, S. J. (2008). *Radiometric calibration methods from image sequences*. The University of North Carolina at Chapel Hill.

Kjaer, K. H., & Ottosen, C. O. (2015). 3D laser triangulation for plant phenotyping in challenging environments. *Sensors, 15*(6), 13533–13547.

Knauer, U., Matros, A., Petrovic, T., Zanker, T., Scott, E. S., & Seiffert, U. (2017). Improved classification accuracy of powdery mildew infection levels of wine grapes by spatial-spectral analysis of hyperspectral images. *Plant Methods, 13*(1), 47.

Kolláth, Z., Cool, A., Jechow, A., Kolláth, K., Száz, D., & Tong, K. P. (2020). Introducing the dark sky unit for multi-spectral measurement of the night sky quality with commercial digital cameras. *Journal of Quantitative Spectroscopy and Radiative Transfer, 253*, 107162.

Konanz, S., Kocsányi, L., & Buschmann, C. (2014). Advanced multi-color fluorescence imaging system for detection of biotic and abiotic stresses in leaves. *Agriculture, 4*(2), 79–95.

Leinonen, I., & Jones, H. G. (2004). Combining thermal and visible imagery for estimating canopy temperature and identifying plant stress. *Journal of Experimental Botany, 55*(401), 1423–1431.

Lenk, S., Chaerle, L., Pfündel, E. E., Langsdorf, G., Hagenbeek, D., Lichtenthaler, H. K., Van Der Straeten, D., & Buschmann, C. (2006). Multispectral fluorescence and reflectance imaging at the leaf level and its possible applications. *Journal of Experimental Botany, 58*(4), 807–814.

Li, D., Quan, C., Song, Z., Li, X., Yu, G., Li, C., & Muhammad, A. (2021). High-throughput plant phenotyping platform (HT3P) as a novel tool for estimating agronomic traits from the lab to the field. *Frontiers in Bioengineering and Biotechnology, 8*(1533).

Li, Y., Sun, J., Wu, X., Chen, Q., Lu, B., & Dai, C. (2019). Detection of viability of soybean seed based on fluorescence hyperspectra and CARS-SVM-AdaBoost model. *Journal of Food Processing and Preservation, 43*(12).

Li, L., Zhang, Q., & Huang, D. (2014). A review of imaging techniques for plant phenotyping. *Sensors, 14*(11), 20078–20111.

Liang, Z., Pandey, P., Stoerger, V., Xu, Y., Qiu, Y., Ge, Y., & Schnable, J. C. (2018). Conventional and hyperspectral time-series imaging of maize lines widely used in field trials. *GigaScience, 7*(2), 1–11.

Lichtenthaler, H. K., & Miehé, J. A. (1997). Fluorescence imaging as a diagnostic tool for plant stress. *Trends in Plant Science, 2*(8), 316–320.

Lillesand, T. M., Kiefer, R. W., & Chipman, J. W. (2004). *Remote sensing and image interpretation* (5th ed.). Wiley.

López-Maestresalas, A., Keresztes, J. C., Goodarzi, M., Arazuri, S., Jarén, C., & Saeys, W. (2016). Non-destructive detection of blackspot in potatoes by Vis-NIR and SWIR hyperspectral imaging. *Food Control, 70*, 229–241.

Lu, B., Dao, P. D., Liu, J., He, Y., & Shang, J. (2020). Recent advances of hyperspectral imaging technology and applications in agriculture. *Remote Sensing, 12*(16), 2659.

Ludovisi, R., Tauro, F., Salvati, R., Khoury, S., Mugnozza Scarascia, G., & Harfouche, A. (2017). UAV-based thermal imaging for high-throughput field phenotyping of black poplar response to drought. *Frontiers in Plant Science, 8*, 1681.

Maimaitijiang, M., Ghulam, A., Sidike, P., Hartling, S., Maimaitiyiming, M., Peterson, K., Shavers, E., Fishman, J., Peterson, J., Kadam, S., Burken, J., & Fritschi, F. (2017). Unmanned aerial system (UAS)-based phenotyping of soybean using multi-sensor data fusion and extreme learning machine. *ISPRS Journal of Photogrammetry and Remote Sensing, 134*, 43–58.

Marchant, J. A., Tillett, R. D., & Brivot, R. (1998). Real-time segmentation of plants and weeds. *Real-Time Imaging, 4*(4), 243–253.

Martinez-Guanter, J., Ribeiro, Á., Peteinatos, G. G., Pérez-Ruiz, M., Gerhards, R., Bengochea-Guevara, J. M., Machleb, J., & Andújar, D. (2019). Low-cost three-dimensional modeling of crop plants. *Sensors, 19*(13), 2883.

Martins, S. M., Brito, G. G., Gonçalves, W. C., Tripode, B. M. D., Lartaud, M., Duarte, J. B., Morello, C. L., & Giband, M. (2020). PhenoRoots: An inexpensive non-invasive phenotyping system to assess the variability of the root system architecture. *Scientia Agricola, 77*(5).

Mastrodimos, N., Lentzou, D., Templalexis, C., Tsitsigiannis, D., & Xanthopoulos, G. (2019). Development of thermography methodology for early diagnosis of fungal infection in table grapes: The case of Aspergillus carbonarius. *Computers and Electronics in Agriculture, 165*, 104972.

Maxwell, K., & Johnson, G. N. (2000). Chlorophyll fluorescence—A practical guide. *Journal of Experimental Botany, 51*(345), 659–668.

Melkus, G., Rolletschek, H., Fuchs, J., Radchuk, V., Grafahrend-Belau, E., Sreenivasulu, N., Rutten, T., Weier, D., Heinzel, N., & Schreiber, F. (2011). Dynamic 13C/1H NMR imaging uncovers sugar allocation in the living seed. *Plant Biotechnology Journal, 9*(9), 1022–1037.

Meyer, G. E., Camargo Neto, J., Jones, D. D., & Hindman, T. W. (2004). Intensified fuzzy clusters for classifying plant, soil, and residue regions of interest from color images. *Computers and Electronics in Agriculture, 42*(3), 161–180.

Mir, R. R., Reynolds, M., Pinto, F., Khan, M. A., & Bhat, M. A. (2019). High-throughput phenotyping for crop improvement in the genomics era. *Plant Science, 282*, 60–72.

Moghimi, A., Yang, C., & Anderson, J. A. (2020). Aerial hyperspectral imagery and deep neural networks for high-throughput yield phenotyping in wheat. *Computers and Electronics in Agriculture, 172*, 105299.

Möller, M., Alchanatis, V., Cohen, Y., Meron, M., Tsipris, J., Naor, A., Ostrovsky, V., Sprintsin, M., & Cohen, S. (2007). Use of thermal and visible imagery for estimating crop water status of irrigated grapevine. *Journal of Experimental Botany, 58*(4), 827–838.

Moradi, A., Oswald, S., Nordmeyer-Massner, J., Pruessmann, K. P., Robinson, B., & Schulin, R. (2010). Analysis of nickel concentration profiles around the roots of the hyperaccumulator plant Berkheya coddii using MRI and numerical simulations. *Plant and Soil, 328*(1), 291–302.

Mwinuka, P. R., Mbilinyi, B. P., Mbungu, W. B., Mourice, S. K., Mahoo, H. F., & Schmitter, P. (2021). The feasibility of hand-held thermal and UAV-based multispectral imaging for canopy water status assessment and yield prediction of irrigated African eggplant (Solanum aethopicum L). *Agricultural Water Management, 245*, 106584.

Naik, H. S., Zhang, J., Lofquist, A., Assefa, T., Sarkar, S., Ackerman, D., Singh, A., Singh, A. K., & Ganapathysubramanian, B. (2017). A real-time phenotyping framework using machine learning for plant stress severity rating in soybean. *Plant Methods, 13*(1), 1–12.

Olsen, J. L., Ceccato, P., Proud, S. R., Fensholt, R., Grippa, M., Mougin, E., Ardö, J., & Sandholt, I. (2013). Relation between seasonally detrended shortwave infrared reflectance data and land surface moisture in semi-arid Sahel. *Remote Sensing, 5*(6), 2898–2927.

Ortiz-Bustos, C. M., Pérez-Bueno, M. L., Barón, M., & Molinero-Ruiz, L. (2016). Fluorescence imaging in the red and far-red region during growth of sunflower plantlets. Diagnosis of the early infection by the parasite Orobanche cumana. *Frontiers in Plant Science, 7*(884), 1–10.

Otsu, N. (1979). A threshold selection method from gray-level histograms. *IEEE Transactions on Systems, Man, and Cybernetics, 9*(1), 62–66.

Pandey, P., Ge, Y., Stoerger, V., & Schnable, J. C. (2017). High throughput in vivo analysis of plant leaf chemical properties using hyperspectral imaging. *Frontiers in Plant Science, 8*, 1348.

Parmley, K. A., et al. (2019). Machine Learning Approach for Prescriptive Plant Breeding. Scientific Reports 9(1), 17132.

Paulus, S., Dupuis, J., Riedel, S., & Kuhlmann, H. (2014). Automated analysis of barley organs using 3D laser scanning: An approach for high throughput phenotyping. *Sensors, 14*(7), 12670–12686.

Pereyra-Irujo, G. A., Gasco, E. D., Peirone, L. S., & Aguirrezábal, L. A. N. (2012). GlyPh: A low-cost platform for phenotyping plant growth and water use. *Functional Plant Biology, 39*(11), 905–913.
Phalempin, M., Lippold, E., Vetterlein, D., & Schlüter, S. (2021). An improved method for the segmentation of roots from X-ray computed tomography 3D images: Rootine v.2. *Plant Methods, 17*(1), 39.
Raj, R., Walker, J. P., Pingale, R., Nandan, R., Naik, B., & Jagarlapudi, A. (2021). Leaf area index estimation using top-of-canopy airborne RGB images. *International Journal of Applied Earth Observation and Geoinformation, 96*, 102282.
Rascher, U., Blossfeld, S., Fiorani, F., Jahnke, S., Jansen, M., Kuhn, A. J., Matsubara, S., Märtin, L. L., Merchant, A., & Metzner, R. (2011). Non-invasive approaches for phenotyping of enhanced performance traits in bean. *Functional Plant Biology, 38*(12), 968–983.
Reid, J. F., & Searcy, S. W. (1987). Vision-based guidance of an agricultural tractor. *IEEE Control Systems Magazine, 7*(2), 39–43.
Rojas-Lima, J. E., Domínguez-Pacheco, A., Hernández-Aguilar, C., Hernández-Simón, L. M., & Cruz-Orea, A. (2021). Statistical methods for the analysis of thermal images obtained from corn seeds. *SN Applied Sciences, 3*(4), 499.
Ryu, J., Hwang, B. G., Kim, Y. X., & Lee, S. J. (2016). Direct observation of local xylem embolisms induced by soil drying in intact Zea mays leaves. *Journal of Experimental Botany, 67*(9), 2617–2626.
Schreiber, U., Schliwa, U., & Bilger, W. (1986). Continuous recording of photochemical and non-photochemical chlorophyll fluorescence quenching with a new type of modulation fluorometer. *Photosynthesis Research, 10*(1–2), 51–62.
Serbin, G., Daughtry, C. S., Hunt, E. R., Brown, D. J., & McCarty, G. W. (2009). Effect of soil spectral properties on remote sensing of crop residue cover. *Soil Science Society of America Journal, 73*(5), 1545–1558.
Serôdio, J., Schmidt, W., Frommlet, J. C., Christa, G., & Nitschke, M. R. (2018). An LED-based multi-actinic illumination system for the high throughput study of photosynthetic light responses. *PeerJ, 6*, e5589.
Shafiekhani, A., Kadam, S., Fritschi, F. B., & DeSouza, G. N. (2017). Vinobot and Vinoculer: Two robotic platforms for high-throughput field phenotyping. *Sensors, 17*(1), 214.
Silván-Cárdenas, J., Corona, N., Pizaña, J., Núñez, J. M., & Madrigal, J. (2015). Geospatial technologies to support coniferous forests research and conservation efforts in Mexico. In R. P. Weber (Ed.), *Old-growth forests and coniferous forests: Ecology, habitat and conservation* (pp. 67–123).
Staton, M. (2017). *What is the relationship between soybean maturity group and yield?* Michigan State University Extension.
Strange, H., Zwiggelaar, R., Sturrock, C., Mooney, S. J., & Doonan, J. H. (2014). Automatic estimation of wheat grain morphometry from computed tomography data. *Functional Plant Biology, 42*(5), 452–459.
Suarez, L., Zhang, P., Sun, J., Wang, Y., Poblete, T., Hornero, A., & Zarco-Tejada, P. J. (2021). Assessing wine grape quality parameters using plant traits derived from physical model inversion of hyperspectral imagery. *Agricultural and Forest Meteorology, 306*, 108445.
Swathandran, S., & Aslam, M. A. M. (2019). Assessing the role of SWIR band in detecting agricultural crop stress: A case study of Raichur district, Karnataka, India. *Environmental Monitoring and Assessment, 191*(7), 442.
Tellaeche, A., Burgos-Artizzu, X. P., Pajares, G., & Ribeiro, A. (2008). A vision-based method for weeds identification through the Bayesian decision theory. *Pattern Recognition, 41*(2), 521–530.
Thenkabail, P. S., & Lyon, J. G. (2016). *Hyperspectral remote sensing of vegetation*. CRC Press.
Thomas, S., Wahabzada, M., Kuska, M. T., Rascher, U., & Mahlein, A.-K. (2017). Observation of plant–pathogen interaction by simultaneous hyperspectral imaging reflection and transmission measurements. *Functional Plant Biology, 44*(1), 23.

Tracy, S. R., Black, C. R., Roberts, J. A., Dodd, I. C., & Mooney, S. J. (2015). Using X-ray computed tomography to explore the role of abscisic acid in moderating the impact of soil compaction on root system architecture. *Environmental and Experimental Botany, 110*, 11–18.

Tracy, S. R., Gómez, J. F., Sturrock, C. J., Wilson, Z. A., & Ferguson, A. C. (2017). Non-destructive determination of floral staging in cereals using X-ray micro computed tomography (μCT). *Plant Methods, 13*(1), 1–12.

Tresch, L., Mu, Y., Itoh, A., Kaga, A., Taguchi, K., Hirafuji, M., Ninomiya, S., & Guo, W. (2019). Easy MPE: Extraction of quality microplot images for UAV-based high-throughput field phenotyping. *Plant Phenomics, 2019*, 2591849.

Vadez, V., Kholová, J., Hummel, G., Zhokhavets, U., Gupta, S., & Hash, C. T. (2015). LeasyScan: A novel concept combining 3D imaging and lysimetry for high-throughput phenotyping of traits controlling plant water budget. *Journal of Experimental Botany, 66*(18), 5581–5593.

Van As, H., & Van Duynhoven, J. (2013). MRI of plants and foods. *Journal of Magnetic Resonance, 229*, 25–34.

Vescovo, L., Wohlfahrt, G., Balzarolo, M., Pilloni, S., Sottocornola, M., Rodeghiero, M., & Gianelle, D. (2012). New spectral vegetation indices based on the near-infrared shoulder wavelengths for remote detection of grassland phytomass. *International Journal of Remote Sensing, 33*(7), 2178–2195.

Vázquez-Arellano, M., Griepentrog, H. W., Reiser, D., & Paraforos, D. S. (2016). 3-D imaging systems for agricultural applications—a review. Sensors, 16(5), 618.

Wang, D., Li, W., Liu, X., Li, N., & Zhang, C. (2020). UAV environmental perception and autonomous obstacle avoidance: A deep learning and depth camera combined solution. *Computers and Electronics in Agriculture, 175*, 105523.

Wang, H., Lin, Y., Wang, Z., Yao, Y., Zhang, Y., & Wu, L. (2017). Validation of a low-cost 2D laser scanner in development of a more-affordable mobile terrestrial proximal sensing system for 3D plant structure phenotyping in indoor environment. *Computers and Electronics in Agriculture, 140*, 180–189.

Wang, Z., Nie, C., Wang, H., Ao, Y., Jin, X., Yu, X., Bai, Y., Liu, Y., Shao, M., Cheng, M., Liu, S., Wang, S., & Tuohuti, N. (2021). Detection and analysis of degree of maize lodging using UAV-RGB image multi-feature factors and various classification methods. *ISPRS International Journal of Geo-Information, 10*(5), 309.

Wang, H., Qian, X., Zhang, L., Xu, S., Li, H., Xia, X., Dai, L., Xu, L., Yu, J., & Liu, X. (2018). A method of high throughput monitoring crop physiology using chlorophyll fluorescence and multispectral imaging. *Frontiers in Plant Science, 9*, 407.

Wang, L., & Qu, J. J. (2007). NMDI: A normalized multi-band drought index for monitoring soil and vegetation moisture with satellite remote sensing. *Geophysical Research Letters, 34*(20).

Windt, C. W., Vergeldt, F. J., De Jager, P. A., & Van As, H. (2006). MRI of long-distance water transport: A comparison of the phloem and xylem flow characteristics and dynamics in poplar, castor bean, tomato and tobacco. *Plant, Cell & Environment, 29*(9), 1715–1729.

Wu, D., Wu, D., Feng, H., Duan, L., Dai, G., Liu, X., Wang, K., Yang, P., Chen, G., Gay, A. P., Doonan, J. H., Niu, Z., Xiong, L., & Yang, W. (2021). A deep learning-integrated micro-CT image analysis pipeline for quantifying rice lodging resistance-related traits. *Plant Communications, 2*(2), 100165.

Xie, C., & Yang, C. (2020). A review on plant high-throughput phenotyping traits using UAV-based sensors. *Computers and Electronics in Agriculture, 178*, 105731.

Yang, W., Feng, H., Zhang, X., Zhang, J., Doonan, J. H., Batchelor, W. D., Xiong, L., & Yan, J. (2020). Crop phenomics and high-throughput phenotyping: Past decades, current challenges, and future perspectives. *Molecular Plant, 13*(2), 187–214.

Yang, C., & Hoffmann, W. C. (2015). Low-cost single-camera imaging system for aerial applicators. *Journal of Applied Remote Sensing, 9*(1), 096064.

Yang, H., Inagaki, T., Ma, T., & Tsuchikawa, S. (2017a). High-resolution and non-destructive evaluation of the spatial distribution of nitrate and its dynamics in spinach (Spinacia oleracea L.) leaves by near-infrared hyperspectral imaging. *Frontiers in Plant Science, 8*, 1937.

Yang, G., Liu, J., Zhao, C., Li, Z., Huang, Y., Yu, H., Xu, B., Yang, X., Zhu, D., Zhang, X., Zhang, R., Feng, H., Zhao, X., Li, Z., Li, H., & Yang, H. (2017b). Unmanned aerial vehicle remote sensing for field-based crop phenotyping: Current status and perspectives. *Frontiers in Plant Science, 8*, 1111.

Yuan, L., Zhang, J., Shi, Y., Nie, C., Wei, L., & Wang, J. (2014). Damage mapping of powdery mildew in winter wheat with high-resolution satellite image. *Remote Sensing, 6*(5), 3611–3623.

Zhang, J., Cheng, T., Guo, W., Xu, X., Qiao, H., Xie, Y., & Ma, X. (2021). Leaf area index estimation model for UAV image hyperspectral data based on wavelength variable selection and machine learning methods. *Plant Methods, 17*(1), 49.

Zhang, Y., Du, J., Wang, J., Ma, L., Lu, X., Pan, X., Guo, X., & Zhao, C. (2018). High-throughput micro-phenotyping measurements applied to assess stalk lodging in maize (Zea mays L.). *Biological Research, 51*, 40.

Zhang, Y., & Zhang, N. (2018). Imaging technologies for plant high-throughput phenotyping: A review. *Frontiers of Agricultural Science and Engineering, 5*(4), 406–419.

Zhao, R., An, L., Song, D., Li, M., Qiao, L., Liu, N., & Sun, H. (2021). Detection of chlorophyll fluorescence parameters of potato leaves based on continuous wavelet transform and spectral analysis. *Spectrochimica Acta - Part A: Molecular and Biomolecular Spectroscopy, 259*, 119768.

Zhao, C., Zhang, Y., Du, J., Guo, X., Wen, W., Gu, S., Wang, J., & Fan, J. (2019). Crop phenomics: Current status and perspectives. *Frontiers in Plant Science, 10*, 714–714.

Zhou, J., Chen, H., Zhou, J., Fu, X., Ye, H., & Nguyen, H. T. (2018). Development of an automated phenotyping platform for quantifying soybean dynamic responses to salinity stress in greenhouse environment. *Computers and Electronics in Agriculture, 151*, 319–330.

Zhou, J., Fu, X., Zhou, S., Zhou, J., Ye, H., & Nguyen, H. T. (2019). Automated segmentation of soybean plants from 3D point cloud using machine learning. *Computers and Electronics in Agriculture, 162*, 143–153.

Zhou, J., Mou, H., Zhou, J., Ali, M. L., Ye, H., Chen, P., & Nguyen, H. T. (2021a). Qualification of soybean responses to flooding stress using UAV-based imagery and deep learning. *Plant Phenomics, 2021*, 9892570.

Zhou, S., Mou, H., Zhou, J., Zhou, J., Ye, H., & Nguyen, H. T. (2021b). Development of an automated plant phenotyping system for evaluation of salt tolerance in soybean. *Computers and Electronics in Agriculture, 182*, 106001.

Zhou, J., Zhou, J., Ye, H., Ali, M. L., Nguyen, H. T., & Chen, P. (2020). Classification of soybean leaf wilting due to drought stress using UAV-based imagery. *Computers and Electronics in Agriculture, 175*, 105576.

Chapter 5
Data-Driven Modeling for Crop Growth in Plant Factories

Zhixian Lin, Shanye Wang, Rongmei Fu, Kuan-Chong Ting, and Tao Lin

5.1 Introduction

World crop production is becoming increasingly threatened by abnormal climates, land shortages, and insufficient labor. According to the United Nations, the world population will increase to nearly 10 billion people by 2050, and 68% of this population will live in cities (United Nations, 2019). There is a high demand for fresh and healthy foods, especially in the urban population. An efficient form of agricultural cultivation is required to address this challenge.

Plant factories, also known as vertical farms, are fast evolving agricultural systems that integrate a variety of modern technologies; they can cultivate crops on multiple layers and achieve high-efficiency and -quality production (Graamans et al., 2018; Kozai, 2013; Kozai et al., 2019). Besides, plant factories are almost completely insulated from the external climate and allow internal environmental factors such as light, temperature, humidity, CO_2 concentration, and nutrient solution to be controlled precisely and automatically; thus, they are rarely constrained by climatic conditions and geographical location. Recently, plant factories have achieved rapid development and created many opportunities for commercial and scientific research. In 2021, the global plant factory market was estimated to be worth USD 121.8 billion, and it is expected to reach USD 172.5 billion by 2026, with a compound annual growth rate of 7.2% (Markets & Markets, 2021). According

Z. Lin · S. Wang · R. Fu · T. Lin (✉)
College of Biosystems Engineering and Food Science, Zhejiang University, Hangzhou, Zhejiang, China
e-mail: lintao1@zju.edu.cn

K.-C. Ting
Department of Agricultural and Biological Engineering, University of Illinois at Urbana-Champaign, Urbana, IL, USA

S. Ma et al. (eds.), *Sensing, Data Managing, and Control Technologies for Agricultural Systems*, Agriculture Automation and Control,
https://doi.org/10.1007/978-3-031-03834-1_5

to data from the last three years, the number of plant factories in commercial production is roughly estimated at over 200 in Japan, about 200 in China, and over 500 around the world (Kozai, 2018; Yang, 2019).

The current problems of plant factory development are the high initial and production costs. According to Kozai et al. (2019), the cost of electricity and labor accounts for more than 50% of a plant factory's production costs. Although the yield of crops grown in plant factories has been greatly improved compared with traditional agricultural methods, the unit production costs are still high. Moreover, the quality of crops often cannot satisfy requirements. The yield and quality of crops are affected by genetic, physical, chemical, and biological environmental factors during crop growth, as well as cultivation methods (Kozai et al., 2019). Theoretically, it is relatively easier to control and improve the yield and quality of crops in plant factories. However, incomplete understanding of the causal relationship between the environment and yield/quality leads to the necessity for improved environmental control and growth prediction in current plant factories.

Data science has played an increasingly important role in plant factories, following the rapid development and application of information technology, intelligent control, sensors, artificial intelligence, and other advanced technologies. Massive crop-related data have surged from equipped sensors and monitoring hardware in plant factories. Digitization and automation by data-driven modeling are becoming increasingly important in reducing production costs and improving the efficiency of plant factories. This chapter describes the applications and prospects of data-driven modeling in plant factories from three perspectives: environmental factor sensing, crop growth monitoring, and crop growth models.

5.2 Environmental Factor Acquisition and Sensing

The environment significantly affects crop quality and yield in plant factories. Therefore, it is necessary to understand the effects of important environmental factors and their measurement methods. The major environmental factors that affect plant growth in plant factories are light, temperature, relative air humidity, carbon dioxide concentration, and air current speed (Ahmed et al., 2020). The effects of environmental factors are shown in Fig. 5.1. The effects of environmental factors on crop growth primarily operate via the stomatal conductance, photosynthetic rate, and transpiration rate. This chapter presents the properties and effects of these environmental factors. In addition, a measurement system for data accumulation for subsequent data-driven modeling is described.

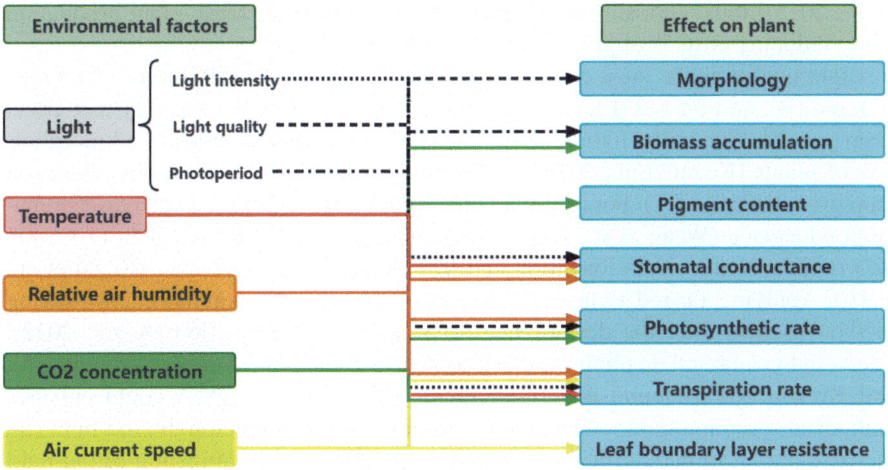

Fig. 5.1 Effects of environmental factors on plants

5.2.1 *Environmental Factors that Affect Plant Growth*

Light

Light is one of the most critical environmental factors involved in plant growth. The most important characteristics of light are light intensity, light quality, and photoperiod. The cost of lighting comprises a large portion of plant factories' expenses, with artificial lighting accounting for ~60% of the total energy cost of a plant factory (Yang, 2019). Understanding and choosing an appropriate light formula can significantly improve the economic benefits.

Current methods of expressing light intensity include illumination (unit: lx), photosynthetically active radiation (PAR: $W \bullet m^{-2}$), and photosynthetic photon flux density (PPFD: $\mu mol\ m^{-2}s^{-1}$). PAR describes the wavelengths of light that sit within the visible range of 400–700 nm (i.e., the light that plants can utilize in photosynthesis). The term PPFD expresses the measurement of PAR. Its value determines the quantity of PAR produced by any one lighting system over a time interval. PPFD describes the number of photons of PAR that reach or pass within a unit time and unit area; it is the most common measure in plant factories. The daily light integral (DLI: $mol\ m^{-2}d^{-1}$) is a better predictor for plant growth (Kozai, 2022); it expresses the light quantum flux of photosynthesis intercepted over a day per unit area. The conversion equations for PPFD and DLI are as follows:

$$\mathrm{DLI} = \mathrm{PPFD}\left(\mu \mathrm{mol} m^{-2} \mathrm{s}^{-1}\right) \cdot \mathrm{Photoperiod}\left(\mathrm{h}\right) \times 3600 \times 10^{-6} \tag{5.1}$$

Light intensity affects the leaf nitrate content, phytochemical accumulation, transpiration rate, and stomatal resistance (Kozai, 2022; Zhang et al., 2018a). Optimum light intensity can promote photosynthesis and further promote plant growth (Bian et al., 2015); thus, choosing a suitable light intensity to increase plant production while reducing costs is of great importance (Kozai, 2022).

Light quality is the most complex light factor. Ultraviolet (UV), blue (B), green (G), red (R), and far-red (FR) are important components of light quality that affect plant growth. Generally, light quality affects both the photosynthesis and morphology of plants (Kozai et al., 2016). Different light qualities have various effects on crop growth (Table 5.1). For example, blue light (420–470 nm) can increase stomatal conductance (Wang et al., 2009), whereas supplemental far-red light (700–800 nm) influences the stem elongation and leaf expansion of the leaves (Kozai et al., 2016). Applying far-red light near the end of the day is a popular energy-saving method used to alter the development of indoor crop production (Kozai, 2022). Green light can regulate photosynthesis and growth in the lower leaves of plants. Li and Kubota (2009) found that supplemental blue light or UV-A could enhance anthocyanin accumulation. Compared with blue light and green light, red light can increase photosynthetic efficiency (Kozai et al., 2016). A combination of red light and blue light can increase the photosynthetic rate and plant biomass (Wang et al., 2009). At present, different ratios of red light and blue light have been used as the core spectrum of light formulas in plant factories.

The photoperiod defines the relative lengths of daytime and nighttime within a 24 h period. According to their photoperiod response types forming flowers, plants can be classified into three types: long-day plants, short-day plants, and day-neutral plants. Zhang et al. (2018a) found that extending the photoperiod could increase carbohydrate production, increase growth, and improve quality. Considering both energy efficiency and plant growth rate, a photoperiod between 16 and 18 h day^{-1} was found to be optimal for lettuce in plant factories (Ahmed et al., 2020).

Table 5.1 Effects of light quality on crop growth

Light	Crop	Effect	Reference
Red	Lettuce	Promote photosynthesis and growth rate	Shimizu et al. (2011)
	Lettuce, kale, and pepper	Stimulate plant height	Naznin et al. (2019)
Blue	Lettuce	Increase anthocyanin content, stomatal conductance, leaf net photosynthetic rate	Bukhov et al. (1995), Goto (2012), Hogewoning et al. (2010)
Red and blue	Lettuce	Improve lettuce growth characteristics and the accumulation of antioxidant phenolic compounds	Son et al. (2016)
Green		Stimulate photosynthesis deep in the canopy providing to carbon gain	Smith et al. (2017)
Far-red	Lettuce	Enhance leaf area	Li and Kubota (2009)
		Stem elongation, leaf expansion	Kozai et al. (2016)
UV	Lettuce	Increases phenolics; accumulates phytochemicals	Bian et al. (2015), Lee et al. (2014)

Temperature

Every plant has three critical temperature levels: lowest, highest, and optimum. When the ambient temperature is below the lowest temperature or higher than the highest temperature, the plant cannot grow normally. Plants grow fastest at their optimum temperature. Temperature has significant effects on photosynthesis, respiration, and transportation, as well as the accumulation of photosynthetic products and the development and growth of various organs. Within a suitable temperature range, all biological processes of a plant increase when the temperature rises (Chowdhury et al., 2021). In addition, temperature is another important factor (besides lighting) that affects energy costs. Therefore, the optimal, crop-specific management of temperature is essential (Yang, 2019).

Relative Air Humidity

Relative air humidity is based upon the maximum amount of water that air can retain at a given temperature and pressure; it represents the degree to which moist air approaches saturation. It is usually expressed as a percentage or ratio of a given water vapor content to the maximum value at a given temperature. Relative air humidity primarily affects the leaf transpiration rate and nutrient uptake (Ahmed et al., 2020; Kozai et al., 2016). It affects the vapor pressure deficit between plant leaves and the surrounding air, which influences plant transpiration and photosynthesis. At a low relative air humidity, the leaves evaporate more water, which results in closed stomata (Ahmed et al., 2020). Therefore, a relatively low relative air humidity will inhibit photosynthesis and plant growth rates (Ryu et al., 2014). In contrast, at a high relative air humidity, the evaporation from leaves is small, and the roots' nutrient solution absorption rate is reduced, which in turn affects the plant's photosynthesis.

Carbon Dioxide Concentration

Carbon dioxide represents the carbon source for plant photosynthesis. Plants absorb CO_2 during photosynthesis and release it during respiration. CO_2 affects several physiological processes involved in plant growth. An excessively high carbon dioxide concentration (or one maintained at a high level for a long time) will cause the stomata to close and the photosynthetic rate to decrease. High carbon dioxide concentrations also reduce the transpiration rate by increasing stomatal resistance. Within an appropriate range, increasing carbon dioxide concentration can increase the photosynthetic rate. Different crops have varied sensitivities to carbon dioxide (Park & Lee, 2001). For example, the maturity and yield of lettuce are sensitive to CO_2.

Air Current Speed

The air current speed is defined as the distance that air travels over a certain period of time (Kozai et al., 2016). It is an important factor in the production design of plant factories. The air current speed primarily affects transpiration rate, photosynthesis, microenvironment, and quality (Kitaya et al., 2000; Yang, 2019). An appropriate air current speed in plant factories can increase the frequency of material exchange between plants and the environment, thereby accelerating the plant transpiration rate. Similarly, carbon dioxide near the plant canopy is constantly renewed, increasing the rate of photosynthesis.

5.2.2 Environmental Data Collection

Sensors

Environmental factor measurement is the basis for digital management and control. Selecting suitable and reliable sensors is highly important for obtaining continuous and reliable data. Light-measurement methods include photometric, quantum, and radiometric methods. Light quality is often measured using a spectroradiometer, which is designed to measure the spectral power distribution of a light source. When selecting a temperature sensor, the operating temperature range, reliability, sensitivity, and stability should be comprehensively considered. Thermistors and thermocouples are commonly used as air temperature measuring devices, and they are ideal for monitoring air temperature at several locations along the growing shelves, both vertically and horizontally. The relative humidity is measured using a hygrometer. Most humidity sensors require regular calibration to ensure accuracy. Nondispersive infrared CO_2 sensors are the most popular CO_2 sensors. The air current speed is typically measured using an anemometer. High sensitivity and small size should be prioritized (Kozai et al., 2016).

The effects of environmental factors are shown in Table 5.2, along with the sensors commonly used to detect them. A better understanding of environmental factors and the suitable selection of sensors can help provide a data basis for the subsequent application of data-driven methods in plant factories, and this can reduce costs.

IoT System

The "Internet of Things (IoT)" refers to the networked interconnection of everyday objects and things to make them individually machine-readable and traceable on the Internet (Group & Lake, 2015; Xia et al., 2012).

The plant factory IoT system has three layers, the application layer, network layer, and perception layer, as shown in Fig. 5.2. The perception layer uses different

Table 5.2 Effects of commonly used sensors for specific environmental factors

Environmental factors	Effect	Sensors
Light intensity	Leaf nitrate content, phytochemical accumulation, transpiration rate, stomatal resistance	Quantum sensor
Light quality	Photosynthesis and morphology	Spectroradiometer
Photoperiod	Carbohydrate	
Temperature	Photosynthesis, respiration, transportation, accumulation of photosynthetic products	Thermistors and thermocouples
Relative air humidity	Leaf transpiration rate and nutrient uptake	Hygrometer
CO_2 concentration	Several physiological processes	NDIR CO_2 sensor
Air current speed	Transpiration rate, photosynthesis, plant microenvironment, and quality	Anemometer

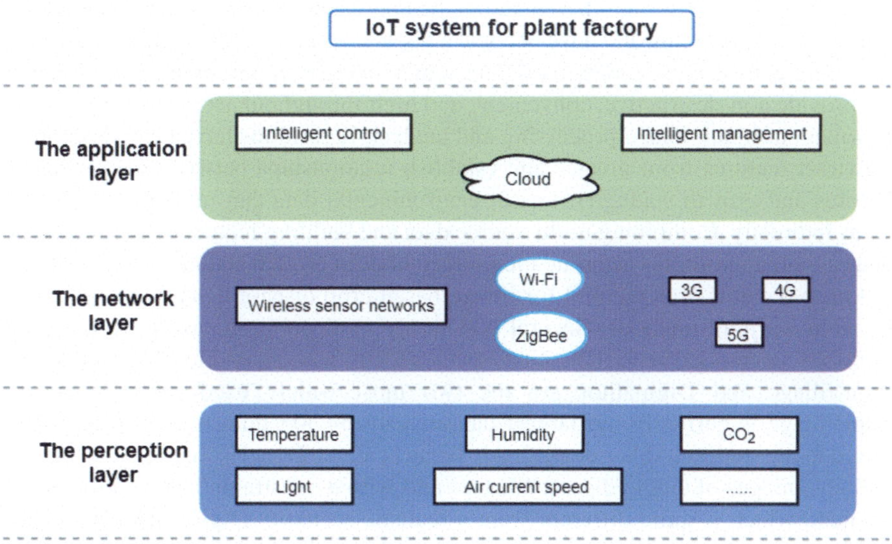

Fig. 5.2 IoT system architecture of plant factories

types of sensors to sense the environmental and physiological information of crops in plant factories. The network layer realizes remote control and management. Wireless sensor networks are an essential component of the network layer. Wireless sensor nodes construct networks by interacting with physical objects and/or their environment and communicating with their neighboring nodes or a gateway. The application layer conducts intelligent management based on the data obtained from the perception layer (Tzounis et al., 2017).

Through the use of the IoT system, the growth environment and physiological information of crops can be sensed, and data can be accumulated for subsequent crop models, in turn reducing unnecessary cost through precise control.

5.3 Crop Growth Monitoring in Plant Factories

Accurate and timely crop growth information can reflect the current growth status and future growth potential of crops, providing important information for crop management decision-making. However, crop growth monitoring in horticulture initially relied upon manual observations and empirical judgment, which was time-consuming, laborious, and destructive. Most commercial plant factories have a large production capacity of over 10,000 individual plants at a time. Operation on such scale is a challenging work for traditional crop monitoring methods (Kozai et al., 2019). Recently, data-driven automatic approaches showed their potential for crop growth monitoring and have become a popular research topic in plant factories.

Images are the most widely used data source in crop growth monitoring. Images can provide non-destructive, convenient, and high-throughput access to crop growth information. Many image processing and analysis algorithms have been developed to extract features from images and establish relationships between the extracted features and growth status. Historical environmental data can reflect crop growth status indirectly. Combining multi-source data and building feature extraction models is a viable option for accurate monitoring of crop growth status.

Image datasets are essential for image-based crop disease detection, especially for deep learning methods. Several commonly used public datasets are listed in Table 5.3, covering disease detection, species identification, and leaf segmentation. PlantVillage and Digipathos are the two most widely used public datasets. PlantVillage is currently the largest dataset with 54,305 images, which has supported many studies (Abbas et al., 2021; Wang et al., 2021b). However, PlantVillage contains images of single diseased leaves and with a constrained acquisition environment, which is quite different from a realistic scenario. Digipathos solves this problem, but the number of pictures is relatively small (2326 images). Therefore, researchers still need to spend a lot of time and effort to create their own datasets in many cases.

Plant factories are equipped with various sensors and cameras that can monitor images and environmental factors (e.g., temperature, humidity, and CO_2), thereby automatically gather data for crop growth monitoring. Data-driven crop growth monitoring methods are widely applied in plant factories (Jiang et al., 2018; Zhang et al., 2020). Based on the purposes of crop growth monitoring, these applications can be divided into growth monitoring measurement and abnormal growth detection. The former can obtain the status of crop growth, which is conducive to environmental management and the construction of crop growth models and the latter can provide early warning of diseases, so that decision makers can implement appropriate management operations timely.

Table 5.3 Plant-related public datasets

Application	Dataset	Crop	Number of images	Reference
Disease detection and identification	PlantVillage	38 categories	54,305	Hughes and Salathe (2015)
	Digipathos	21 plant species	2326	Barbedo et al. (2018)
	PlantDoc	13 plant species	2598	Singh et al. (2020)
	CMTL	311 host species	12,290	Lee et al. (2021)
Species identification	Swedish leaf dataset	15 species	1125	Soderkvist (2001)
	Flavia	32 species	1907	Wu et al. (2007)
	ImageCLEF11/12	71 tree species	6436/11,572	Müller et al. (2010)
	Leafsnap	185 tree species	23,147	Kumar et al. (2012)
	Oxford flower 17	17 flower species	1360	Nilsback and Zisserman (2006)
	Oxford flower 102	102 flower species	8189	Nilsback and Zisserman (2008)
Leaf segmentation	CVPPP	Tobacco, arabidopsis	810	Minervini et al. (2016)

5.3.1 Growth Monitoring Measurement

Crop growth can be defined as an increment of biomass or morphological size (Bakker et al., 1995). Monitoring crop growth by measuring morphological changes is an effective and straightforward method. Crop morphological changes can be described by many quantitative parameters, such as leaf area, height, width, plant weight, leaf shape changes, internode distance, and fruit counts. Generally, traditional measurement methods are mainly based on observations and manual measurement. For example, researchers use the grid method, gravimetric method (Ross et al., 2000), and leaf area meter (Igathinathane et al., 2008) to measure the leaf area, manually use a ruler to measure the height, and directly weigh the product to obtain the weight. These ways are simple but time-consuming and laborious, as well as destructive to crops.

Non-contact and automated methods can better meet the needs of real-time monitoring of plant factory production. Images are typical non-contact data widely used in crop monitoring. Invisible spectral bands such as near-infrared, red-edge, and ultraviolet can be used for camera imaging benefiting from the development of spectroscopic techniques. The fluorescent signature of chlorophyll can be used in imaging to detect nutrients. Depth cameras can also obtain distance information between the camera and imaged surface. Data collection methods such as laser scanning, structured light, time of flight, structure from motion, and stereo vision

Table 5.4 Summary of image-based crop growth monitoring

Parameter	Data	Analytical method	Reference
Leaf area	RGB images	Traditional image processing	Lü et al. (2010)
Leaf area, height, volume, diameter	Stereo vision	Image processing, geometric feature calculation	Yeh et al. (2014)
Leaf area, height	RGB-depth images	Learning- and model-free 3D point clouds	Yang et al. (2020)
Leaf area	RGB images	Deep learning framework (ANN)	Wijaya et al. (2020)
Height	Stereo vision	Depth perception, time series estimation	Nugroho et al. (2020)
Weight and growth rate	RGB images	Image segmentation framework in deep learning (R-CNN)	Reyes-Yanes et al. (2020)
Weight	RGB images	Colored 3D point clouds	Mortensen et al. (2018)
Weight	RGB images	Deep learning framework (CNN)	Zhang et al. (2020)

can be used for 3D reconstruction of crops. Image-based methods can reduce hardware costs, achieve real-time monitoring, and better meet the requirements of plant factories. Many image-based studies have been carried out in crop growth monitoring (Table 5.4).

There are two common pipelines for image-based crop morphological parameter measurement: the first is to perform image processing, feature extraction, and parameter regression analysis on data; the other is an end-to-end approach, using deep learning for pattern recognition (Fig. 5.3). Image processing includes preprocessing and automatic segmentation of crop organs, involving traditional image processing methods and deep learning methods. Feature extraction means obtaining relevant parameters for regression, such as height and width relative to leaf area. Finally, mathematical algorithms such as support vector machines, random forests, multiple linear regression, and various deep learning frameworks are used for regression analysis to get target parameters. The end-to-end method based on deep learning omits the above steps, does not need to manually design rules, and can automatically mine the potential features of the data. However, the disadvantage is that deep learning requires higher quality and quantity of the dataset for model training. Compared with the two-dimensional method, the three-dimensional method makes up for the lack of depth information, can obtain accurate distance and intuitive shape, and has certain advantages in estimating plant growth and biomass (Vandenberghe et al., 2018). Generally, researchers do not apply these methods individually but apply similar methods to the extraction of target parameters according to requirements or combine different strategies to improve the accuracy and robustness of the algorithm.

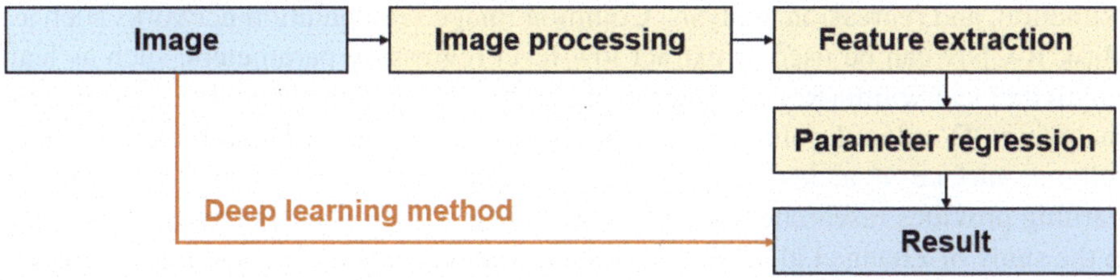

Fig. 5.3 Two pipelines of image-based crop morphological parameter measurement

Leaf Area

Leaf area measurements can be divided into three dimensions: single leaf area, single plant leaf area, and population leaf area measurements (Leroy et al., 2007). Techniques used for the non-contact method include coefficient regression (Cristofori et al., 2008) and image processing (Hajjdiab & Obaid, 2010). The coefficient regression method uses easily measurable parameters (such as leaf length and leaf width) to construct mathematical equations for the leaf area. Combining various strategies under the image processing framework for segmentation and calculation can well improve the accuracy of pixel statistics and achieve algorithms with better accuracy and robustness (Lü et al., 2010). Using three-dimensional methods such as stereo vision or 3D point cloud can represent leaves with complex shapes to obtain more morphological parameters, including leaf area, height, volume, and diameter (Yeh et al., 2014). The obtained crop morphological data combined with time series can also be used to construct the crop growth curve.

Plant Height

Previous studies have implemented non-contact automatic measurement for plant height, using laser scanning or LiDAR (Hoffmeister et al., 2016), ultrasonic sensors (Chang et al., 2017), stereo vision (Nugroho et al., 2020), or high-resolution RGB image methods that combine motion algorithms. For crops with prominent stalks such as rice and corn, plant height is an easy-to-judge key parameter. Researchers have developed algorithms and automatic height measurement systems based on depth perception methods such as stereo vision (Constantino et al., 2015; Shrestha et al., 2021). For other crops with shorter stems, plant height can be extracted as a secondary parameter along with other parameters (Yang et al., 2020; Yeh et al., 2014).

Weight

As an important approach for tracking crop growth, the non-contact measurement of fresh weight is to establish the relationship between morphological features and fresh weight. It generally covers three steps, including image processing, feature extraction, and regression analysis. Common image segmentation networks such as Mask R-CNN can be used to extract low-level regression parameters, such as leaf length and leaf width (Reyes-Yanes et al., 2020). Three-dimensional means such as analyzing 3D colored point clouds can also extract regression parameters such as volume, surface area, leaf coverage, and height (Mortensen et al., 2018). Deep learning provides researchers with an end-to-end way to obtain plant fresh weight. In the study of Zhang et al. (2020), CNN showed its convenience and high accuracy for estimating leaf area, leaf dry weight, and leaf fresh weight. The model also had great generalization in dealing with images from different seasons and growth stages.

5.3.2 Abnormal Growth Detection

Crops in plant factories have a lower probability of being infected by soil diseases and pests than greenhouses or fields. However, the relatively high humidity and temperature of plant factories provide a suitable living environment for crop diseases. Bacteria and viruses in the air can infect hydroponic crops through the nutrient solution. Most crop diseases generate apparent symptoms such as yellow leaves, rusty brown spots, peeling, dehydration, and scorching. Using computer vision can offer high feasibility and accuracy for plant disease observation and diagnosis, significantly reducing labor costs and improving automation and intelligent management. Abnormal growth detection in plant factories can be divided into disease and pest detection and nutritional stress detection. The former is caused by exogenous pathogens, and the latter is caused by inappropriate nutrients or environmental factors.

Disease and Pest Detection

Hydroponic crops are susceptible to diseases caused by bacteria and viruses. The temperature and humidity environments in plant factories are also very suitable for the breeding of pests such as aphids, mealworms, scale insects, and leaf rolls. Accurate identification and classification of crop diseases with numerous symptoms and complex causes require considerable expertise. For operators and staff who lack professional knowledge, a rapid and convenient method is needed to provide early disease warnings. Deep learning has shown its remarkable effects on image recognition and classification, and it has been verified in numerous fields of research (Table 5.5).

The deep learning network for plant disease detection can be roughly divided into three categories: classification network, target detection network, and segmentation network. According to the specific tasks, the classification network method can be subdivided into three sub-categories: using the network as a feature extractor, directly using the network for classification, and using the network for lesion location (Liu & Wang, 2021). Most existing crop diseases and insect pests classification

Table 5.5 Summary of disease and pest detection models

Approach	Model	Crop	Dataset	Functionality	Reference
Classification network	AlexNet, GoogLeNet, ResNet	Tomato	5550 images	Classify leaf diseases	Zhang et al. (2018b)
	VGG-16, VGG-19, ResNet, InceptionV3	Tomato	2681 images	Classify leaf diseases	Ahmad et al. (2020)
	DenseNet	Tomato	PlantVillage	Classify leaf diseases	Abbas et al. (2021)
Detection network	Faster R-CNN, YOLO	Lettuce	873 images	Identify diseases	Pratama et al. (2020)
	DBA_SDD	14 varieties	PlantVillage	Identify diseases, classify disease degree	Wang et al. (2021b)
	YOLOv4-tiny	5 varieties	1500 images	Identify diseases in real time	Mohandas et al. (2021)
Segmentation network	Mask R-CNN	Strawberry	2500 images	Segment diseases instance	Afzaal et al. (2021)
	DeepLabV3+	Lettuce	500 images	Segment abnormal leaves	Wu et al. (2021)
	DeepLabV3+, U-Net	Cucumber	1000 images	Segment leaves and disease spots	Wang et al. (2021a)

procedures use CNNs with hybrid structures, such as AlexNet, VGGNet, ResNet, GoogLeNet, EfficientNet, HRNet, and DenseNet. Researchers can also design specific network structures based on actual problems. In the face of intra- and inter-class differences in complex scenes, the three main object detection networks, Faster R-CNN, SSD, and R-FCN can effectively identify nine types of tomato leaf diseases (Fuentes et al., 2017). The detection network can be divided into two- and one-stage networks. The two-stage network is represented by faster R-CNN (Ren et al., 2015), which first generates a proposed region that may contain lesions and then further performs target screening. The first-level network is represented by SSD (Wang et al., 2021b) and You Only Look Once (YOLO) (Mohandas et al., 2021), and it directly uses the features extracted from the network to predict the location and category of the lesion. Region-level or pixel-level segmentation of diseased parts is usually more conducive to disease identification. The general network can be roughly divided into FCNs (Long et al., 2015) and Mask R-CNNs (Afzaal et al., 2021). As new semantic segmentation networks, DeepLabV3+ and U-Net have also achieved good results in the severity classification of cucumber petioles (Wang et al., 2021a).

Crop pests and diseases detection based on images and deep learning has developed rapidly, with broad development prospects and great potential. However, in the process of moving towards agricultural application, the application in real situations is still facing many challenges. There are still some problems that need to be solved, such as the dataset being limited to the current laboratory environment, the disease data (negative samples) being insufficient, and prior knowledge may be required to guide network learning and other difficulties.

Nutrient Stress Detection

Crops are supplied with nutrients via hydroponic technology in plant factories. When the nutrient supply is insufficient or excessive, it inhibits crop growth and causes a phenomenon known as nutrient stress (Tian et al., 2021). Crops often show several typical symptoms under stress conditions for certain nutrients, such as nitrogen, phosphorus, and potassium. For example, an increase in light intensity and a rapid growth rate will lead to the loss of calcium ions in lettuce leaves, thereby significantly increasing the ratio of lettuce edge burning (Sago, 2016). The increased irradiation time of red and blue LED light can cause photo-oxidation damage to lettuce (Shao et al., 2020).

There are several diagnostic methods for assessing crop nutritional status, including crop sap analysis (Farneselli et al., 2014), chlorophyll fluorescence diagnostics (Simko et al., 2015), and spectral analysis (Eshkabilov et al., 2021). Researchers also combined machine learning with image analysis, chlorophyll fluorescence, and hyperspectral technology to develop powerful methods for the early detection of biotic stress (Behmann et al., 2015).

Combining computer vision technologies to identify crop nutritional stress and automatically diagnose nutritional status opens up a new approach for the non-destructive detection and rapid diagnosis of crop nutritional stress. However, such approach still needs further improvements owing to the similarities between diseases and nutritional stress symptoms. Zhang et al. (2012) performed ground measurements on the hyperspectral reflectance of inoculated yellow rust and nutrient stress treatments to detect and distinguish wheat yellow rust and nutrient stress. Shimamura et al. (2019) perform binary discrimination of tipburn occurrence on lettuce using convolutional neural networks and identify early symptoms of tipburn. Gozzovelli et al. (2021) proposed a lettuce tip burn detection method using a Wasserstein generative adversarial network to overcome the problems of sample data imbalance, and they combined this with YOLOv2 and U-net to perform region screening at healthy and burned edge regions, respectively. The result showed that their model could accurately locate and detect the tipburn stress of the crop canopy. The above studies show that image-based nutrient stress recognition has both methodological convenience and achievability, but the specific application still needs to be promoted.

5.4 Development and Application of Crop Growth Model

The concept of system analysis is widely used in agriculture research and production, which generates a large number of conceptual models to help understand agricultural production, and promotes the quantitative research of agricultural systems (Ting et al., 2016). Among these models, the crop growth model has developed rapidly and is gradually being applied to agricultural research and production.

Crop growth can be defined as an increase in biomass or morphological size (Bakker et al., 1995). Quantitative research on the growth, physiology, and ecological processes of crops can better understand the dynamic responses of crops to the environment. The crop growth model performs dynamic quantitative analysis and growth simulation research on the crop and its ecological environment factors to achieve precise environmental control and thereby improve crop yield and quality.

Based on the different dynamics and timescales, the crop growth simulation in agriculture can be divided into instantaneous and long-term state simulations (Fig. 5.4). The instantaneous state includes fast crop dynamics (i.e., photosynthesis, respiration, and transpiration) and the associated microenvironment dynamics. Simulation of such states is essential for real-time crop regulation. The output of long-term state simulation arises from the long-term accumulation of crop growth, which is reflected in yield as fresh weight, dry weight, and fruit weight, and in

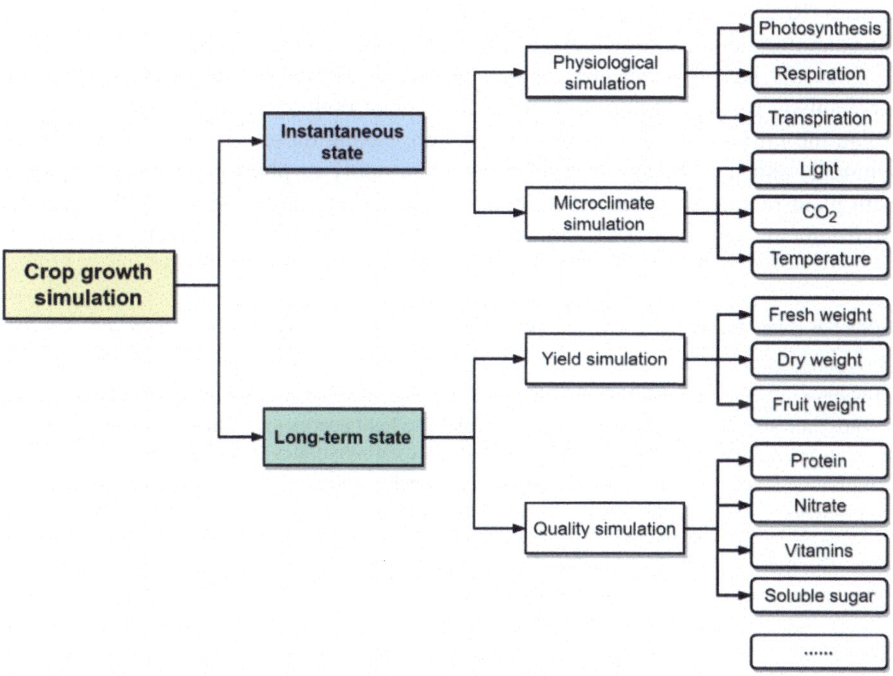

Fig. 5.4 The main research contents in crop growth simulation

quality as the accumulation of protein, nitrates, vitamins, and so on. Long-term state simulations are generally used to predict yield and quality, as well as to assess the influencing factors thereof.

Plant factories, as modern agricultural production systems equipped with various sensors and actuators, are suitable for the research and application of crop growth models. The crop growth model also plays a vital role in the cultivation management optimization of plant factories. Recently, a number of crop growth models have been developed, which can be divided into two categories (according to methodology): mechanistic models and data-driven models.

5.4.1 Mechanistic Models

The mechanistic model, also known as the knowledge-driven or explanatory model, primarily relies upon existing domain knowledge (both theoretical and empirical). A mechanistic model describes a system that uses mathematical language, as well as its theorems and symbols. To study basic crop growth principles, it is necessary to understand the fundamental physiological processes underlying it. The basic physiological processes of crops that closely determine their growth are photosynthesis, respiration, and transpiration (Table 5.6). These physiological processes are affected by various environmental factors and are characterized by numerous interactions (Rodríguez et al., 2015). The mechanistic model is an important support tool for obtaining detailed knowledge of these effects and interactions.

Over the past few decades, researchers have conducted numerous studies on mechanistic models (Table 5.7). Some European countries and the United States began the initial crop growth model research, basing these upon greenhouse environments and then further expanding them to apply to plant factories. Compared with plant factories, greenhouses involve more energy exchanges with the outside environment, and their microclimates have been extensively researched to model

Table 5.6 Fundamental physiological processes are closely associated with crop growth

Physiological processes	Definition	Impact on crops
Photosynthesis	A series of reactions that use light energy to assimilate CO_2 to synthesize carbohydrates	Directly affect growth and yield
Respiration (dark respiration)	A form of respiration in which carbon dioxide is released without the presence of light	Provide chemical energy and reduce power
Respiration (Photorespiration)	Takes place where carbon dioxide is given out, which is the opposite process of photosynthesis	Waste energy, maintain the normal physiology
Transpiration	Drives xylem transpiration stream to drive root uptake of water and nutrients	Maintain favorable water and nutritional status

Table 5.7 Summary of mechanistic models

Model	Crop	Application	Reference
TOMGRO	Tomato	Simulate the development and yield of tomatoes	Jones et al. (1991)
LETSGROW	Lettuce	Simulate hydroponic lettuce production under a range of environmental conditions	Both (1995)
HORTISIM	Tomato and sweet pepper	Simulate the growth of crop in response to climate in greenhouses for greenhouse environmental control and management	Gijzen et al. (1997)
TOMSIM	Tomato	Simulate dry matter production of tomato	Heuvelink (1997)
NiCoLet	Lettuce	Determine optimal climatic control strategies to prevent high nitrate concentration	Seginer et al. (1997)
TOMPOUSSE	Tomato	Simulate weekly prediction of greenhouse tomato	Abreu et al. (1998)
Vanthoor	Tomato	Guide the design of greenhouses for a broad range of climatic and economic conditions by climate-yield models	Vanthoor et al. (2011b)

greenhouse crop growth and development. The dynamic behavior of a microclimate represents a combination of physical processes involving energy transfer (radiation and heat) and mass balance (water vapor flux and CO_2 concentration). Below are several representative models and studies.

TOMGRO (Jones et al., 1991) is a physiological model of tomato crop development and yield. It uses the source-sink theory, which focuses on the relationship between tomato growth and greenhouse environmental factors such as solar radiation, temperature, and CO_2. TOMGRO describes the tomato plant via seven basic physiological crop factors: the number of leaves, number of main stem segments, number of fruits, dry weights of leaves plus petioles, dry weights of main stem segments, dry weights of fruits, and areas of leaves. TOMGRO is one of the most remarkable models of tomato growth and yield in greenhouses, and it has had an extensive influence on crop growth and simulation research (Bacci et al., 2012; Jones et al., 1999; Wang et al., 2013). HORTISIM (Gijzen et al., 1997) is a popular growth model that provides effective strategies for greenhouse environmental control and management. It can simulate the growth of several crops, including tomato, cucumber, and sweet pepper. HORTISIM is combined with seven sub-models (weather, greenhouse climate, soil, crop, greenhouse manager, soil manager, and crop manager) as well as a simulation process manager. This structure allows it to adapt to different crops by setting different model configurations, and it is widely used in growth simulation, yield prediction, and environmental control strategy optimization (Heuvelink et al., 2000; Marcelis et al., 2000). NiCoLet (Seginer et al., 1997) is a two-state-variable model that describes the nitrate concentration in lettuce when the nitrate supply is unlimited. The NiCoLet model predicts growth and the dynamic fluctuations of soluble carbon and nitrate based on the mechanisms of photosynthesis and maintenance of turgor pressure. Researchers have conducted extensive research using this model. Ioslovich et al. (2002) modified the original model to predict lettuce's growth rate and nitrate content in a sparse canopy. Mathieu et al. (2006) modified a NiCoLet model and evaluated both short- and long-term changes in lettuce growth and nitrate accumulation. The Vanthoor model (Vanthoor et al., 2011a, b) is a well-designed model that features two components: a greenhouse climate model and a crop yield model. The greenhouse climate model describes the greenhouse microclimate for a wide range of designs and climates. It consists of a set of first-order differential equations to ensure that it can be combined with a crop yield model. Its crop yield model employs the buffer theory and produces two crop growth status results (leaf area index and dry matter) (Vanthoor et al., 2011c). Although this structure makes its variables relatively easy to obtain, it struggles to reflect the physiological characteristics of the crop intuitively, and many of its variables are more affected by microclimate parameters than crop status ones.

The mechanistic model can provide highly reliable predictions when crop growth is normal (compared with situations used for model calibration). However, owing to the complex interactions between crop growth physiological processes, it remains difficult to establish suitable growth models for a broader range of applications (Chang et al., 2021; Medina-Ruíz, 2011). For example, TOMGRO is more suitable for crop management optimization, whereas Vanthoor can be better applied to greenhouse regulation. Certain researchers have focused on optimizing or modifying the model parameters to improve the model response (Boote et al., 2012; Ioslovich et al., 2002). Following the improvement of computing abilities, resources, and experience sharing between plant scientists, mathematicians, and computer scientists, numerous recent studies have incorporated multiple mechanistic models to exploit the advantages of different models and improve performance or applicability (Lin et al., 2019; Mathieu et al., 2006; Zhao et al., 2019).

Another weakness of the mechanistic model is that it requires many state variables, input variables, and parameters, making it difficult for users to implement these models in practical applications due to the complicated calibration process (Boote et al., 2012). The environmental controllability and enclosed characteristics of plant factories considerably simplify the application of the mechanistic model (e.g., solar radiation and external climate can be neglected). However, the application of mechanistic models in plant factories is relatively rare compared to greenhouses, and most of these play the role of sub-models in environmental optimization (Xu et al., 2018, 2021). With the rapid development of plant factories and mathematical modeling techniques, crop growth mechanistic models are expected to play a more important role in the future.

5.4.2 Data-Driven Models

The data-driven model, also called the statistical or empirical model, can extract mathematical relationships between crop status (yield or growth states) and management variables (environmental factors or control strategies) from the gathered data (Fan et al., 2015). Therefore, the quality and quantity of data directly affect the model's performance. Data collection has been a significant challenge of data-driven model research in the past. Researchers have to record data manually, which is time-consuming and laborious, and the data quality is unreliable. Nowadays, with the development of sensors and IoT technology in agriculture, data have become more available, yielding more studies on data-driven crop models (Table 5.8).

Typically, the data-driven model workflow involves three processes, data collection, model building, and model application. Data collection forms the basis for building a data-driven model, and it significantly affects model performance. After organization and preprocessing, the data are used as inputs and outputs for the

Table 5.8 Summary of data-driven crop models in horticulture

Approach	Model	Crop	Data	Functionality	Reference
Regression analysis	Linear regression	Strawberry	Plant variables (manual measured) and environmental parameters	Prediction of the strawberry growth and fruit yield	Dong et al. (2018)
	Partial least squares regression	Snap bean	Plant variables (manual measured) and hyperspectral images	Yield modeling of snap bean	Hassanzadeh et al. (2020)
Machine learning	ANN	Lettuce	Lettuce height and environmental parameters	Prediction of lettuce growth in plant factory	Rizkiana et al. (2021)
	Neural fuzzy	Lettuce	Environmental parameters and image of lettuce	Prediction of lettuce growth, harvest day, and quality in greenhouse	Chang et al. (2021)
Deep learning	EFuNN	Tomato	Yield and environmental parameters	Prediction of tomato yield in greenhouse	Qaddoum et al. (2013)
	Due Attention-LSTM	Tomato	Plant variables (manual measured) and environment parameters	Prediction of tomato yield in plant house and interpretation on how factor effect yield	De Alwis et al. (2019)
	LSTM	Tomato	Stem diameter values and environment parameters	Prediction of tomato yield in greenhouse	Alhnaity et al. (2020)
	TCN and RNN	Tomato	Yield and daily recorded environmental parameters	Prediction of tomato yield in greenhouse	Gong et al. (2021)
	LSTM-based encoder-decoder	Tomato	Plant stem diameter and environment parameters	Prediction of tomato stem diameter in greenhouse	Alhnaity et al. (2021)

model training process. The input data are fed through a data-driven algorithm to correlate with the output data and the results of this correlation are used to adjust the model. In general, the model's performance can be measured by the accuracy of the training or validation datasets. After model building, the trained model can be used to classify, predict, or cluster new samples (testing data), using the experience obtained during the model-building process. Regression analysis is one of the most widely used methods of crop model construction, which can quantitatively describe the relationships between independent and dependent variables. Data-driven regression does not impose domain-based knowledge but only considers datasets in the form $(X_i, y_i)_{1 \leq i \leq N}$, where X_i represents the multidimensional input variables and y_i represents the response variable (Domijan et al., 2006). Classical statistical regression methods include linear regression, logistic regression, polynomial regression, and stepwise regression. Linear and polynomial regression models are straightforward, and their coefficients can directly reflect each variable's interpretation, which performs well in simple situations (Sim et al., 2020). However, the assumption of linear or polynomial relationships between crop growth and environmental factors is not always valid, owing to the nonlinear dynamic processes involved in the physical and biological systems of plant factories.

Machine learning has attracted increasing attention in recent years. Approaches such as support vector machines, decision trees, and artificial neural networks (ANNs) have been applied to tackle complex problems in agricultural systems (Liakos et al., 2018). These approaches can extract functional patterns for crop growth and development simulations from the high-dimensional environmental and management data. Many studies have been conducted using machine learning models, and these have shown that the error in the regression problem can be reduced with sufficient data (Chen & Cournède, 2018; Rizkiana et al., 2021; Zaidi et al., 1999).

Deep learning has emerged alongside high-performance computing and big data technologies to create new techniques for crop modeling in the field of multidisciplinary agricultural technology. As an end-to-end network approach, deep learning models can extract and automatically organize intricate relationships from high-dimensional data, using multiple levels of representation (LeCun et al., 2015). Neural networks have been widely applied in agricultural fields for a long time (Kocian et al., 2020; Seginer, 1997; Zaidi et al., 1999); they use a simple structure to combine input and output variables to indicate correlation, and they can describe complex systems using limited parameters. With strong feature extraction and learning abilities, deep learning models can achieve relatively high performances even for complex problems (Deng & Yu, 2014). Another advantage of deep learning models is that they can be easily integrated with other algorithms based on their hierarchical structure. Qaddoum et al. (2013) proposed a deep learning model that integrated fuzzy and neural networks. The model achieved a high prediction accuracy for weekly tomato production. A combination of ANNs and image processing methods was proposed by Zaborowicz et al. (2017) to analyze the quality of tomatoes.

The temporal-sequential deep learning model has shown its potential in crop growth modeling. Its temporal architecture renders it inherently suitable for crop growth process simulations. Recurrent neural networks (RNNs), as the most popular temporal-sequential model, are specifically designed to learn sequential relationships by explicitly linking adjacent observations (Werbos, 1990). As a variant of an RNN, long short-term memory (LSTM) has been successfully applied in crop yield prediction. Alhnaity et al. (2020) employed an LSTM model to predict tomato growth and yield in greenhouse environments. They fed former yield, growth, stem diameter values, and microclimate conditions into the model. The results showed that LSTM provided a superior performance to SVR and random forest regression. Gong et al. (2021) developed a greenhouse crop yield prediction model by combining two state-of-the-art networks for temporal sequence processing: a temporal convolutional network and an RNN. The proposed model achieved a higher yield prediction performance than traditional machine learning methods and other classical deep neural networks.

Compared with mechanistic models, data-driven models can solve complex problems particularly well, especially when there is sufficient data. Generally, data-driven models have more complex structures and more parameters, which are learned from data through their own method of computing—no human help is required. All data-driven models are to some extent black boxes. For data-driven models like linear regression, the models are relatively well understood and interpretable. The interpretation becomes more difficult for support vector machines or random forest models. For deep neural networks with millions of parameters, the interpretation becomes extremely difficult. Despite the increased utilization of data-driven models, there is still a lack of sufficient techniques to explain and interpret the behavior and decisions of these models. Recently, domain knowledge (such as physiological principles and cultivation experience) has been used to improve the performance and interpretability of data-driven crop models (Worrall et al., 2021; Yin et al., 2021). Crop growth models integrating data-driven and mechanistic will have more significant potential in the future and will perform better in real application scenarios.

5.5 Summary

In this chapter, we have introduced the applications and prospects of data science in plant factories. The data-driven modeling approach can describe the dynamic process of crop growth and its interaction with the environment and then provide support for plant factories' decision-making. Massive data have been available with the application of sensor and IoT technologies. The automated measurement procedures of environmental factors in plant factories are summarized. Monitoring the crop growth status can provide information regarding the variables that affect crop growth and development, thereby providing information for cultivation management. Based on environmental factors and crop growth data, crop growth models can quantitatively simulate the growth processes of crops and their responses to the environment. We have summarized the characteristics and contributions of data-driven modeling to the main production processes in plant factories. Plant factories are emerging and promising crop production systems. High costs constrain the development and business viability of plant factories. We believe that further applications of data-driven modeling can effectively reduce the production costs of plant factories and lead to successful commercial production.

References

Abbas, A., Jain, S., Gour, M., & Vankudothu, S. (2021). Tomato plant disease detection using transfer learning with C-GAN synthetic images. *Computers and Electronics in Agriculture, 187*, 106279. https://doi.org/10.1016/j.compag.2021.106279

Abreu, P., Meneses, J. F., & Gary, C. (1998). Tompousse, a model of yield prediction for tomato crops: Calibration study for unheated plastic greenhouses. *Electronic Information in Horticulture, 519*, 141–150.

Afzaal, U., Bhattarai, B., Pandeya, Y. R., & Lee, J. (2021). An instance segmentation model for strawberry diseases based on mask R-CNN. *Sensors, 21*(19), 6565. https://doi.org/10.3390/s21196565

Ahmad, I., Hamid, M., Yousaf, S., Shah, S. T., & Ahmad, M. O. (2020). Optimizing pretrained convolutional neural networks for tomato leaf disease detection. *Complexity, 2020*, 1–6. https://doi.org/10.1155/2020/8812019

Ahmed, H. A., Yu-Xin, T., & Qi-Chang, Y. (2020). Optimal control of environmental conditions affecting lettuce plant growth in a controlled environment with artificial lighting: A review. *South African Journal of Botany, 130*, 75–89. https://doi.org/10.1016/j.sajb.2019.12.018

Alhnaity, B., Kollias, S., Leontidis, G., Jiang, S., Schamp, B., & Pearson, S. (2021). An autoencoder wavelet based deep neural network with attention mechanism for multi-step prediction of plant growth. *Information Sciences, 560*, 35–50. https://doi.org/10.1016/j.ins.2021.01.037

Alhnaity, B., Pearson, S., Leontidis, G., & Kollias, S. (2020). Using deep learning to predict plant growth and yield in greenhouse environments. *Acta Horticulturae, 1296*, 425–432. https://doi.org/10.17660/ActaHortic.2020.1296.55

Bacci, L., Battista, P., & Rapi, B. (2012). Evaluation and adaptation of TOMGRO model to Italian tomato protected crops. *New Zealand Journal of Crop and Horticultural Science, 40*(2), 115–126.

Bakker, J. C., Bot, G. P. A., Challa, H., & van de Braak, N. J. (1995). *Greenhouse climate control: An integrated approach*. Wageningen Academic Publishers.

Barbedo, J., Vieira Koenigkan, L., Almeida Halfeld-Vieira, B., Veras Costa, R., Lima Nechet, K., Vieira Godoy, C., Lobo Junior, M., Rodrigues Alves Patricio, F., Talamini, V., Gonzaga Chitarra, L., Alves Santos Oliveira, S., Nakasone Ishida, A. K., Cunha Fernandes, J. M., Teixeira Santos, T., Rossi Cavalcanti, F., Terao, D., & Angelotti, F. (2018). Annotated plant pathology databases for image-based detection and recognition of diseases. *IEEE Latin America Transactions, 16*(6), 1749–1757. https://doi.org/10.1109/TLA.2018.8444395

Behmann, J., Mahlein, A.-K., Rumpf, T., Römer, C., & Plümer, L. (2015). A review of advanced machine learning methods for the detection of biotic stress in precision crop protection. *Precision Agriculture, 16*(3), 239–260. https://doi.org/10.1007/s11119-014-9372-7

Bian, Z. H., Yang, Q. C., & Liu, W. K. (2015). Effects of light quality on the accumulation of phytochemicals in vegetables produced in controlled environments: A review. *Journal of the Science of Food and Agriculture, 95*(5), 869–877. https://doi.org/10.1002/jsfa.6789

Boote, K. J., Rybak, M. R., Scholberg, J. M., & Jones, J. W. (2012). Improving the CROPGRO-tomato model for predicting growth and yield response to temperature. *HortScience, 47*(8), 1038–1049.

Both, A. J. (1995). *Dynamic simulation of supplemental lighting for greenhouse hydroponic lettuce production*. Rutgers University. https://doi.org/10.13140/RG.2.2.11209.29282

Bukhov, N. G., Drozdova, I. S., & Bondar, V. V. (1995). Light response curves of photosynthesis in leaves of sun-type and shade-type plants grown in blue or red light. *Journal of Photochemistry and Photobiology B: Biology, 30*(1), 39–41. https://doi.org/10.1016/1011-1344(95)07124-K

Chang, C.-L., Chung, S.-C., Fu, W.-L., & Huang, C.-C. (2021). Artificial intelligence approaches to predict growth, harvest day, and quality of lettuce (Lactuca sativa L.) in a IoT-enabled greenhouse system. *Biosystems Engineering, 212*, 77–105. https://doi.org/10.1016/j.biosystemseng.2021.09.015

Chang, Y. K., Zaman, Q. U., Rehman, T. U., Farooque, A. A., Esau, T., & Jameel, M. W. (2017). A real-time ultrasonic system to measure wild blueberry plant height during harvesting. *Biosystems Engineering, 157*, 35–44. https://doi.org/10.1016/j.biosystemseng.2017.02.004

Chen, X., & Cournéde, P.-H. (2018). Model-driven and data-driven approaches for crop yield prediction: Analysis and comparison. *International Journal of Mathematical and Computational Sciences, 11*(7), 334–342.

Chowdhury, M., Kiraga, S., Islam, M. N., Ali, M., Reza, M. N., Lee, W.-H., & Chung, S.-O. (2021). Effects of temperature, relative humidity, and carbon dioxide concentration on growth and glucosinolate content of kale grown in a plant factory. *Food, 10*(7), 1524. https://doi.org/10.3390/foods10071524

Constantino, K. P., Gonzales, E. J., Lazaro, L. M., Serrano, E. C., & Samson, B. P. (2015). *Plant height measurement and tiller segmentation of rice crops using image processing* (Vol. 3, p. 6). Springer.

Cristofori, V., Fallovo, C., Gyves, E. M., Rivera, C. M., Bignami, C., & Rouphael, Y. (2008). Non-destructive, analogue model for leaf area estimation in persimmon (Diospyros kaki L.f.) based on leaf length and width measurement. *European Journal of Horticultural Science, 73*, 216–221.

De Alwis, S., Zhang, Y., Na, M., & Li, G. (2019). Duo attention with deep learning on tomato yield prediction and factor interpretation. In A. C. Nayak & A. Sharma (Eds.), *PRICAI 2019: Trends in artificial intelligence* (pp. 704–715). Springer International Publishing. https://doi.org/10.1007/978-3-030-29894-4_56

Deng, L., & Yu, D. (2014). Deep learning: Methods and applications. *Foundations and Trends in Signal Processing, 7*(3–4), 197–387.

Domijan, K., Jorgensen, M., & Reid, J. (2006). Semi-mechanistic modelling in nonlinear regression: A case study. *Australian & New Zealand Journal of Statistics, 48*(3), 373–392.

Dong, Q., Sun, Q., Hu, Y., Xu, Y., & Qu, M. (2018). Data-driven horticultural crop model for optimal fertigation management - A methodology description. *IFAC-PapersOnLine, 51*(17), 472–476. https://doi.org/10.1016/j.ifacol.2018.08.167

Eshkabilov, S., Lee, A., Sun, X., Lee, C. W., & Simsek, H. (2021). Hyperspectral imaging techniques for rapid detection of nutrient content of hydroponically grown lettuce cultivars. *Computers and Electronics in Agriculture, 181*, 105968. https://doi.org/10.1016/j.compag.2020.105968

Fan, X.-R., Kang, M.-Z., Heuvelink, E., de Reffye, P., & Hu, B.-G. (2015). A knowledge-and-data-driven modeling approach for simulating plant growth: A case study on tomato growth. *Ecological Modelling, 312*, 363–373. https://doi.org/10.1016/j.ecolmodel.2015.06.006

Farneselli, M., Tei, F., & Simonne, E. (2014). Reliability of petiole sap test for N nutritional status assessing in processing tomato. *Journal of Plant Nutrition, 37*(2), 270–278. https://doi.org/10.1080/01904167.2013.859696

Fuentes, A., Yoon, S., Kim, S., & Park, D. (2017). A robust deep-learning-based detector for real-time tomato plant diseases and pests recognition. *Sensors, 17*(9), 2022. https://doi.org/10.3390/s17092022

Gijzen, H., Heuvelink, E., Challa, H., Marcelis, L. F. M., Dayan, E., Cohen, S., & Fuchs, M. (1997). HORTISIM: A model for greenhouse crops and greenhouse climate. *Modelling Plant Growth, 456*, 441–450.

Gong, L., Yu, M., Jiang, S., Cutsuridis, V., & Pearson, S. (2021). Deep learning based prediction on greenhouse crop yield combined TCN and RNN. *Sensors, 21*(13), 4537. https://doi.org/10.3390/s21134537

Goto, E. (2012). Plant production in a closed plant factory with artificial lighting. *Acta Horticulturae, 956*, 37–49. https://doi.org/10.17660/ActaHortic.2012.956.2

Gozzovelli, R., Franchetti, B., Bekmurat, M., & Pirri, F. (2021). Tip-burn stress detection of lettuce canopy grown in Plant Factories. *IEEE/CVF International Conference on Computer Vision Workshops (ICCVW), 2021*, 1259–1268. https://doi.org/10.1109/ICCVW54120.2021.00146

Graamans, L., Baeza, E., van den Dobbelsteen, A., Tsafaras, I., & Stanghellini, C. (2018). Plant factories versus greenhouses: Comparison of resource use efficiency. *Agricultural Systems, 160*, 31–43. https://doi.org/10.1016/j.agsy.2017.11.003

Group, S. M. A., & Lake, V. (2015). Internet of Things (IoT): A literature review. *Journal of Computer and Communications, 3*(5), 164. https://doi.org/10.4236/jcc.2015.35021

Hajjdiab, H., & Obaid, A. (2010). A vision-based approach for nondestructive leaf area estimation. *Environmental Science and Information Application Technology, 3*, 53–56. https://doi.org/10.1109/ESIAT.2010.5568973

Hassanzadeh, A., van Aardt, J., Murphy, S. P., & Pethybridge, S. J. (2020). Yield modeling of snap bean based on hyperspectral sensing: A greenhouse study. *Journal of Applied Remote Sensing, 14*(2), 024519.

Heuvelink, E. (1997). *TOMSIM: A dynamic simulation model for tomato crop growth and development*. In ISHS Second Int. Symp. on Models for Plant Growth, Env. Control and Farm Management in Protected Cultivation, Wageningen, The Netherlands.

Heuvelink, E., Lee, J. H., Buiskool, R. P. M., & Ortega, L. (2000). Light on cut chrysanthemum: Measurement and simulation of crop growth and yield. *International ISHS Symposium on Artificial Lighting, 580*, 197–202.

Hoffmeister, D., Waldhoff, G., Korres, W., Curdt, C., & Bareth, G. (2016). Crop height variability detection in a single field by multi-temporal terrestrial laser scanning. *Precision Agriculture, 17*(3), 296–312. https://doi.org/10.1007/s11119-015-9420-y

Hogewoning, S. W., Trouwborst, G., Maljaars, H., Poorter, H., van Ieperen, W., & Harbinson, J. (2010). Blue light dose–responses of leaf photosynthesis, morphology, and chemical composition of Cucumis sativus grown under different combinations of red and blue light. *Journal of Experimental Botany, 61*(11), 3107–3117. https://doi.org/10.1093/jxb/erq132

Hughes, D. P., & Salathe, M. (2015). *An open access repository of images on plant health to enable the development of mobile disease diagnostics*. https://arxiv.org/abs/1511.08060v2

Igathinathane, C., Chennakesavulu, B., Manohar, K., Womac, A. R., & Pordesimo, L. O. (2008). Photovoltaic leaf area meter development and testing. *International Journal of Food Properties, 11*(1), 53–67. https://doi.org/10.1080/10942910600954739

Ioslovich, I., Seginer, I., & Baskin, A. (2002). SE—Structures and environment: fitting the nicolet lettuce growth model to plant-spacing experimental data. *Biosystems Engineering, 83*(3), 361–371. https://doi.org/10.1006/bioe.2002.0130

Jiang, J., Kim, H.-J., & Cho, W.-J. (2018). On-the-go image processing system for spatial mapping of lettuce fresh weight in plant factory. *IFAC-PapersOnLine, 51*(17), 130–134. https://doi.org/10.1016/j.ifacol.2018.08.075

Jones, J., Dayan, E., Allen, L., Keulen, & Challa, H. (1991). A dynamic tomato growth and yield model (TOMGRO). *Transactions of ASAE, 34*, 2. https://doi.org/10.13031/2013.31715

Jones, J. W., Kenig, A., & Vallejos, C. E. (1999). Reduced state–variable tomato growth model. *Transactions of ASAE, 42*(1), 255.

Kitaya, Y., Tsuruyama, J., Kawai, M., Shibuya, T., & Kiyota, M. (2000). Effects of air current on transpiration and net photosynthetic rates of plants in a closed plant production system. In C. Kubota & C. Chun (Eds.), *Transplant production in the 21st century: Proceedings of the international symposium on transplant production in closed system for solving the global issues on environmental conservation, food, resources and energy* (pp. 83–90). Springer. https://doi.org/10.1007/978-94-015-9371-7_13

Kocian, A., Massa, D., Cannazzaro, S., Incrocci, L., Di Lonardo, S., Milazzo, P., & Chessa, S. (2020). Dynamic Bayesian network for crop growth prediction in greenhouses. *Computers and Electronics in Agriculture, 169*, 105167. https://doi.org/10.1016/j.compag.2019.105167

Kozai, T. (2013). Resource use efficiency of closed plant production system with artificial light: Concept, estimation and application to plant factory. *Proceedings of the Japan Academy, Series B, 89*(10), 447–461. https://doi.org/10.2183/pjab.89.447

Kozai, T. (2018). *Smart plant factory: The Next generation indoor vertical farms*. Springer. https://doi.org/10.1007/978-981-13-1065-2

Kozai, T. (2022). Chapter 8—Balances and use efficiencies of CO2, water, and energy. In T. Kozai, G. Niu, & J. Masabni (Eds.), *Plant factory basics, applications and advances* (pp. 129–151). Academic Press. https://doi.org/10.1016/B978-0-323-85152-7.00003-3

Kozai, T., Niu, G., & Takagaki, M. (2019). *Plant factory: An indoor vertical farming system for efficient quality food production*. Academic Press.

Kozai, T., Niu, G., & Takagaki, M. (2016). *Plant factory: An indoor vertical farming system for efficient quality food production*. Academic Press.

Kumar, N., Belhumeur, P. N., Biswas, A., Jacobs, D. W., Kress, W. J., Lopez, I. C., & Soares, J. V. B. (2012). Leafsnap: A computer vision system for automatic plant species identification. In A. Fitzgibbon, S. Lazebnik, P. Perona, Y. Sato, & C. Schmid (Eds.), *Computer vision – ECCV 2012* (Vol. 7573, pp. 502–516). Springer. https://doi.org/10.1007/978-3-642-33709-3_36

LeCun, Y., Bengio, Y., & Hinton, G. (2015). Deep learning. *Nature, 521*(7553), 436–444. https://doi.org/10.1038/nature14539

Lee, M.-J., Son, J. E., & Oh, M.-M. (2014). Growth and phenolic compounds of Lactuca sativa L. grown in a closed-type plant production system with UV-A, -B, or -C lamp. *Journal of the Science of Food and Agriculture, 94*(2), 197–204. https://doi.org/10.1002/jsfa.6227

Lee, S. H., Goeau, H., Bonnet, P., & Joly, A. (2021). Conditional multi-task learning for plant disease identification. In *2020 25th international conference on pattern recognition (ICPR)* (pp. 3320–3327). ICPR. https://doi.org/10.1109/ICPR48806.2021.9412643

Leroy, C., Saint-André, L., & Auclair, D. (2007). Practical methods for non-destructive measurement of tree leaf area. *Agroforestry Systems, 71*(2), 99–108. https://doi.org/10.1007/s10457-007-9077-2

Li, Q., & Kubota, C. (2009). Effects of supplemental light quality on growth and phytochemicals of baby leaf lettuce. *Environmental and Experimental Botany, 67*(1), 59–64. https://doi.org/10.1016/j.envexpbot.2009.06.011

Liakos, K. G., Busato, P., Moshou, D., Pearson, S., & Bochtis, D. (2018). Machine learning in agriculture: A review. *Sensors, 18*(8), 2674. https://doi.org/10.3390/s18082674

Lin, D., Wei, R., & Xu, L. (2019). An integrated yield prediction model for greenhouse tomato. *Agronomy, 9*(12), 873. https://doi.org/10.3390/agronomy9120873

Liu, J., & Wang, X. (2021). Plant diseases and pests detection based on deep learning: A review. *Plant Methods, 17*(1), 22. https://doi.org/10.1186/s13007-021-00722-9

Long, J., Shelhamer, E., & Darrell, T. (2015). *Fully convolutional networks for semantic segmentation*. arXiv:1411.4038.

Lü, C., Ren, H., Zhang, Y., & Shen, Y. (2010). Leaf area measurement based on image processing. *International Conference on Measuring Technology and Mechatronics Automation, 2*, 580–582. https://doi.org/10.1109/ICMTMA.2010.141

Marcelis, L. F. M., Maas, F. M., & Heuvelink, E. (2000). The latest developments in the lighting technologies in Dutch horticulture. *International ISHS Symposium on Artificial Lighting, 580*, 35–42.

Markets and Markets. (2021). *Plant factory market by growing system (soil-based, non-soil-based, and hybrid), facility type (greenhouses, indoor farms, other facility types), light type, crop type (vegetables, fruits, flowers & ornamentals), and region—global forecast to 2026* (AGI 7962).

Markets and Markets. https://www.marketsandmarkets.com/Market-Reports/plant-factory-market-199919959.html

Mathieu, J., Linker, R., Levine, L., Albright, L., Both, A. J., Spanswick, R., Wheeler, R., Wheeler, E., deVilliers, D., & Langhans, R. (2006). Evaluation of the nicolet model for simulation of short-term hydroponic lettuce growth and nitrate uptake. *Biosystems Engineering, 95*(3), 323–337. https://doi.org/10.1016/j.biosystemseng.2006.07.006

Medina-Ruíz, C. A. (2011). Mathematical modeling on tomato plants: A review. *African Journal of Agricultural Reseearch, 6*, 33. https://doi.org/10.5897/AJARX11.001

Minervini, M., Fischbach, A., Scharr, H., & Tsaftaris, S. A. (2016). Finely-grained annotated datasets for image-based plant phenotyping. *Pattern Recognition Letters, 81*, 80–89. https://doi.org/10.1016/j.patrec.2015.10.013

Mohandas, A., Anjali, M. S., & Rahul Varma, U. (2021). Real-time detection and identification of plant leaf diseases using YOLOv4-tiny. In *2021 12th international conference on computing communication and networking technologies (ICCCNT)* (pp. 1–5). ICCCNT. https://doi.org/10.1109/ICCCNT51525.2021.9579783

Mortensen, A. K., Bender, A., Whelan, B., Barbour, M. M., Sukkarieh, S., Karstoft, H., & Gislum, R. (2018). Segmentation of lettuce in coloured 3D point clouds for fresh weight estimation. *Computers and Electronics in Agriculture, 154*, 373–381. https://doi.org/10.1016/j.compag.2018.09.010

Müller, H., Clough, P., Deselaers, T., & Caputo, B. (2010). *ImageCLEF: Experimental evaluation in visual information retrieval* (Vol. 32). Springer. https://doi.org/10.1007/978-3-642-15181-1

Naznin, M. T., Lefsrud, M., Gravel, V., & Azad, M. O. K. (2019). Blue light added with red LEDs enhance growth characteristics, pigments content, and antioxidant capacity in lettuce, spinach, kale, basil, and sweet pepper in a controlled environment. *Plants, 8*(4), 93. https://doi.org/10.3390/plants8040093

Nilsback, M.-E., & Zisserman, A. (2006). A visual vocabulary for flower classification. *IEEE Computer Society Conference on Computer Vision and Pattern Recognition, 2*, 1447–1454. https://doi.org/10.1109/CVPR.2006.42

Nilsback, M.-E., & Zisserman, A. (2008). Automated flower classification over a large number of classes. *Computer Vision Graphics Image Processing, 2008*, 722–729. https://doi.org/10.1109/ICVGIP.2008.47

Nugroho, A. P., Fadilah, M. A. N., Wiratmoko, A., Azis, Y. A., Efendi, A. W., Sutiarso, L., & Okayasu, T. (2020). Implementation of crop growth monitoring system based on depth perception using stereo camera in plant factory. *IOP Conference Series: Earth and Environmental Science, 542*(1), 012068. https://doi.org/10.1088/1755-1315/542/1/012068

Park, M. H., & Lee, Y. B. (2001). Effects of CO_2 concentration, light intensity and nutrient level on growth of leaf lettuce in a plant factory. *Acta Horticulturae, 548*, 377–384. https://doi.org/10.17660/ActaHortic.2001.548.43

Pratama, I. Y., Wahab, A., Alaydrus, M. (2020). Deep learning for assessing unhealthy lettuce hydroponic using convolutional neural network based on faster R-CNN with inception V2. In *2020 Fifth international conference on informatics and computing (ICIC)* (pp. 1–6). ICIC. https://doi.org/10.1109/ICIC50835.2020.9288554

Qaddoum, K., Hines, E. L., & Iliescu, D. D. (2013). Yield prediction for tomato greenhouse using EFuNN. *International Scholarly Research Notices, 2013*, 430986.

Ren, S., He, K., Girshick, R., & Sun, J. (2015). Faster R-CNN: Towards real-time object detection with region proposal networks. *Advances in Neural Information Processing Systems, 28*, 38046.

Reyes-Yanes, A., Martinez, P., & Ahmad, R. (2020). Real-time growth rate and fresh weight estimation for little gem romaine lettuce in aquaponic grow beds. *Computers and Electronics in Agriculture, 179*, 105827. https://doi.org/10.1016/j.compag.2020.105827

Rizkiana, A., Nugroho, A. P., Salma, N. M., Afif, S., Masithoh, R. E., Sutiarso, L., & Okayasu, T. (2021). Plant growth prediction model for lettuce (Lactuca sativa.) in plant factories using artificial neural network. *IOP Conference Series: Earth and Environmental Science, 733*(1), 012027. https://doi.org/10.1088/1755-1315/733/1/012027

Rodríguez, F., Berenguel, M., Guzmán, J. L., & Ramírez-Arias, A. (2015). *Modeling and control of greenhouse crop growth*. Springer International Publishing. https://doi.org/10.1007/978-3-319-11134-6

Ross, J., Ross, V., & Koppel, A. (2000). Estimation of leaf area and its vertical distribution during growth period. *Agricultural and Forest Meteorology, 101*(4), 237–246. https://doi.org/10.1016/S0168-1923(00)00089-7

Ryu, D. K., Kang, S. W., Ngo, V. D., Chung, S. O., Choi, J. M., Park, S. U., & Kim, S. J. (2014). Control of temperature, humidity, and CO_2 concentration in small-sized experimental plant factory. *Acta Horticulturae*. https://doi.org/10.17660/ActaHortic.2014.1037.59

Sago, Y. (2016). Effects of light intensity and growth rate on tipburn development and leaf calcium concentration in butterhead lettuce. *HortScience, 51*(9), 1087–1091. https://doi.org/10.21273/HORTSCI10668-16

Seginer, I. (1997). Some artificial neural network applications to greenhouse environmental control. *Computers and Electronics in Agriculture, 18*(2–3), 167–186.

Seginer, I., Straten, G., & Buwalda, F. (1997). Nitrate concentration in greenhouse lettuce: A modeling study. *International Society for Horticultural Science, 456*, 189–198.

Shao, M., Liu, W., Zha, L., Zhou, C., Zhang, Y., & Li, B. (2020). Differential effects of high light duration on growth, nutritional quality, and oxidative stress of hydroponic lettuce under red and blue LED irradiation. *Scientia Horticulturae, 268*, 109366. https://doi.org/10.1016/j.scienta.2020.109366

Shimamura, S., Uehara, K., & Koakutsu, S. (2019). *Automatic identifcation of plant physiological disorders in plant factory using convolutional neural networks*. The Society of Digital Information and Wireless Communications.

Shimizu, H., Saito, Y., Nakashima, H., Miyasaka, J., & Ohdoi, K. (2011). Light environment optimization for lettuce growth in plant factory. *IFAC Proceedings Volumes, 44*(1), 605–609. https://doi.org/10.3182/20110828-6-IT-1002.02683

Shrestha, D., Steward, B. L., & Kaspar, T. (2021). *Determination of early stage corn plant height using stereo-vision*.

Sim, H. S., Kim, D. S., Ahn, M. G., Ahn, S. R., & Kim, S. K. (2020). *Prediction of strawberry growth and fruit yield based on environmental and growth data in a greenhouse for soil cultivation with applied autonomous facilities*. Korean Society for Horticultural Science.

Simko, I., Jimenez-Berni, J. A., & Furbank, R. T. (2015). Detection of decay in fresh-cut lettuce using hyperspectral imaging and chlorophyll fluorescence imaging. *Postharvest Biology and Technology, 106*, 44–52. https://doi.org/10.1016/j.postharvbio.2015.04.007

Singh, D., Jain, N., Jain, P., Kayal, P., Kumawat, S., & Batra, N. (2020). PlantDoc: A dataset for visual plant disease detection. In *Proceedings of the 7th ACM IKDD CoDS and 25th COMAD* (pp. 249–253). ACM. https://doi.org/10.1145/3371158.3371196

Smith, H. L., McAusland, L., & Murchie, E. H. (2017). Don't ignore the green light: Exploring diverse roles in plant processes. *Journal of Experimental Botany, 68*(9), 2099–2110. https://doi.org/10.1093/jxb/erx098

Soderkvist, O. J. O. (2001). *Computer vision classifcation of leaves from Swedish trees*.

Son, K.-H., Jeon, Y.-M., & Oh, M.-M. (2016). Application of supplementary white and pulsed light-emitting diodes to lettuce grown in a plant factory with artificial lighting. *Horticulture, Environment, and Biotechnology, 57*(6), 560–572. https://doi.org/10.1007/s13580-016-0068-y

Tian, Z., Ma, W., Yang, Q., & Duan, F. (2021). Application status and challenges of machine vision in plant factory—A review. *Information Processing in Agriculture*. https://doi.org/10.1016/j.inpa.2021.06.003

Ting, K. C., Lin, T., & Davidson, P. C. (2016). Integrated urban controlled environment agriculture systems. In T. Kozai, K. Fujiwara, & E. S. Runkle (Eds.), *LED lighting for urban agriculture* (pp. 19–36). Springer. https://doi.org/10.1007/978-981-10-1848-0_2

Tzounis, A., Katsoulas, N., Bartzanas, T., & Kittas, C. (2017). Internet of Things in agriculture, recent advances and future challenges. *Biosystems Engineering, 164*, 31–48. https://doi.org/10.1016/j.biosystemseng.2017.09.007

United Nations. (2019). *World urbanization prospects: The 2018 revision*. UN.

Vandenberghe, B., Depuydt, S., & Messem, A. V. (2018). *How to make sense of 3D representations for plant phenotyping: A compendium of processing and analysis techniques*. OSF Preprints.

Vanthoor, B. H. E., de Visser, P. H. B., Stanghellini, C., & van Henten, E. J. (2011a). A methodology for model-based greenhouse design: Part 2, description and validation of a tomato yield model. *Biosystems Engineering, 110*(4), 378–395. https://doi.org/10.1016/j.biosystemseng.2011.08.005

Vanthoor, B. H. E., Stanghellini, C., van Henten, E. J., & de Visser, P. H. B. (2011b). A methodology for model-based greenhouse design: Part 1, a greenhouse climate model for a broad range of designs and climates. *Biosystems Engineering, 110*(4), 363–377. https://doi.org/10.1016/j.biosystemseng.2011.06.001

Vanthoor, B. H. E., Van Henten, E. J., Stanghellini, C., & De Visser, P. H. B. (2011c). A methodology for model-based greenhouse design: Part 3, sensitivity analysis of a combined greenhouse climate-crop yield model. *Biosystems Engineering, 110*(4), 396–412.

Wang, C., Du, P., Wu, H., Li, J., Zhao, C., & Zhu, H. (2021a). A cucumber leaf disease severity classification method based on the fusion of DeepLabV3+ and U-Net. *Computers and Electronics in Agriculture, 189*, 106373. https://doi.org/10.1016/j.compag.2021.106373

Wang, H., Gu, M., Cui, J., Shi, K., Zhou, Y., & Yu, J. (2009). Effects of light quality on CO2 assimilation, chlorophyll-fluorescence quenching, expression of Calvin cycle genes and carbohydrate accumulation in Cucumis sativus. *Journal of Photochemistry and Photobiology B: Biology, 96*(1), 30–37. https://doi.org/10.1016/j.jphotobiol.2009.03.010

Wang, J., Yu, L., Yang, J., & Dong, H. (2021b). DBA_SSD: A novel end-to-end object detection algorithm applied to plant disease detection. *Information, 12*(11), 474. https://doi.org/10.3390/info12110474

Wang, W., Zhang, M., Liu, C., Li, M., & Liu, G. (2013). Real-time monitoring of environmental information and modeling of the photosynthetic rate of tomato plants under greenhouse conditions. *Applied Engineering in Agriculture, 29*(5), 783–792.

Werbos, P. J. (1990). Backpropagation through time: What it does and how to do it. *Proceedings of the IEEE, 78*(10), 1550–1560. https://doi.org/10.1109/5.58337

Wijaya, R., Hariono, B., Saputra, T. W., & Rukmi, D. L. (2020). Development of plant monitoring systems based on multi-camera image processing techniques on hydroponic system. *IOP Conference Series: Earth and Environmental Science, 411*(1), 012002. https://doi.org/10.1088/1755-1315/411/1/012002

Worrall, G., Rangarajan, A., & Judge, J. (2021). Domain-guided machine learning for remotely sensed in-season crop growth estimation. *ArXiv:2106.13323 [Cs]*. http://arxiv.org/abs/2106.13323

Wu, S. G., Bao, F. S., Xu, E. Y., Wang, Y.-X., Chang, Y.-F., & Xiang, Q.-L. (2007). A leaf recognition algorithm for plant classification using probabilistic neural network. In *2007 IEEE international symposium on signal processing and information technology* (pp. 11–16). IEEE. https://doi.org/10.1109/ISSPIT.2007.4458016

Wu, Z., Yang, R., Gao, F., Wang, W., Fu, L., & Li, R. (2021). Segmentation of abnormal leaves of hydroponic lettuce based on DeepLabV3+ for robotic sorting. *Computers and Electronics in Agriculture, 190*, 106443. https://doi.org/10.1016/j.compag.2021.106443

Xia, F., Yang, L. T., Wang, L., & Vinel, A. (2012). Internet of things. *International Journal of Communication Systems, 25*(9), 1101–1102. https://doi.org/10.1002/dac.2417

Xu, D., Ahmed, H. A., Tong, Y., Yang, Q., & van Willigenburg, L. G. (2021). Optimal control as a tool to investigate the profitability of a Chinese plant factory—Lettuce production system. *Biosystems Engineering, 208*, 319–332. https://doi.org/10.1016/j.biosystemseng.2021.05.014

Xu, D., Du, S., & van Willigenburg, G. (2018). Adaptive two time-scale receding horizon optimal control for greenhouse lettuce cultivation. *Computers and Electronics in Agriculture, 146*, 93–103. https://doi.org/10.1016/j.compag.2018.02.001

Yang, Q. C. (2019). *Plant factory*. Tsinghua Unveristy Press.

Yang, S., Zheng, L., Gao, W., Wang, B., Hao, X., Mi, J., & Wang, M. (2020). An efficient processing approach for colored point cloud-based high-throughput seedling phenotyping. *Remote Sensing, 12*(10), 1540. https://doi.org/10.3390/rs12101540

Yeh, Y.-H. F., Lai, T.-C., Liu, T.-Y., Liu, C.-C., Chung, W.-C., & Lin, T.-T. (2014). An automated growth measurement system for leafy vegetables. *Biosystems Engineering, 117*, 43–50. https://doi.org/10.1016/j.biosystemseng.2013.08.011

Yin, X., Struik, P. C., & Goudriaan, J. (2021). On the needs for combining physiological principles and mathematics to improve crop models. *Field Crops Research, 271*, 108254. https://doi.org/10.1016/j.fcr.2021.108254

Zaborowicz, M., Boniecki, P., Koszela, K., Przybylak, A., & Przybył, J. (2017). Application of neural image analysis in evaluating the quality of greenhouse tomatoes. *Scientia Horticulturae, 218*, 222–229.

Zaidi, M. A., Murase, H., & Honami, N. (1999). Neural network model for the evaluation of lettuce plant growth. *Journal of Agricultural Engineering Research, 74*(3), 237–242. https://doi.org/10.1006/jaer.1999.0452

Zhang, X., He, D., Niu, G., Yan, Z., & Song, J. (2018a). Effects of environment lighting on the growth, photosynthesis, and quality of hydroponic lettuce in a plant factory. *International Journal of Agricultural and Biological Engineering, 11*(2), 33–40. https://doi.org/10.25165/ijabe.v11i2.3420

Zhang, J., Pu, R., & Huang, W. (2012). Using in-situ hyperspectral data for detecting and discriminating yellow rust disease from nutrient stresses. *Field Crops Research, 134*, 165–174. https://doi.org/10.1016/j.fcr.2012.05.011

Zhang, K., Wu, Q., Liu, A., & Meng, X. (2018b). Can deep learning identify tomato leaf disease? *Advances in Multimedia, 2018*, 1–10. https://doi.org/10.1155/2018/6710865

Zhang, L., Xu, Z., Xu, D., Ma, J., Chen, Y., & Fu, Z. (2020). Growth monitoring of greenhouse lettuce based on a convolutional neural network. *Horticulture Research, 7*(1), 124. https://doi.org/10.1038/s41438-020-00345-6

Zhao, C., Liu, B., Xiao, L., Hoogenboom, G., Boote, K. J., Kassie, B. T., Pavan, W., Shelia, V., Kim, K. S., Hernandez-Ochoa, I. M., Wallach, D., Porter, C. H., Stockle, C. O., Zhu, Y., & Asseng, S. (2019). A simple crop model. *European Journal of Agronomy, 104*, 97–106. https://doi.org/10.1016/j.eja.2019.01.009

Chapter 6
Data-Driven Modeling for Crop Mapping and Yield Estimation

Xingguo Xiong, Qiyu Tian, Luis F. Rodriguez, and Tao Lin

6.1 Introduction

Increasing food production requires the sustainable management of agricultural activities, and there is a clear demand for the monitoring of crop growth statuses across different locations and environmental contexts (Dong et al., 2016; Weiss et al., 2020). Satellite-based crop mapping represents an essential tool for agricultural resource monitoring, and it can provide important information for acreage surveys, yield estimation, and water management (Löw et al., 2013). Reliable yield estimation is used to understand how crop yields respond to various environmental factors (Lobell, 2013). It offers critical information for decision-making in agricultural insurance and agriculture-related policies (Lobell et al., 2015).

Satellite remote sensing offers the unique advantage of mapping crop types and monitoring crop yields across different spatiotemporal scales because of its wide spatial coverage and high temporal resolution (Hu et al., 2019; Song et al., 2017). Most moderate spatial resolution (10–1000 m) earth observation data (e.g., Sentinel, Landsat, MODIS) are free of charge to the public and are produced daily on average (Rußwurm & Körner, 2019). The reflectance spectra of crops vary with crop growth, directly describing the phenological transitions of vegetation. These vegetation-characteristic phenological transitions have been applied to improve crop mapping and yield estimation (Cao et al., 2021; Dong et al., 2016; Rußwurm & Körner,

X. Xiong · Q. Tian · T. Lin (✉)
College of Biosystems Engineering and Food Science, Zhejiang University, Hangzhou, Zhejiang, China
e-mail: lintao1@zju.edu.cn

L. F. Rodriguez
Department of Agricultural and Biological Engineering, University of Illinois at Urbana-Champaign, Urbana, IL, USA

S. Ma et al. (eds.), *Sensing, Data Managing, and Control Technologies for Agricultural Systems*, Agriculture Automation and Control,
https://doi.org/10.1007/978-3-031-03834-1_6

2018a). Several useful vegetation indices have been designed to reduce the dimensionality of multiple spectral bands and to characterize the biophysical properties of crops. For example, the normalized difference vegetation index (NDVI) and enhanced vegetation index (EVI) exhibit an obvious temporal pattern in summer crops, with an increasing trend before the heading phase followed by a rapid decrease (Dong et al., 2016). In addition, decades' worth of meteorological datasets has been constructed around the world (Abatzoglou et al., 2018). Multiple agro-meteorological indices have been calculated to better capture the impacts of extreme meteorological stress on crop yields (Zhang et al., 2014).

Crop mapping and yield estimation have achieved comparable successes by combining training data and advanced computer vision techniques (Burke et al., 2021). Numerous efforts have been made in agricultural surveys to collect reference data. Crop yield is updated annually in many regions, and crop-type maps are generated regionally. For example, the Cropland Data Layer and the Crop Inventory dataset were created by the United States Department of Agriculture and Agriculture and Agri-Food Canada, respectively (Song et al., 2017). For these reference data, a variety of data-driven methods have been applied to classify crops and estimate crop yields. Linear regression is employed for estimating yields based on a set of predictor variables, such as vegetation indexes or meteorological factors. Non-parametric machine learning better captured the nonlinear relationships between input predictor and output response; it also exhibited a robust performance for high-dimensional data. Deep learning has been introduced to learn features and generate predictions from labeled data in an end-to-end fashion.

This chapter reviews a variety of data-driven approaches that have been applied for crop mapping and yield estimation, with a focus on machine learning and deep learning approaches. The workflow for data analysis is summarized as follows: data collection, data preprocessing, model development, model evaluation, and model improvement. The application of these approaches in crop mapping and yield estimation is described. The following section pays particular attention to data-driven approaches that use algorithms from machine learning and deep learning for crop mapping and yield estimation. Key constraints on future progress are explored, and possible future research directions are highlighted.

6.2 Workflow of Data-Driven Modeling Analysis

Data-driven modeling analysis was implemented by a standardized workflow (Fig. 6.1); this includes the processes of data collection, data preprocessing, model development, model evaluation, and model improvement. These processes are broadly applied for crop mapping and yield estimation.

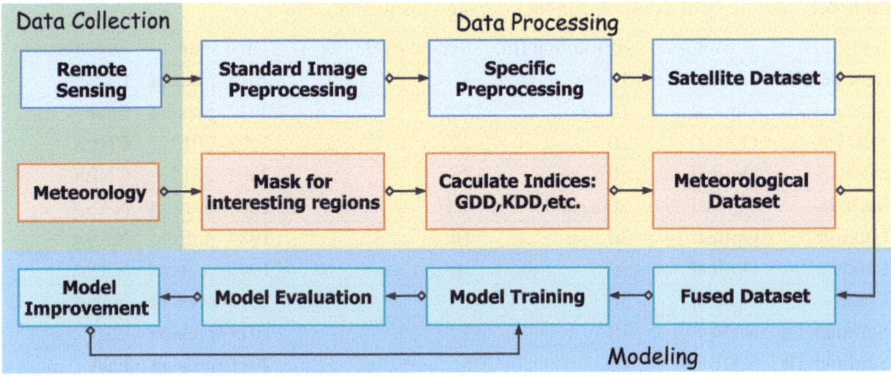

Fig. 6.1 Workflow of data-driven modeling analysis

6.2.1 Data Collection

Satellite remote sensing observations and meteorological records are used as primary data sources because of their ability to represent crop growth characteristics from various perspectives. The main methods of collecting the desired data include experimental collection, random sampling, historical data retrieval, and online public datasets. Other measurement values can be supplemented by expert opinions and domain information.

Satellite Remote Sensing Data

Satellite remote sensing has been used as a nondestructive tool to identify crop growth statuses. It offers systematic and periodic crop information at a large spatial scale, which facilitates the representation of spatiotemporal heterogeneity within an area of interest. Crop monitoring using satellite remote sensing has been widely addressed in multiple specific applications, including crop mapping and yield estimation (Weiss et al., 2020). Moderate spatial resolution (10–1000 m) data are being captured by an increasing number of satellite missions (Table 6.1), and these data are broadly used for crop cultivar retrieval (Ma et al., 2017). These satellite data are publicly available, mainly from the Moderate-resolution Imaging Spectroradiometer (MODIS), Landsat, and Sentinel.

Meteorological Data

Crop yields are affected by many variables, including seed quality, meteorological factors, and water usage (Guo & Xue, 2012). Meteorological effects often interact in nonlinear ways; this can both positively and negatively influence gross primary production and crop yields (Lischeid et al., 2022). Meteorological data are retrieved

Table 6.1 Summary of available satellite remote sensing data source

Name	Sensor type	Resolution (m)	Repeat cycle (days)	Time span	Organization
Sentinel-2A	Optical	10-60	5	2015–present	ESA
Sentinel-2B	Optical	10–60	5	2017–present	ESA
SPOT 4	Optical	20	26	1998–2013	CNES
SPOT 5	Optical	20	26	2002–2015	CNES
MODIS	Optical	500/1000	8/16	2000–present	NASA
Landsat 5	Optical	30	16	1984–present	NASA
Landsat 7	Optical	30	16	1999–present	NASA
Landsat 8	Optical	30	16	2013–present	NASA
Sentinel-1A	SAR	10	5	2014–present	ESA
Sentinel-1B	SAR	10	5	2016–present	ESA

SPOT systeme probatoire d'observation de la terre; *MODIS* moderate-resolution imaging spectro-radiometer, *SAR* synthetic aperture radar, *ESA* European Space Agency, *CNES* Centre National d'Etudes Spatiales, *NASA* National Aeronautics and Space Administration

Table 6.2 Summary of meteorological factors

Meteorological factors	Variables	Meaning	Type
Temperature	Tmin	Daily minimum temperature	Raw
	Tmax	Daily maximum temperature	Raw
Water supply	Pre	Daily total precipitation	Raw
	Pdsi	Palmer drought severity index	Calculated
Water demand	Pet	Evapotranspiration	Calculated
	Vap	Vapor pressure	Calculated
	Vpd	Vapor pressure deficit	Calculated
Land surface temperature	NGDD	Normal growing degree days	Calculated
	HKDD	Hot killing degree days	Calculated
	CKDD	Cold killing degree days	Calculated

Note: "Raw" means raw data, "Calculated" means data values obtained from the predefined equation

from meteorological observation stations or remote sensing-derived products. Several frequently used variables include precipitation, air temperature, relative humidity, and wind speed (Table 6.2).

6.2.2 Data Preprocessing

Data preprocessing is necessary to remove noise and redundant information before model development in crop mapping and yield estimation; this includes cloud removal in optical images and pixel-level speckle filtering in synthetic aperture radar (SAR) images. In addition, basic statistical metrics (e.g., variance, standard deviation, and quantile) are often used to detect outliers. The outliers are removed based on predefined thresholds that are set based upon these metrics. Filtering algorithms, such as the Lee filter, are also utilized to eliminate speckle noise in SAR imagery.

Remote sensing data are preprocessed in two major steps. First, standard image preprocessing is carried out, comprising radiometric calibration, atmospheric correction, and geometric correction. For SAR images, additional steps are necessary, including thermal noise removal and terrain correction. Fortunately, available remote sensing products such as Sentinel-2 L2A are ready-to-use products that have been preprocessed by these standard operations. After the first step, specific preprocessing is used for crop mapping and yield estimation. Satellite image data are generally processed using the following steps to construct a continuous time series. Resampling and coordinate system transformations are applied to ensure alignment with a geo-referenced raster; pixels of interest are extracted using existing land cover and land use products; image noise is removed in images. For SAR images, speckles in the image are filtered using suitable filters, such as the averaged structure Lee filter (Wei et al., 2021). For optical images, because of cloud coverage, images are screened by the cloud coverage percentage or by visual inspection (Vuolo et al., 2018). A cloud removal algorithm is applied to eliminate small clouds and obtain cloud-free reflectance values; the missing values (attributable to cloud contamination) are reconstructed by generating median or mean composites within a predefined time span. Interpolation and filtering algorithms represent another possible approach (Gumma et al., 2014). The combination of linear interpolation and Savitzky–Golay filter has been used in many multi-temporal studies (Chen et al., 2021; Liu et al., 2020). Vegetation indices sensitive to various biophysical properties are calculated from the row reflectance band for further analysis (Table 6.3). Meteorological data are often obtained from an already processed dataset, in which outliers have been removed and missing data have been filled. Agro-meteorological indices can be calculated to better capture the impacts of extreme stress (Cao et al., 2021). Both remote sensing data and meteorological data are aggregated into a specific spatial and temporal scale to assist data analysis at various levels.

Table 6.3 Indexes derived from remote sensing and meteorological data

Names	Index type	Characteristics	References
NDVI	Vegetation index	Vegetation status	Rouse et al. (1974)
EVI	Vegetation index	Vegetation status, canopy structure	Huete et al. (2002)
LSWI	Vegetation index	Water content, residue cover	Xiao et al. (2002)
NDTI	Vegetation index	Nonphotosynthetic components, residue cover	Van Deventer et al. (1997)
NDWI	Vegetation index	Water content	Gao (1996)
GDD	Meteorological index	Crop growth	McMaster and Wilhelm (1997)
KDD	Meteorological index	Crop growth	Yan and Hunt (1999)

LSWI land surface water index, *NDTI* normalized difference tillage index, *NDWI* normalized difference water index, *GDD* growing degree days, *KDD* killing degree days

Google Earth Engine (GEE) and Python are the two most commonly used tools for remote sensing and agro-meteorological data preprocessing.

Google Earth Engine (GEE)

GEE is a cloud-based platform for global-scale geospatial analysis; it is designed to help not only traditional remote sensing scientists but also users who lack the technical ability to use supercomputing resources (Gorelick et al., 2017). The GEE consists of a data catalog and a high-performance parallel computation service. The data catalog houses a large publicly available geospatial dataset repository with a capacity of several petabytes. Optical sensors such as Landsat and MODIS are the most frequently used data sources, followed by Sentinel-2 and Sentinel-1 (Tamiminia et al., 2020).

Users can access and analyze data using an Internet-accessible application programming interface (API) and an associated online interactive development environment (IDE) (Gorelick et al., 2017). The online IDE (also called the Earth Engine Code Editor) is often used to send interactive or batch queries. Queries are constructed by combining operations from a client library with more than 800 functions. These functions range from simple mathematical operations to complex geostatistical computations, machine learning, and image processing (Table 6.4).

In the past few years, the GEE has been applied to many high-impact societal issues, including deforestation, food security, and climate monitoring (Gorelick et al., 2017). The 349 papers that have employed GEE in a wide variety of remote sensing applications are comprehensively summarized (Tamiminia et al., 2020). These applications can be classified into 11 groups. Crop mapping is the most

Table 6.4 Built-in functions in the GEE platform

Package	Capabilities	Package	Capabilities
Image	Band composites, registration	Reducer	Image composition, image aggregation
	Masking, mosaicking, clipping		Statistics of images and features
	Edge and texture extraction		Rasterizing and vectorization
	Spatial operation		Resampling and reducing resolution
Image collection	Filtering, reducing, mapping	Charts	Time-series visualization
	Composing and mosaicking		Histograms
	Iteration	Machine learning	Random forest
Feature, feature collection	Filtering, reducing		K-means clustering

common object of study (including vegetation and rice paddies), followed by agricultural monitoring.

Python

Python is a useful tool for big data analysis, owing to its easy readability and statistical analysis capacities. It offers many well-tested analytical libraries. A Python API package called "ee" has been developed based on GEE. All functions in the client library are accessible through this package, and queries can be constructed in the local IDE to analyze large-scale geospatial data. In addition, a large number of libraries include packages such as numerical computing, data analysis, statistical analysis, visualization, and machine learning (Table 6.5). For example, NumPy is used for scientific computing and basic and advanced array operations, and Matplotlib can generate data visualizations such as two- or three-dimensional diagrams and graphs. Big datasets can be easily processed and analyzed using the Python library, with the help of powerful local computing resources.

6.2.3 Model Development

Models are built to fit the relationships between predictor variables (e.g., vegetation index, meteorological index) and response variables (e.g., crop type and crop yield). The trained model can predict the response variable from the predictor variables. For feature importance analysis, the model helps to elucidate whether the predictor variables are important in explaining the response variable, as well as how each component of the predictor variable affects the response variable. No one method dominates over all other methods and all possible datasets, and a suitable algorithm should be selected prior to model construction (James et al., 2013).

Table 6.5 Commonly used Python libraries

Category	Package	Functionality	References
Data processing	NumPy	Scientific computing and array operation	Harris et al. (2020)
	SciPy	Linear algebra, integration, optimization, and statistics	Virtanen et al. (2020)
	Pandas	Data analysis and manipulation	McKinney (2010)
Modeling	SciKit-Learn	Machine learning and data mining	Pedregosa et al. (2011)
	PyTorch	Deep learning	Paszke et al. (2019)
	TensorFlow	Machine learning and deep learning	Abadi et al. (2016)
Data visualization	Matplotlib	Data visualization	Hunter (2007)
	Seaborn	Visualizing statistical models	Waskom (2021)

Three data-driven methods are widely used in crop mapping and yield estimation: linear regression, non-parametric machine learning, and deep learning. Linear regression is a straightforward method for predicting quantitative responses regarding a set of predictor variables. The coefficients of the model are estimated by minimizing the least-squares criterion (James et al., 2013). Variants of linear regression [e.g., ridge regression and least absolute shrinkage and selection operator (LASSO)] are fitted to address the problem of collinearity between variables, thereby facilitating robust prediction performance. A penalty term is added to the loss function to regularize or shrink the coefficients. Non-parametric machine learning methods such as random forest (RF) or support vector machines (SVMs) are not constrained by assumptions such as the normal distribution of input data (Löw et al., 2013). The RF classifier is composed of many decision-tree classifiers. Each tree is trained using a bootstrap or bagging strategy (Breiman, 1996). Input features are randomly selected from the total feature set to build each tree in the forest. A majority vote strategy is employed for the results of all trees, and the model output is calculated accordingly (Breiman, 2001). Meanwhile, the variable importance can be calculated using RF. SVM can separate different class distributions in high-dimensional feature spaces, which are mapped using nonlinear kernel functions (Vapnik, 1999). An optimal hyper-plane is fitted to maximize the interval between the plane and the closest training data. The deep learning model architecture comprises a series of connected units that facilitate hierarchical feature representation exclusively from data (Zhong et al., 2019a, b). These units learn features and perform classification in an end-to-end fashion without the need for manual feature engineering. Developed from multilayer perceptron (MLP), the deep learning model contains an input layer, hidden layers, and an output layer. An activation function is applied to the output of the hidden layer to increase the nonlinearity; this output is then used as the input of the next layer. The models are trained iteratively using forward and backward propagation. Weight decay and dropout are used to tackle the problems of over-fitting (Zhang et al., 2021). Under increasing numbers of layers, the parameters must be updated using more labeled data.

6.2.4 Model Evaluation and Improvement

Qualitative and quantitative indicators are used to evaluate model output. For yield estimation analysis, the performance of the model is typically assessed using two indicators: the root mean squared error (RMSE) and the R^2 statistic. The RMSE is small if the predicted value is close to the observed one; it is large if the predicted and observed values differ significantly. R^2 takes a value between 0 and 1. It measures the proportion of variability in the response variable, which can be explained using predictor variables (James et al., 2013). The error matrix plays an important role in the accuracy assessment in crop-type mapping studies. The quantitative accuracy metrics computed from the error matrix include the overall accuracy, F1-score, producer's accuracy, and user's accuracy. The overall accuracy and

F1-score are comprehensive measures that characterize the accuracy of a given class using a single value. The user's accuracy represents the proportion of correctly predicted samples in all predicted samples of the target class. The producer's accuracy quantifies the accuracy of the correctly identified samples in the reference samples.

The model is improved and updated according to the aforementioned evaluation indicators. At this stage, if the expected accuracy is not achieved, the model is redeveloped by parameter tuning or selecting a more suitable one, and another model evaluation is performed to obtain the best results for the given target.

6.3 Application in Crop Mapping and Yield Estimation

6.3.1 Crop Mapping

Timely and reliable crop mapping is important for monitoring crop growth and achieving sustainable food security (Wang et al., 2019; Xu et al., 2020). Crop type information has traditionally been obtained from field surveys and censuses (Wang et al., 2019). Satellite observations can provide crop growth information on a large spatial scale, which helps to reduce the burden of field data collection (Song et al., 2017). Crop-type maps are created using features derived from remote sensing observations and advanced data analysis technologies (Wang et al., 2019). These technologies include threshold-based methods and emerging data-driven methods. The former has been developed using domain expert knowledge and applied at various temporal and spatial scales; the latter allows for automated analysis of satellite imagery with ground truth data, and these latter methods can be categorized into two groups: machine learning-based and deep learning-based.

Threshold-Based Method

The vegetation index time-series curves are closely related to crop phenology (Sakamoto et al., 2010). Individual crop phenologies allow multi-temporal remote sensing data to effectively characterize and extract target crops (Pan et al., 2012).

Phenological indexes are manually developed from time-series profiles or specific periods in which phenological characteristics are inherently exhibited, and a set of thresholds at which these indexes realize robust crop discrimination are manually determined (Table 6.6). Temporal matching techniques calculate the global temporal similarity index to quantify the similarity between the target crop and other land covers. Distance metrics (which measure the time-series profile similarity) are calculated, and a suitable threshold is set for these distance metrics, using ground truth data or agricultural statistical data (Dong et al., 2020; Gumma et al., 2015; Sun et al., 2012). However, time-series profiles of vegetation indices (VIs) typically exhibit intra-class variability due to variations in crop growth, climate conditions, and plant structures (Qiu et al., 2015; Wardlow et al., 2007).

Table 6.6 Vegetation indices developed for crop classification

Spectral response	Phenological stage	Band	Sensors	Crops	References
LSWI + a > EVI/NDVI, 0< a < 0.2	Flooding/transplanting stage	VI	MODIS	Rice	Xiao et al. (2006)
LSWI + a > EVI/NDVI, 0< a < 0.2	Flooding/transplanting stage	VI	MODIS	Rice	Dong et al. (2016)
The minima and maxima of the EVI	Flooding/transplanting, heading stage	VI	MODIS	Rice	Boschetti et al. (2017)
The adjusted kappa of NDVI in key growth stage	Jointing, heading, milking stages	VI	MODIS	Multiple crops	Massey et al. (2017)
EVI in transition date of phenological stages	Jointing, heading, milking stages	VI	MODIS	Corn, soybean	Zhong et al. (2016)
Peaks and troughs in EVI	Seeding, tillering, heading and harvest stage	VI	MODIS	Winter wheat	Qiu et al. (2017)

VI vegetation index
Note: "Multiple crops" means more than three types of crops

Phenological metrics derived from specific growth stages have been developed and widely applied for crop mapping (Dong et al., 2016; Xiao et al., 2005, 2006; Zhang et al., 2015). Flooding and rice transplanting signals are key features for identifying rice paddy fields (Dong et al., 2016). The relationship between LSWI and NDVI (EVI) has been effectively explored and captured to identify flooding/transplanting signals. In cross-year corn and soybean mapping, phenological indices have been obtained by fitting a predefined curve to the EVI time series (Zhong et al., 2016). The EVI amplitude, changing rate, and transition dates pertaining to the start, rapid growth, and end of the entire growing season were calculated. A set of universal decision rules was built using substantial expert input and intensive image interpretation (Zhong et al., 2016). Winter crops (e.g., winter wheat) have distinctive phenological features due to their long over-winter periods. Seeding date, first peak date, second peak date, and harvest date represent the four key phenological stages used for winter wheat identification. The NDVI (EVI) in the four stages is calculated to establish a specific index for cross-region and cross-year winter wheat mapping (Pan et al., 2012; Qiu et al., 2017; Qu et al., 2021).

Phenological index-based methods depend on expert experience and local agricultural knowledge. These distinctive indexes help realize crop mapping across years and at a large spatial scale (Dong et al., 2016; Xiao et al., 2006). However, these algorithms have been applied using limited vegetation indices, primarily NDVI and EVI derived from optical images. It is difficult for humans to use multi-temporal, multi-spectral, and multimodal data in combination. The process of phenological index construction (to account for complex factors such as inter-class similarity and intra-class variability) is overwhelming to human capabilities in the big data era. This is often the case for multi-temporal classification and multi-source data fusion (Zhong et al., 2019a, b).

Emerging Data-Driven Methods

Data-driven models using big labeled datasets have been developed to achieve better results without manual intervention (Xu et al., 2020; Zhong et al., 2019a, b). Efforts have been made in data-driven crop mapping using different data sources from optical and microwave remote sensing (Table 6.7).

The spectral reflectivity of optical images is useful for describing growth cycles and phenological characteristics, which is helpful for training robust data-driven models in large-scale crop mapping. Images of key phenological stages are utilized, though they are limited by the low temporal resolution of remote sensing data or the cloud contamination. The spectral response in the key phenological stage has been used to generate descriptive statistics for specific crops (Mathur & Foody, 2008). For example, smoothed NDVIs of weeks 21 and 35 were generated for training a decision tree (DT) model (Howard & Wylie, 2014). Depending on the spectral features of the target crop in certain phases, a variety of regional-scale crop cover products can be provided using machine learning methods (Ozdogan & Gutman, 2008). Information for the multi-temporal crop has been explicitly explored to improve crop mapping at larger spatiotemporal scales. Traditionally, each time step in the VI series is treated as an independent dimension, and classical machine learning methods such as DT, RF, and SVM are employed (Wardlow & Egbert, 2008; Zheng et al., 2015). Temporal features derived from seasonal statistics and predefined equations improve classification accuracies compared to the original VI values (Song et al., 2017; Wardlow et al., 2007; Zhong et al., 2014). These features directly represent crop progress and growing conditions, which include dates of phenological transition stages and the rate of vegetative development (Zhong et al., 2016). Spectral features, such as the NDVI and EVI for vegetation depiction, NDWI for water detection, and the LSWI which is sensitive to soil moisture, have been calculated at every time step (Pelletier et al., 2016). In the same way, large numbers of spatial features (e.g., geometric and texture) have been utilized (Pelletier et al., 2016). Variants of convolutional neural networks (CNNs) and recurrent neural networks (RNNs) have also been developed to identify crop growth patterns and classify crops from raw spectral reflectance (Ndikumana et al., 2018; Rußwurm & Körner, 2018b; Zhong et al., 2019a, b).

Microwave remote sensing data (e.g., SAR imagery) are useful for crop mapping, because they offer all-weather acquisition capabilities (Zhao et al., 2020). SAR data contain contextual information regarding crop growth, which can be extracted from the phase and intensity of single, dual, and polarimetric modes. Handcrafted features derived from SAR data have been fed into machine learning models to achieve crop classification (McNairn et al., 2009). Deep learning models have been employed to process the row backscatter coefficient time series and achieve high classification accuracies (Du et al., 2019; Wei et al., 2021). However, SAR signals can easily become saturated by soil roughness and vegetation growth, which can limit the application of SAR data (Adrian et al., 2021; Huang et al., 2018).

Optical and SAR images are often fused to remove noise generated by clouds and speckles, as well as to provide complementary information regarding the same

Table 6.7 Data sources utilized in data-driven crop mapping

Sensor type	Sensors	Input information	Classifier	Objectives	Reference
Optical	MODIS	NDVI at every time step	TML	Corn and soybean	Wardlow and Egbert (2008)
	Landsat	NDVI at every time step	TML	Multiple crops	Zheng et al. (2015)
	MODIS	Parameters of the predefined equation of EVI	TML	Corn and soybean	Zhong et al. (2016)
	Landsat	Every 10 percentile of row band and VI value	TML	Soybean	Song et al. (2017)
	Landsat	Time series of EVI	DL	Multiple crops	Zhong et al. (2019a, b)
	Sentinel-2	Time series of row band	DL	Multiple crops	Rußwurm and Korner (2017)
SAR	ALOS PALSAR	Polarimetric feature derived from VV, etc. at key stages	TML	Multiple crops	McNairn et al. (2009)
	Sentinel-1	Time series of row VV and VH	DL	Rice	Wei et al. (2021)
	ZY-3, sentinel-1	Time series of row VV and VH	DL	Rice, rape, cotton	Zhou et al. (2019)
	Sentinel-1	Time series of row VV and VH	DL	Multiple crops	Ndikumana et al. (2018)
Optical + SAR	Sentinel-1, sentinel-2	Harmonic regression parameters of optical time-series data, the monthly composition of SAR data	TML	Multiple crops	Pott et al. (2021)
	Sentinel-1, sentinel-2, landsat 8	Row band, VI, VV, and VH of each time step	TML	Multiple crops	Blickensdörfer et al. (2022)
	Sentinel-1, sentinel-2	Time series of row band, VI and VV, VH	DL	Multiple crops	Ienco et al. (2019)
	Sentinel-1, sentinel-2	Time series of VV, VH, and VI	DL	Multiple crops	Zhao et al. (2020)
	Sentinel-1, sentinel-2	Mean, median, and standard deviation of VV of objects	DL	Multiple crops	Garioud et al. (2021)

VV vertical transmit and vertical receive, *VI* vegetation index, *ALOS PALSAR* advanced land observing satellite phased array type L-band synthetic aperture radar, *ZY-3* Chinese resources satellite three, *TML* typical machine learning, *DL* deep learning

crop, for better crop classification. The temporal resolution and spatial coverage of optical sensors are strongly affected by cloud cover, resulting in significant missing information (Garioud et al., 2021). Although SAR data are unaffected by clouds, speckles in SAR are generally serious, causing difficulties in image interpretation. In addition, both the geometric imaging pattern and physical radiation mechanism of these two sensors differ (Li et al., 2020). The complementarity of SAR and optical images facilitates the reconstruction of missing information (Garioud et al., 2021). A direct SAR-optical fusion strategy was developed by stacking features extracted from SAR and optical imagery in a high-dimensional feature space. All features are fed into a typical machine learning algorithm such as RF, which has been demonstrated as suitable for large-scale crop mapping tasks (Blickensdörfer et al., 2022; Pott et al., 2021). Studies have also been devoted to achieving data fusion through deep learning methods to reconstruct NDVI time series that can be used for crop monitoring and crop classification. An integrated deep learning network has also been used to construct a relationship between the backscatter coefficient time series from Sentinel-1 and the NDVI time series calculated from Sentinel-2. This network comprised a one-dimensional convolutional neural network and long short-term memory (LSTM) layers (Ienco et al., 2019; Zhao et al., 2020).

Various classification algorithms have also been developed for data-driven crop mapping; machine learning and deep learning play an important role in processing multi-temporal remote sensing data and facilitating crop classification across regions and years (Table 6.8).

Temporal, spectral, and spatial crop features extracted from multi-temporal data have been fed into typical machine-learning classifiers to utilize data efficiently. In these studies, temporal features from the VI time series were extracted by a function or a set of functions. Multi-spectral and high-spatial information embedded in multi-temporal satellite images is also explored. Thus, dozens or even hundreds of features have been applied to depict crop properties from many perspectives. All the features were fed into machine learning models, including RF and SVM. These classifiers were originally designed to process high-dimensional data and were employed to support crop mapping in large-scale and multi-class scenarios. However, large feature sizes produce a heavy computational burden in big data analysis. With limited training data, classification accuracy can degrade when the number of features increases (Hughes, 1968; Löw et al., 2013). As a result, various dimension reduction strategies have been introduced to achieve equivalent or better accuracies and reduce the computational cost without discarding the main information in the full feature space. Existing dimension reduction methods in crop mapping are divided into two categories: feature subset selection and feature contribution quantification. The former typically divides the full feature into several homogeneous feature subsets, and a few labeled data are used to evaluate the impact of these subsets on classification (Hu et al., 2019; Löw et al., 2013); subsets with higher classification accuracies are typically selected. The latter method computes each feature's importance score using RF. Features with scores above a certain threshold are selected and combined to accomplish the subsequent classification task (Pelletier et al., 2016; Zhong et al., 2014). In practical scenarios, it is difficult to produce effective feature representations that rely on human experience and expertise. Feature extraction manually is also labor-intensive and time-consuming. An ideal feature extractor should be trained to automatically capture the crop growth patterns and act as a human to recognize crop types (Zhong et al., 2019a, b).

Table 6.8 Data-driven algorithms for crop mapping

Approaches	Classifier	Band	Feature extraction	Feature selection	Objectives	Reference
Machine learning	SVM	VI	Handcrafted 2D features	Feature subset selection	Corn	Hu et al. (2019)
	RF	OB, VI, PV	Handcrafted 2D features	Feature contribution quantification	Corn and soybean	Zhong et al. (2014)
	RF	OB, VI	Handcrafted 2D features	Feature contribution quantification	Multiple crops	Pelletier et al. (2016)
	SVM	OB, VI	Handcrafted 3D features	Feature subset selection	Multiple crops	Löw et al. (2013)
	RF	OB, VI, PM	Handcrafted 3D features	Feature contribution quantification	Multiple crops	You et al. (2017)
Deep learning	ConvLSTM	OB	Automatic feature learning	N/A	Multiple crops	Rußwurm and Körner (2018a)
	AtLSTM	OB	Automatic feature learning	N/A	Winter wheat	Hu et al. (2019)
	ConvLSTM	OB, VI	Automatic feature learning	N/A	Multiple crops	Rußwurm and Körner (2019)
	AtLSTM	OB	Automatic feature learning	N/A	Corn and soybean	Xu et al. (2020)
	ConvLSTM	OB	Automatic feature learning	N/A	Multiple crops	de Macedo et al. (2020)
	Self-attention	OB	Automatic feature learning 1	N/A	Multiple crops	Rußwurm and Körner (2019)

AtLSTM attention-based long short-term memory, *ConvLSTM* convolutional long short-term memory, *OB* optical bands, *PM* polarization modes, *PV* phenological variables
Notes: 2D features mean features extracted from temporal and spectral dimensions; 3D features spatial, spectral, and temporal dimensions; "N/A" means not exist

Deep learning models show significant potential in crop feature representation for remotely sensed time series; they are marked by complete data-driven feature extraction and crop classification processes, without requiring extra feature selection. In multi-temporal corn and soybean classification, a one-dimensional convolutional neural network (Conv1D) was applied to capture the temporal pattern or shape of the VI series. Conv1D layers can be stacked such that the lower layers focus on local features, and the upper layers summarize more general patterns to a larger extent. The optimal Conv1D model was demonstrated to be superior to popular classifiers such as XGBoost, RF, and SVM (Zhong et al., 2019a, b). In addition to CNNs, RNNs are specialized for sequential data analysis. The original backscatter coefficient time series was used as the input to train an RNN to perform multi-class crop classification in southeastern France, and it outperformed the classical approaches (Ndikumana et al., 2018). Variants of RNNs, such as LSTM, have been successfully applied in crop mapping. LSTM is employed to extract temporal characteristics from a dense time series of optical or SAR remote sensing observations (Rußwurm & Korner, 2017; Zhou et al., 2019). With the exception of widely used temporal properties, the considerable information exhibited by multi-spectral and high spatial resolution data has not been adequately explored. The integration of this information is expected to resist the intra-class variability and inter-class similarity. Three-dimensional convolutional layers were used for the winter wheat mapping. Features were learned simultaneously along spatial and temporal dimensions (Zhong et al., 2019a, b). In addition, multiple optical band time series were stacked to train one LSTM, which learned temporal relationships from multi-temporal-spectral optical data (Rußwurm & Korner, 2017). The newly developed self-attention mechanism has also shown a strong crop mapping performance (Rußwurm & Körner, 2019). In practice, a CNN or LSTM alone may not be adequate for multi-dimensional feature representation. Some studies have incorporated attention mechanisms into LSTM layers; here, multi-temporal and multi-spectral patterns were automatically learned to help cross-region corn and soybean mapping (Xu & Cheng, 2021). Moreover, a CNN and LSTM were integrated to represent spatial-temporal-spectral features. This model first extracted temporal features and then learned the spatial context of the extracted temporal features for crop mapping (Yang et al., 2021). It was demonstrated that the coupled network could outperform an individual CNN or LSTM.

6.3.2 Yield Estimation

Process-Driven Mechanism Methods

Process-driven mechanism methods (crop models) predict crop yield using mechanistic crop growth equations. These equations originated from the canopy photosynthesis theory, and they represent crop phenology, organ growth, dry matter accumulation, water balance, nutrient dynamics, conventional management, and environmental stress (Loomis & Williams, 1963; Monsi, 1953). The crop models were normally developed in the following steps: theoretical development, single-crop model development, and crop model system development. Many theoretical studies were conducted to build single-crop models (Childs et al., 1977; De Vries & Van Laar, 1982). However, it is difficult for these models to apply in practical scenes due to inconsistent input data. Therefore, crop model systems were established by unifying various input data.

Crop model systems can be classified into two categories, depending on the driving factors of crop growth (Table 6.9). One category [e.g., CERES (Basso et al.,

Table 6.9 Process-driven crop simulation models

Driving	Crop model	Crop	Simulated process	Reference
Light	CERES	Cereal	Common processes	Basso et al. (2016)
	WOFOST	Cereal	Common processes, stress of diseases, pests and weeds	Du et al. (2019)
	APSIM	Cereal, beans, cash crop	Common processes, soil salinization, residue decomposition	Holzworth et al. (2014)
Water	AquaCrop	Cereal, tubers vegetables, fruits	Common processes, canopy coverage	Steduto et al. (2009)

Notes: Common processes include phenology, organ growth, dry matter accumulation, water balance, nutrient dynamics, conventional management (sow, harvest, irrigation, fertilization), and environmental stress (temperature, water, nutrient)

2016), WOFOST (Du et al., 2019), and APRSIM (Holzworth et al., 2014)] is driven by light utilization. The other category [e.g., AquaCrop (Steduto et al., 2009)] is driven by water and often applied in drought regions. AquaCrop model simulates the crop growth process using water productivity and estimates crop yield through the transpiration and the harvest index.

The process-driven mechanism methods are fully interpretable and designed for simulating crop growth and yield response, but they are not efficient in capturing several effects (e.g., management, crop varieties, irrigation, and agronomic technologies) because of limited temporal and spatial relevant information. In addition, several parameters in crop models are often set empirically, which constrain the model performance.

Data-Driven Statistical Methods

Data-driven statistical methods quantify the linear or nonlinear relationship between various predictor variables and crop yields, and have become the dominant technique for estimating crop yields. These variables are often retrieved from climatic data, soil maps, crop progress reports, and satellite observations. Climatic data are publicly available from site scales to global scales. Soil maps provide the soil properties of crop growing environments. Crop progress reports are released by the relevant national department and contain the growth states of planted crops. Satellites observe the crops growth from a remote perspective based on optical reflection. Various vegetation indices (VIs) are derived from satellite observations and are widely used in crop yield estimation because of their spatiotemporal coverage. According to the data source used for yield estimation, data-driven statistical methods can be divided into single-source and multi-source yield estimation.

Single-Source Yield Estimation

The vegetation indices (VIs) are typically used as the primary predictor variable for remote-sensing-based yield estimation, and they can explain crop yield variation to some extent (Table 6.10). It is assumed that there is a direct relationship between VIs and canopy biomass, and an indirect relationship between biomass and the final yield (Sakamoto, 2020). The VIs are calculated from raw spectral bands. Several widely used VIs include NDVI, EVI, and NDWI (Xue & Su, 2017). NDVI is sensitive to chlorophyll, while EVI is more responsive to canopy structure changes (e.g., leaf area index, canopy type, plant orientation, and canopy structure). Moreover, EVI is less sensitive to different soil backgrounds (Bidoglio et al., 2002). Based on these VIs, linear regression models are employed. New indicators derived from different VIs have been designed to combine their respective advantages on capturing biological crop growth status. For example, combined with growth stages, NDVI- and EVI-based growth metrics have been designed and applied to estimate soybean yields in the Mississippi Delta of the United States (Shammi & Meng, 2021). Compared with traditional linear regression models, the emergence of powerful nonlinear fitting methods (e.g., random forest) make it possible to build models using raw spectral data. CNN and LSTM were adopted to automatically learn high-order nonlinear properties from the raw optical bands. A Gaussian process was added to clarify the spatiotemporal structure of the model that eliminated the spatial correlation error and improved the model accuracy (You et al., 2017).

Compared to traditional yield estimation patterns based on crop models or on the compilation of the survey information, VIs-based yield estimation is a unique means to provide crop status information over large areas with regular revisits, and allows deriving spatially explicit and temporally resolved maps of production and yield

Table 6.10 Remote sensing indices for yield estimation

Approaches	Input	Sensors	Method	Objectives	Ref
Traditional regression	NDVI	MODIS	Linear regression	Spring wheat etc.	Mkhabela (2011)
	EVI	MODIS	Linear regression	Rice	Song et al. (2017)
	EVI2, NDWI	MODIS	Linear regression	Maize, soybean	Bolton and Friedl (2013)
	NDVI- and -EVI-based metrics	MODIS	Linear regression	Soybean	Shammi and Meng (2021)
Deep learning	Raw spectral bands	MODIS	CNN, LSTM, GP	Soybean	You et al. (2017)
	NDVI	MODIS	Ridge regression, LSTM	Soybean	Wang et al. (2019)

NDVI normalized difference vegetation index, *EVI* enhanced vegetation index, *EVI2* enhanced vegetation index-2, *NDWI* normalized difference water index, *CNN* convolutional neural networks, *LSTM* long short-term memory, *GP* Gaussian process

(Mateo-Sanchis et al., 2019). However, previous studies have demonstrated that VI-based yield estimation methods tended to underestimate yield (Bolton & Friedl, 2013). The variability of VIs could not always represent the yield loss caused by environmental stress (Sakamoto, 2020).

The meteorological factors are applied in data-driven yield estimation to improve estimation accuracy (Table 6.11). Temperature and precipitation are the two key meteorological factors that affect crop yields. When the temperature exceeds the range denoted by the highest and lowest temperatures, crops begin to suffer damage and even death. Raw temperature variables include daily average temperature (T-average), daily minimum temperature (T-min), and daily maximum temperature (T-max) (Lobell & Burke, 2010). Daily precipitation is used as another variable to represent rainfall. Moderate precipitation improves the physiological activities of plants. In addition, solar radiation and air humidity are also used as predictors (Landau et al., 2000; Lischeid et al., 2022). GDD and KDD derived from raw temperature data represent the accumulated effect of temperature on the specific crop (Lin et al., 2020). The supply and demand between precipitation and the required water of crops are often expressed by the vapor pressure deficit (VPD) and potential evapotranspiration (ETP) (Gornott & Wechsung, 2016; Jiang et al., 2020).

Crop yields can be estimated from climatic variables over the entire growing season or for a specific phenological period, as crop growth stages differ in their sensitivity to climate (Gornott & Wechsung, 2016). Various statistical learning and recently developed deep learning models help to quantify the relationship between meteorological factors and crop yields. Multiple linear regression (MLR) is a common method for yield estimations based on multiple meteorological variables (Lobell & Burke, 2010). Collinearity problems among various meteorological variables may exist, which adversely affects the yield estimation accuracy. To select the

Table 6.11 Meteorological variables used for yield estimation

Input type	Climate variable	Model	References
Raw data	T-average, Pre	Multiple linear regression	Lobell and Burke (2010)
Raw data	T-min, T-max, pre, radiation	Multiple linear regression	Landau et al. (2000)
Raw data	T-min, T-max, pre-, radiation, wind speed, air humidity	Random forest, support vector machine models	Lischeid et al. (2022)
Processed index	GDD, KDD, VPD	Long short-term memory networks	Jiang et al. (2020)
Processed index	GDD, KDD, pre	Attention-based long short-term memory networks	Lin et al. (2020)
Processed index	ETP, SRT, pre	Separate time series model, panel data model, random coefficient model	Gornott and Wechsung (2016)

T-average daily average temperature, *T-min* daily minimum temperature, *T-max* daily maximum temperature, *Pre* daily precipitation, *GDD* growing degree days, *KDD* killing degree days, *VPD* vapor pressure deficit, *ETP* potential evapotranspiration optimal variables, stepwise linear regression is introduced and often combined with MLR (Qian et al., 2009). LASSO is another model to address the problem of collinearity and to identify important features. LSTM has become popular in recent years as it has the advantages of capturing temporal features and enabling a powerful nonlinear fitting capacity (Lin et al., 2020).

Multi-Source Yield Estimations

The VIs are directly related to the amount of vegetation, and yield variation induced by environmental stress is correlated with environmental and meteorological variables from air temperature, precipitation, solar radiation, and soil moisture. Thus, multi-source data need to be considered in yield estimation models. Commonly used data sources include meteorological data, soil data, remote sensing observations, and management data. Combining satellite observations and meteorology is the primary method of multi-source yield estimation (Li et al., 2019; Peng et al., 2018). On this basis, soil and management data are integrated in certain cases (Kern et al., 2018; Ma et al., 2021).

Machine learning and emerging deep learning models facilitate the accuracy of multi-source yield estimation. Because the yield distribution varies to a large extent, it is difficult to establish a general model for crop yield estimation using multi-source data. Traditional regression methods [e.g., MLR, stepwise regression, and LASSO regression (Mueller et al., 2012)] are often adopted to establish a linear or nonlinear relationship based on remote sensing, soil, and management measurements. The newly developed nonlinear fitting models (e.g., machine learning and deep learning) can effectively capture the complex relationship between yields and multimodal characteristics. SVR was used to extract features that significantly affect crop growth and could not be quantified ever (Kuwata & Shibasaki, 2015). RF is used to identify the most important meteorological and soil properties drivers of temporal and spatial variability of yield (Brinkhoff & Robson, 2021; Luciano et al., 2021). Certain research also performed feature selection based on the trained DNN model, which successfully decreased the dimension of the input space without significant drop in the prediction accuracy (Khaki & Wang, 2019). Long short-term memory (LSTM) neural network with an attention mechanism (ALSTM) was proposed to assign attention to the key parts of the input sequence that affect the target vectors so that the crop yield features can be accurately extracted (Tian et al., 2021). In addition, certain hybrid models were designed. The CNN-RNN model was designed to capture the time dependencies of environmental factors and the genetic improvement of seeds over time without having their genotype information (Khaki et al., 2020).

6.4 Opportunities and Challenges

Challenges still remain when using data-driven methods to generate crop-type maps and estimate crop yields. First, data-driven methods are heavily dependent on labeled data. These models have achieved only limited accuracy in regions where abundant labeled data are unavailable (Defourny et al., 2019). Second, data-driven models are often considered "black boxes," offering limited interpretability (Xu et al., 2021).

6.4.1 Absent Label Data

While satellite imagery and environmental data are now abundant, the scarcity of labeled data makes the training of data-driven models difficult. The extrapolation of data-driven methods outside of the training region can be compromised by variations in crop growth. Two methods have been very successful in other domains (i.e., land cover and land use mapping), and they may be able to address the problem to some extent:

One-Class Classification

One-class reference data are used to reduce the redundant effort required for non-crop type collection, and one-class classifiers that only require the reference data of the target class are effective methods in this respect. These classifiers can be divided into positive classifiers (using only one-class labels) and positive and unlabeled classifiers (which require additional randomly selected unlabeled data).

Few-Shot Learning

The cross-regional and cross-year transfer of deep models trained with a small amount of labeled data remains a hot topic in land cover and land use mapping. Three strategies have been developed for few-shot learning. The first is self-supervised learning that uses unlabeled data to learn representations valuable to the downstream tasks via a predefined pre-text task. Active learning has been used to actively label a smaller sample set that represents the overall data distribution; this reduces the total annotation effort required to produce a desirable mapping accuracy. Croplands across regions differ, though they still share seasonal and spectral characteristics. Meta-learning can capture these similarities and variations in data, using a unique mechanism that teaches deep models to learn how to learn from data.

6.4.2 "Black Box" Problem of Machine Learning Methods

The lack of interpretability is regarded as another drawback of these high-performance approaches. Interpreting deep learning approaches is crucial for verifying their reliability in both multi-temporal crop mapping and yield estimation.

Most existing data-driven models have not been well integrated with agricultural knowledge or experience. When dealing with sparse data distributions, this raises doubts regarding the reliability of the model. One way to solve these problems is to integrate agricultural knowledge and experience into a machine learning model to constrain and optimize the model.

6.5 Summary

In this chapter, we introduced various data-driven models applied to crop mapping and yield estimation. We summarized the data analysis workflow, with a focus on commonly used data sources, data preprocessing methods, model development processes, model evaluation metrics, and model improvements. Then, we reviewed conventional methods and data-driven models in crop mapping and yield estimation, focusing on the newly developed machine learning and deep learning models. Opportunities and challenges in the application of data-driven models for crop mapping and yield estimation were discussed.

References

Abadi, M., Agarwal, A., Barham, P., Brevdo, E., Chen, Z., Citro, C., Corrado, G. S., Davis, A., Dean, J., & Devin, M. (2016). Tensorflow: Large-scale machine learning on heterogeneous distributed systems. *ArXiv Preprint ArXiv:1603.04467*.

Abatzoglou, J. T., Dobrowski, S. Z., Parks, S. A., & Hegewisch, K. C. (2018). TerraClimate, a high-resolution global dataset of monthly climate and climatic water balance from 1958–2015. *Scientific Data, 5*(1), 1–12.

Adrian, J., Sagan, V., & Maimaitijiang, M. (2021). Sentinel SAR-optical fusion for crop type mapping using deep learning and Google Earth Engine. *ISPRS Journal of Photogrammetry and Remote Sensing, 175*, 215–235. https://doi.org/10.1016/j.isprsjprs.2021.02.018

Basso, B., Liu, L., & Ritchie, J. T. (2016). A comprehensive review of the CERES-wheat, -maize and -rice models' performances. In *Advances in Agronomy* (Vol. 136). Elsevier Inc.. https://doi.org/10.1016/bs.agron.2015.11.004

Bidoglio, G., De Plano, A., Avogadro, A., & Murray, C. N. (2002). Migration behaviour and chemical speciation of Np and Am under nuclear waste repository conditions. *Inorganica Chimica Acta, 95*(1), 1–3. https://doi.org/10.1016/S0020-1693(00)85959-9

Blickensdörfer, L., Schwieder, M., Pflugmacher, D., Nendel, C., Erasmi, S., & Hostert, P. (2022). Mapping of crop types and crop sequences with combined time series of Sentinel-1, Sentinel-2 and Landsat 8 data for Germany. *Remote Sensing of Environment, 269*, 112831. https://doi.org/10.1016/j.rse.2021.112831

Bolton, D. K., & Friedl, M. A. (2013). Forecasting crop yield using remotely sensed vegetation indices and crop phenology metrics. *Agricultural and Forest Meteorology, 173*, 74–84. https://doi.org/10.1016/j.agrformet.2013.01.007

Boschetti, M., Busetto, L., Manfron, G., Laborte, A., Asilo, S., Pazhanivelan, S., & Nelson, A. (2017). PhenoRice: A method for automatic extraction of spatio-temporal information on rice crops using satellite data time series. *Remote Sensing of Environment, 194*, 347–365.

Breiman, L. (1996). Bagging predictors. *Machine Learning, 24*(2), 123–140.

Breiman, L. (2001). Random forests. *Machine Learning, 45*(1), 5–32.

Brinkhoff, J., & Robson, A. J. (2021). Block-level macadamia yield forecasting using spatio-temporal datasets. *Agricultural and Forest Meteorology, 303*, 108369. https://doi.org/10.1016/j.agrformet.2021.108369

Burke, M., Driscoll, A., Lobell, D. B., & Ermon, S. (2021). Using satellite imagery to understand and promote sustainable development. *Science, 371*(6535), 8628. https://doi.org/10.1126/science.abe8628

Cao, J., Zhang, Z., Tao, F., Zhang, L., Luo, Y., Zhang, J., Han, J., & Xie, J. (2021). Integrating multi-source data for rice yield prediction across china using machine learning and deep learning approaches. *Agricultural and Forest Meteorology, 297*, 108275. https://doi.org/10.1016/j.agrformet.2020.108275

Chen, X., Feng, L., Yao, R., Wu, X., Sun, J., & Gong, W. (2021). Prediction of maize yield at the city level in China using multi-source data. *Remote Sensing, 13*(1), 146. https://doi.org/10.3390/rs13010146

Childs, S. W., Gilley, J. R., & Splinter, W. E. (1977). A simplified model of corn growth under moisture stress. *Transactions of the ASAE, 20*(5), 858–865.

de Macedo, M. M. M., Mattos, A. B., & Oliveira, D. A. B. (2020). Generalization of Convolutional LSTM Models for Crop Area Estimation. *IEEE Journal of Selected Topics in Applied Earth Observations and Remote Sensing, 13*, 1134–1142.

De Vries, F. W. T. P., & Van Laar, H. H. (1982). Simulation of growth processes and the model BACROS. In *Simulation of plant growth and crop production* (pp. 114–135). Pudoc.

Defourny, P., Bontemps, S., Bellemans, N., Cara, C., Dedieu, G., Guzzonato, E., Hagolle, O., Inglada, J., Nicola, L., & Rabaute, T. (2019). Near real-time agriculture monitoring at national scale at parcel resolution: Performance assessment of the Sen2-Agri automated system in various cropping systems around the world. *Remote Sensing of Environment, 221*, 551–568.

Dong, J., Fu, Y., Wang, J., Tian, H., Fu, S., Niu, Z., Han, W., Zheng, Y., Huang, J., & Yuan, W. (2020). Early-season mapping of winter wheat in China based on Landsat and Sentinel images. *Earth System Science Data, 12*(4), 3081–3095. https://doi.org/10.5194/essd-12-3081-2020

Dong, J., Xiao, X., Menarguez, M. A., Zhang, G., Qin, Y., Thau, D., Biradar, C., & Moore, B., III. (2016). Mapping paddy rice planting area in northeastern Asia with Landsat 8 images, phenology-based algorithm and Google Earth Engine. *Remote Sensing of Environment, 185*, 142–154.

Du, Z., Yang, J., Ou, C., & Zhang, T. (2019). Smallholder crop area mapped with a semantic segmentation deep learning method. *Remote Sensing, 11*(7), 888.

Gao, B.-C. (1996). NDWI—A normalized difference water index for remote sensing of vegetation liquid water from space. *Remote Sensing of Environment, 58*(3), 257–266.

Garioud, A., Valero, S., Giordano, S., & Mallet, C. (2021). Recurrent-based regression of Sentinel time series for continuous vegetation monitoring. *Remote Sensing of Environment, 263*, 112419. https://doi.org/10.1016/j.rse.2021.112419

Gorelick, N., Hancher, M., Dixon, M., Ilyushchenko, S., Thau, D., & Moore, R. (2017). Google earth engine: Planetary-scale geospatial analysis for everyone. *Remote Sensing of Environment, 202*, 18–27. https://doi.org/10.1016/j.rse.2017.06.031

Gornott, C., & Wechsung, F. (2016). Agricultural and forest meteorology statistical regression models for assessing climate impacts on crop yields: A validation study for winter wheat and silage maize in Germany. *Agricultural and Forest Meteorology, 217*, 89–100. https://doi.org/10.1016/j.agrformet.2015.10.005

Gumma, M. K., Mohanty, S., Nelson, A., Arnel, R., Mohammed, I. A., & Das, S. R. (2015). Remote sensing based change analysis of rice environments in Odisha, India. *Journal of Environmental Management, 148*, 31–41.

Gumma, M. K., Thenkabail, P. S., Maunahan, A., Islam, S., & Nelson, A. (2014). Mapping seasonal rice cropland extent and area in the high cropping intensity environment of Bangladesh using MODIS 500m data for the year 2010. *ISPRS Journal of Photogrammetry and Remote Sensing, 91*, 98–113. https://doi.org/10.1016/j.isprsjprs.2014.02.007

Guo, W. W., & Xue, H. (2012). An incorporative statistic and neural approach for crop yield modelling and forecasting. *Neural Computing and Applications, 21*(1), 109–117.

Harris, C. R., Millman, K. J., van der Walt, S. J., Gommers, R., Virtanen, P., Cournapeau, D., Wieser, E., Taylor, J., Berg, S., & Smith, N. J. (2020). Array programming with NumPy. *Nature, 585*(7825), 357–362.

Holzworth, D. P., Huth, N. I., deVoil, P. G., Zurcher, E. J., Herrmann, N. I., McLean, G., Chenu, K., van Oosterom, E. J., Snow, V., Murphy, C., Moore, A. D., Brown, H., Whish, J. P. M., Verrall, S., Fainges, J., Bell, L. W., Peake, A. S., Poulton, P. L., Hochman, Z., & Keating, B. A. (2014). APSIM - Evolution towards a new generation of agricultural systems simulation. *Environmental Modelling and Software, 62*, 327–350. https://doi.org/10.1016/j.envsoft.2014.07.009

Howard, D. M., & Wylie, B. K. (2014). Annual crop type classification of the US Great Plains for 2000 to 2011. *Photogrammetric Engineering & Remote Sensing, 80*(6), 537–549.

Hu, Q., Sulla-Menashe, D., Xu, B., Yin, H., Tang, H., Yang, P., & Wu, W. (2019). A phenology-based spectral and temporal feature selection method for crop mapping from satellite time series. *International Journal of Applied Earth Observation and Geoinformation, 80*, 218–229. https://doi.org/10.1016/j.jag.2019.04.014

Huang, Y., Chen, Z., Yu, T., Huang, X., & Gu, X. (2018). Agricultural remote sensing big data: Management and applications. *Journal of Integrative Agriculture, 17*(9), 1915–1931. https://doi.org/10.1016/s2095-3119(17)61859-8

Huete, A., Didan, K., Miura, T., Rodriguez, E. P., Gao, X., & Ferreira, L. G. (2002). Overview of the radiometric and biophysical performance of the MODIS vegetation indices. *Remote Sensing of Environment, 83*(1–2), 195–213.

Hughes, G. (1968). On the mean accuracy of statistical pattern recognizers. *IEEE Transactions on Information Theory, 14*(1), 55–63.

Hunter, J. D. (2007). Matplotlib: A 2D graphics environment. *Computing in Science & Engineering, 9*(3), 90–95.

Ienco, D., Interdonato, R., Gaetano, R., & Ho Tong Minh, D. (2019). Combining Sentinel-1 and Sentinel-2 satellite image time series for land cover mapping via a multi-source deep learning architecture. *ISPRS Journal of Photogrammetry and Remote Sensing, 158*, 11–22. https://doi.org/10.1016/j.isprsjprs.2019.09.016

James, G., Witten, D., Hastie, T., & Tibshirani, R. (2013). *An introduction to statistical learning* (Vol. 112). Springer.

Jiang, H., Hu, H., Zhong, R., Xu, J., Xu, J., Huang, J., Wang, S., Ying, Y., & Lin, T. (2020). A deep learning approach to conflating heterogeneous geospatial data for corn yield estimation: A case study of the US Corn Belt at the county level. *Glob Chang Biol, 26*(3), 1754–1766. https://doi.org/10.1111/gcb.14885

Kern, A., Barcza, Z., Marjanović, H., Árendás, T., Fodor, N., Bónis, P., Bognár, P., & Lichtenberger, J. (2018). Statistical modelling of crop yield in Central Europe using climate data and remote sensing vegetation indices. *Agricultural and Forest Meteorology, 260*, 300–320. https://doi.org/10.1016/j.agrformet.2018.06.009

Khaki, S., & Wang, L. (2019). Crop yield prediction using deep neural networks. *Frontiers in Plant Science, 10*(May), 1–10. https://doi.org/10.3389/fpls.2019.00621

Khaki, S., Wang, L., & Archontoulis, S. V. (2020). A CNN-RNN framework for crop yield prediction. *Frontiers in Plant Science, 10*, 1–14. https://doi.org/10.3389/fpls.2019.01750

Kuwata, K., & Shibasaki, R. (2015). Estimating crop yields with deep learning and remotely sensed data. *International Geoscience and Remote Sensing Symposium (IGARSS), 2015*, 858–861. https://doi.org/10.1109/IGARSS.2015.7325900

Landau, S., Mitchell, R. A. C., Barnett, V., Colls, J. J., Craigon, J., & Payne, R. W. (2000). A parsimonious, multiple-regression model of wheat yield response to environment. *Agricultural and Forest Meteorology, 101*(2–3), 151–166. https://doi.org/10.1016/S0168-1923(99)00166-5

Li, W., Dong, R., Fu, H., Wang, J., Yu, L., & Gong, P. (2020). Integrating Google Earth imagery with Landsat data to improve 30-m resolution land cover mapping. *Remote Sensing of Environment, 237*, 111563.

Li, Y., Guan, K., Yu, A., Peng, B., Zhao, L., Li, B., & Peng, J. (2019). Toward building a transparent statistical model for improving crop yield prediction: Modeling rainfed corn in the U.S. *Field Crops Research, 234*, 55–65. https://doi.org/10.1016/j.fcr.2019.02.005

Lin, T., Zhong, R., Wang, Y., Xu, J., Jiang, H., Xu, J., Ying, Y., Rodriguez, L., Ting, K. C., & Li, H. (2020). DeepCropNet: a deep spatial-temporal learning framework for county-level corn yield estimation. *Environmental Research Letters, 15*, 3. https://doi.org/10.1088/1748-9326/ab66cb

Lischeid, G., Webber, H., Sommer, M., Nendel, C., & Ewert, F. (2022). Machine learning in crop yield modelling: A powerful tool, but no surrogate for science. *Agricultural and Forest Meteorology, 312*, 108698. https://doi.org/10.1016/j.agrformet.2021.108698

Liu, L., Xiao, X., Qin, Y., Wang, J., Xu, X., Hu, Y., & Qiao, Z. (2020). Mapping cropping intensity in China using time series Landsat and Sentinel-2 images and Google Earth Engine. *Remote Sensing of Environment, 239*, 111624. https://doi.org/10.1016/j.rse.2019.111624

Lobell, D. B. (2013). The use of satellite data for crop yield gap analysis. *Field Crops Research, 143*, 56–64. https://doi.org/10.1016/j.fcr.2012.08.008

Lobell, D. B., & Burke, M. B. (2010). On the use of statistical models to predict crop yield responses to climate change. *Agricultural Forest Meteorology, 150*(11), 1443–1452.

Lobell, D. B., Thau, D., Seifert, C., Engle, E., & Little, B. (2015). A scalable satellite-based crop yield mapper. *Remote Sensing of Environment, 164*, 324–333. https://doi.org/10.1016/j.rse.2015.04.021

Loomis, R. S., & Williams, W. A. (1963). Maximum crop productivity: An extimate 1. *Crop Science, 3*(1), 67–72.

Löw, F., Michel, U., Dech, S., & Conrad, C. (2013). Impact of feature selection on the accuracy and spatial uncertainty of per-field crop classification using support vector machines. *ISPRS Journal of Photogrammetry and Remote Sensing, 85*, 102–119. https://doi.org/10.1016/j.isprsjprs.2013.08.007

Luciano, A. C. S., Picoli, M. C. A., Duft, D. G., Rocha, J. V., Leal, M. R. L. V., & le Maire, G. (2021). Empirical model for forecasting sugarcane yield on a local scale in Brazil using Landsat imagery and random forest algorithm. *Computers and Electronics in Agriculture, 184*, 106063. https://doi.org/10.1016/j.compag.2021.106063

Ma, L., Li, M., Ma, X., Cheng, L., Du, P., & Liu, Y. (2017). A review of supervised object-based land-cover image classification. *ISPRS Journal of Photogrammetry and Remote Sensing, 130*, 277–293. https://doi.org/10.1016/j.isprsjprs.2017.06.001

Ma, Y., Zhang, Z., Kang, Y., & Özdoğan, M. (2021). Corn yield prediction and uncertainty analysis based on remotely sensed variables using a Bayesian neural network approach. *Remote Sensing of Environment, 259*, 112408. https://doi.org/10.1016/j.rse.2021.112408

Massey, R., Sankey, T. T., Congalton, R. G., Yadav, K., Thenkabail, P. S., Ozdogan, M., & Sánchez Meador, A. J. (2017). MODIS phenology-derived, multi-year distribution of conterminous U.S. crop types. *Remote Sensing of Environment, 198*, 490–503.

Mateo-Sanchis, A., Piles, M., Munoz-Mari, J., Adsuara, J. E., Perez-Suay, A., & Camps-Valls, G. (2019). Synergistic integration of optical and microwave satellite data for crop yield estimation. *Remote Sens Environ, 234*, 111460. https://doi.org/10.1016/j.rse.2019.111460

Mathur, A., & Foody, G. M. (2008). Crop classification by support vector machine with intelligently selected training data for an operational application. *International Journal of Remote Sensing, 29*(8), 2227–2240.

McKinney, W. (2010). Data structures for statistical computing in python. *Proceedings of the Python in Science, 445*, 51–56.

McMaster, G. S., & Wilhelm, W. W. (1997). Growing degree-days: One equation, two interpretations. *Agricultural and Forest Meteorology, 87*(4), 291–300.

McNairn, H., Shang, J., Jiao, X., & Champagne, C. (2009). The contribution of ALOS PALSAR multipolarization and polarimetric data to crop classification. *IEEE Transactions on Geoscience and Remote Sensing, 47*(12), 3981–3992.

Mkhabela, M. S., Bullock, P., Raj, S., Wang, S., & Yang, Y. (2011). Crop yield forecasting on the Canadian Prairies using MODIS NDVI data. *Agricultural and Forest Meteorology, 151*(3), 385–393. https://doi.org/10.1016/j.agrformet.2010.11.012.

Monsi, M. (1953). Uber den Lichtfaktor in den Pflanzen-gesellschaften und seine Bedeutung fur die Stoffproduktion. *Jap Journ Bot, 14*, 22–52.

Mueller, N. D., Gerber, J. S., Johnston, M., Ray, D. K., Ramankutty, N., & Foley, J. A. (2012). Closing yield gaps through nutrient and water management. *Nature, 490*(7419), 254–257. https://doi.org/10.1038/nature11420

Ndikumana, E., Ho Tong Minh, D., Baghdadi, N., Courault, D., & Hossard, L. (2018). Deep recurrent neural network for agricultural classification using multitemporal SAR Sentinel-1 for Camargue, France. *Remote Sensing, 10*(8), 1217.

Ozdogan, M., & Gutman, G. (2008). A new methodology to map irrigated areas using multitemporal MODIS and ancillary data: An application example in the continental US. *Remote Sensing of Environment, 112*(9), 3520–3537. https://doi.org/10.1016/j.rse.2008.04.010

Pan, Y., Li, L., Zhang, J., Liang, S., Zhu, X., & Sulla-Menashe, D. (2012). Winter wheat area estimation from MODIS-EVI time series data using the crop proportion phenology index. *Remote Sensing of Environment, 119*, 232–242. https://doi.org/10.1016/j.rse.2011.10.011

Paszke, A., Gross, S., Massa, F., Lerer, A., Bradbury, J., Chanan, G., Killeen, T., Lin, Z., Gimelshein, N., & Antiga, L. (2019). Pytorch: An imperative style, high-performance deep learning library. *Advances in Neural Information Processing Systems, 32*, 8026–8037.

Pedregosa, F., Varoquaux, G., Gramfort, A., Michel, V., Thirion, B., Grisel, O., Blondel, M., Prettenhofer, P., Weiss, R., & Dubourg, V. (2011). Scikit-learn: Machine learning in Python. *The Journal of Machine Learning Research, 12*, 2825–2830.

Pelletier, C., Valero, S., Inglada, J., Champion, N., & Dedieu, G. (2016). Assessing the robustness of Random Forests to map land cover with high resolution satellite image time series over large areas. *Remote Sensing of Environment, 187*, 156–168. https://doi.org/10.1016/j.rse.2016.10.010

Peng, B., Guan, K., Pan, M., & Li, Y. (2018). Benefits of seasonal climate prediction and satellite data for forecasting U.S. maize yield. *Geophysical Research Letters, 45*(18), 9662–9671. https://doi.org/10.1029/2018gl079291

Pott, L. P., Amado, T. J. C., Schwalbert, R. A., Corassa, G. M., & Ciampitti, I. A. (2021). Satellite-based data fusion crop type classification and mapping in Rio Grande do Sul, Brazil. *ISPRS Journal of Photogrammetry and Remote Sensing, 176*, 196–210. https://doi.org/10.1016/j.isprsjprs.2021.04.015

Qian, B., De Jong, R., Warren, R., Chipanshi, A., & Hill, H. (2009). Statistical spring wheat yield forecasting for the Canadian prairie provinces. *Agricultural and Forest Meteorology, 149*(6–7), 1022–1031. https://doi.org/10.1016/j.agrformet.2008.12.006

Qiu, B., Li, W., Tang, Z., Chen, C., & Qi, W. (2015). Mapping paddy rice areas based on vegetation phenology and surface moisture conditions. *Ecological Indicators, 56*, 79–86.

Qiu, B., Luo, Y., Tang, Z., Chen, C., Lu, D., Huang, H., Chen, Y., Chen, N., & Xu, W. (2017). Winter wheat mapping combining variations before and after estimated heading dates. *ISPRS Journal of Photogrammetry and Remote Sensing, 123*, 35–46.

Qu, C., Li, P., & Zhang, C. (2021). A spectral index for winter wheat mapping using multi-temporal Landsat NDVI data of key growth stages. *ISPRS Journal of Photogrammetry and Remote Sensing, 175*, 431–447. https://doi.org/10.1016/j.isprsjprs.2021.03.015

Rouse, J. W., Haas, R. H., Schell, J. A., & Deering, D. W. (1974). Monitoring vegetation systems in the Great Plains with ERTS. *NASA Special Publication, 351*(1974), 309.

Rußwurm, M., & Korner, M. (2017). Temporal vegetation modelling using long short-term memory networks for crop identification from medium-resolution multi-spectral satellite images. In *Proceedings of the IEEE conference on computer vision and pattern recognition workshops* (pp. 11–19). IEEE.

Rußwurm, M., & Körner, M. (2018a). *Convolutional LSTMs for cloud-robust segmentation of remote sensing imagery*. ArXiv Preprint ArXiv:1811.02471.

Rußwurm, M., & Körner, M. (2018b). Multi-temporal land cover classification with sequential recurrent encoders. *ISPRS International Journal of Geo-Information, 7*(4), 129.

Rußwurm, M., & Körner, M. (2019). *Self-attention for raw optical satellite time series classification*. ArXiv Preprint ArXiv:1910.10536.

Sakamoto, T. (2020). Incorporating environmental variables into a MODIS-based crop yield estimation method for United States corn and soybeans through the use of a random forest regression algorithm. *ISPRS Journal of Photogrammetry and Remote Sensing, 160*, 208–228. https://doi.org/10.1016/j.isprsjprs.2019.12.012

Sakamoto, T., Wardlow, B. D., Gitelson, A. A., Verma, S. B., Suyker, A. E., & Arkebauer, T. J. (2010). A two-step filtering approach for detecting maize and soybean phenology with time-series MODIS data. *Remote Sensing of Environment, 114*(10), 2146–2159.

Shammi, S. A., & Meng, Q. (2021). Use time series NDVI and EVI to develop dynamic crop growth metrics for yield modeling. *Ecological Indicators, 121*, 107124. https://doi.org/10.1016/j.ecolind.2020.107124

Song, X.-P., Potapov, P. V., Krylov, A., King, L., Di Bella, C. M., Hudson, A., Khan, A., Adusei, B., Stehman, S. V., & Hansen, M. C. (2017). National-scale soybean mapping and area estimation in the United States using medium resolution satellite imagery and field survey. *Remote Sensing of Environment, 190*, 383–395.

Steduto, P., Hsiao, T. C., Raes, D., & Fereres, E. (2009). Aquacrop-the FAO crop model to simulate yield response to water: I. Concepts and underlying principles. *Agronomy Journal, 101*(3), 426–437. https://doi.org/10.2134/agronj2008.0139s

Sun, H., Xu, A., Lin, H., Zhang, L., & Mei, Y. (2012). Winter wheat mapping using temporal signatures of MODIS vegetation index data. *International Journal of Remote Sensing, 33*(16), 5026–5042.

Tamiminia, H., Salehi, B., Mahdianpari, M., Quackenbush, L., Adeli, S., & Brisco, B. (2020). Google Earth Engine for geo-big data applications: A meta-analysis and systematic review. *ISPRS Journal of Photogrammetry and Remote Sensing, 164*, 152–170. https://doi.org/10.1016/j.isprsjprs.2020.04.001

Tian, H., Wang, P., Tansey, K., Han, D., Zhang, J., Zhang, S., & Li, H. (2021). A deep learning framework under attention mechanism for wheat yield estimation using remotely sensed indices in the Guanzhong Plain, PR China. *International Journal of Applied Earth Observation and Geoinformation, 102*(17), 102375. https://doi.org/10.1016/j.jag.2021.102375

Van Deventer, A. P., Ward, A. D., Gowda, P. H., & Lyon, J. G. (1997). Using thematic mapper data to identify contrasting soil plains and tillage practices. *Photogrammetric Engineering and Remote Sensing, 63*, 87–93.

Vapnik, V. (1999). *The nature of statistical learning theory*. Springer Science & Business Media.

Virtanen, P., Gommers, R., Oliphant, T. E., Haberland, M., Reddy, T., Cournapeau, D., Burovski, E., Peterson, P., Weckesser, W., & Bright, J. (2020). SciPy 1.0: Fundamental algorithms for scientific computing in Python. *Nature Methods, 17*(3), 261–272.

Vuolo, F., Neuwirth, M., Immitzer, M., Atzberger, C., & Ng, W.-T. (2018). How much does multi-temporal Sentinel-2 data improve crop type classification? *International Journal of Applied Earth Observation and Geoinformation, 72*, 122–130.

Wang, S., Azzari, G., & Lobell, D. B. (2019). Crop type mapping without field-level labels: Random forest transfer and unsupervised clustering techniques. *Remote Sensing of Environment, 222*, 303–317. https://doi.org/10.1016/j.rse.2018.12.026

Wardlow, B. D., & Egbert, S. L. (2008). Large-area crop mapping using time-series MODIS 250 m NDVI data: An assessment for the U.S. Central Great Plains. *Remote Sensing of Environment, 112*(3), 1096–1116. https://doi.org/10.1016/j.rse.2007.07.019

Wardlow, B. D., Egbert, S. L., & Kastens, J. H. (2007). Analysis of time-series MODIS 250 m vegetation index data for crop classification in the US Central Great Plains. *Remote Sensing of Environment, 108*(3), 290–310.
Waskom, M. L. (2021). Seaborn: Statistical data visualization. *Journal of Open Source Software, 6*(60), 3021.
Wei, P., Chai, D., Lin, T., Tang, C., Du, M., & Huang, J. (2021). Large-scale rice mapping under different years based on time-series Sentinel-1 images using deep semantic segmentation model. *ISPRS Journal of Photogrammetry and Remote Sensing, 174*, 198–214.
Weiss, M., Jacob, F., & Duveiller, G. (2020). Remote sensing for agricultural applications: A meta-review. *Remote Sensing of Environment, 236*, 111402.
Xiao, X., Boles, S., Frolking, S., Li, C., Babu, J. Y., Salas, W., & Moore, B., III. (2006). Mapping paddy rice agriculture in South and Southeast Asia using multi-temporal MODIS images. *Remote Sensing of Environment, 100*(1), 95–113.
Xiao, X., Boles, S., Frolking, S., Salas, W., Moore Iii, B., Li, C., He, L., & Zhao, R. (2002). Landscape-scale characterization of cropland in China using vegetation and landsat TM images. *International Journal of Remote Sensing, 23*(18), 3579–3594.
Xiao, X., Boles, S., Liu, J., Zhuang, D., Frolking, S., Li, C., Salas, W., & Moore, B., III. (2005). Mapping paddy rice agriculture in southern China using multi-temporal MODIS images. *Remote Sensing of Environment, 95*(4), 480–492.
Xu, J., Yang, J., Xiong, X., Li, H., Huang, J., Ting, K. C., Ying, Y., & Lin, T. (2021). Towards interpreting multi-temporal deep learning models in crop mapping. *Remote Sensing of Environment, 264*, 112599.
Xu, J., Zhu, Y., Zhong, R., Lin, Z., Xu, J., Jiang, H., Huang, J., Li, H., & Lin, T. (2020). DeepCropMapping: A multi-temporal deep learning approach with improved spatial generalizability for dynamic corn and soybean mapping. *Remote Sensing of Environment, 247*, 111946.
Xu, S., & Cheng, J. (2021). A new land surface temperature fusion strategy based on cumulative distribution function matching and multiresolution Kalman filtering. *Remote Sensing of Environment, 254*, 112256. https://doi.org/10.1016/j.rse.2020.112256
Xue, J., & Su, B. (2017). Significant remote sensing vegetation indices: A review of developments and applications. *Journal of Sensors, 2017*, 1353691. https://doi.org/10.1155/2017/1353691
Yan, W., & Hunt, L. A. (1999). An equation for modelling the temperature response of plants using only the cardinal temperatures. *Annals of Botany, 84*(5), 607–614.
Yang, H., Pan, B., Li, N., Wang, W., Zhang, J., & Zhang, X. (2021). A systematic method for spatio-temporal phenology estimation of paddy rice using time series Sentinel-1 images. *Remote Sensing of Environment, 259*, 112394. https://doi.org/10.1016/j.rse.2021.112394
You, J., Li, X., Low, M., Lobell, D., & Ermon, S. (2017). Deep Gaussian process for crop yield prediction based on remote sensing data. *Proceedings of the AAAI Conference on Artificial Intelligence, 31*, 1.
Zhang, A., Lipton, Z. C., Li, M., & Smola, A. J. (2021). Dive into deep learning. ArXiv Preprint ArXiv:2106.11342.
Zhang, J., Feng, L., & Yao, F. (2014). Improved maize cultivated area estimation over a large scale combining MODIS–EVI time series data and crop phenological information. *ISPRS Journal of Photogrammetry and Remote Sensing, 94*, 102–113. https://doi.org/10.1016/j.isprsjprs.2014.04.023
Zhang, L., Zhang, Q., Zhang, L., Tao, D., Huang, X., & Du, B. (2015). Ensemble manifold regularized sparse low-rank approximation for multiview feature embedding. *Pattern Recognition, 48*(10), 3102–3112. https://doi.org/10.1016/j.patcog.2014.12.016
Zhao, W., Qu, Y., Chen, J., & Yuan, Z. (2020). Deeply synergistic optical and SAR time series for crop dynamic monitoring. *Remote Sensing of Environment, 247*, 111952. https://doi.org/10.1016/j.rse.2020.111952
Zheng, B., Myint, S. W., Thenkabail, P. S., & Aggarwal, R. M. (2015). A support vector machine to identify irrigated crop types using time-series Landsat NDVI data. *International Journal of Applied Earth Observation and Geoinformation, 34*, 103–112.

Zhong, L., Gong, P., & Biging, G. S. (2014). Efficient corn and soybean mapping with temporal extendability: A multi-year experiment using Landsat imagery. *Remote Sensing of Environment, 140*, 1–13. https://doi.org/10.1016/j.rse.2013.08.023

Zhong, L., Hu, L., & Zhou, H. (2019a). Deep learning based multi-temporal crop classification. *Remote Sensing of Environment, 221*, 430–443. https://doi.org/10.1016/j.rse.2018.11.032

Zhong, L., Hu, L., Zhou, H., & Tao, X. (2019b). Deep learning based winter wheat mapping using statistical data as ground references in Kansas and northern Texas. *US. Remote Sensing of Environment, 233*, 111411. https://doi.org/10.1016/j.rse.2019.111411

Zhong, L., Yu, L., Li, X., Hu, L., & Gong, P. (2016). Rapid corn and soybean mapping in US Corn Belt and neighboring areas. *Scientific Reports, 6*(1), 1–14.

Zhou, Y., Luo, J., Feng, L., & Zhou, X. (2019). DCN-based spatial features for improving parcel-based crop classification using high-resolution optical images and multi-temporal SAR data. *Remote Sensing, 11*(13), 1619.

Chapter 7
Artificial Intelligence for Image Processing in Agriculture

Shih-Fang Chen and Yan-Fu Kuo

7.1 Introduction

Humans create tools and solve problems based on accumulated experience, hence the term human intelligence. Machines learn through algorithms created by humans and then find the best solutions using accumulated training. This is the intelligence of machines, also known as artificial intelligence (AI). AI mimics human intelligence to solve problems.

The development of AI can be traced back to the 1950s. Since then, it had been through several AI winters, i.e., low interest, low funding, and low research volume in the field. The recent rise of AI can be attributed to the breakthrough of its subdomain, machine learning, in weak AI. Weak AI, also known as narrow AI, can make judgments close to humans and even compete with domain experts in target-specific tasks, such as computer Go and autonomous driving.

Data-driven machine learning has accelerated the application of AI in various fields due to the success of deep learning. Take the aforementioned computer Go as an example, AlphaGo (Silver et al., 2016) used deep convolutional neural network (CNN) to defeat world Go champion, Ke Jie. Similar frameworks also produce amazing predictive accuracy in other applications, such as ImageNet (Krizhevsky et al., 2012) in image recognition, AlphaFold (Senior et al., 2020) in three-dimensional structure prediction of protein, and AtomNet (Wallach et al., 2015) in drug design. In addition to CNNs, other deep learning algorithms, such as recurrent neural network (RNN), have also revolutionized the processing of serial data,

S.-F. Chen (✉) · Y.-F. Kuo
Department of Biomechatronics Engineering, National Taiwan University, Taipei, Taiwan
e-mail: sfchen@ntu.edu.tw

S. Ma et al. (eds.), *Sensing, Data Managing, and Control Technologies for Agricultural Systems*, Agriculture Automation and Control,
https://doi.org/10.1007/978-3-031-03834-1_7

including speech recognition and machine translation. The success of integrating various AI applications into our daily lives is beyond imagination.

As the algorithms mature and computing hardware efficiency grows, deep learning spreads in various applications, such as smart or intelligent cities, manufacturing, finance, medicine, and agriculture. For a long time, agriculture has faced the challenges of manpower shortage, extreme climate, and insufficient food. AI is expected to reverse those challenges. Some crucial projects include the development of field robots to gradually replace human labor, drones for field monitoring and early detection, and the use of machine learning and big data to build models for crop yield prediction or disease and insect disaster impact assessment.

This chapter introduces the evolution of AI in computer vision and cognition and explains the basic concept of neural networks. It also presents the various changes and impacts of the deep convolutional networks in computer vision and its application in agriculture, forestry, fisheries, and animal husbandry. The impact of this wave of AI on human life and economic development is promising. As big data, high-efficiency algorithms, and hardware acceleration become more available and integrate more seamlessly, users can easily enjoy the convenience and wonder of AI by using appropriate devices.

7.2 From AI to Image Recognition

To understand how humans solve problems, one should first comprehend the most common ways people perceive environmental information in their daily lives. Junichi Nomura, the author of "The Secrets of Colors," mentioned that vision accounts for 87% of the functional proportion of the five human senses. Animals detect natural enemies by vision to avoid all kinds of dangers. Humans use vision to drive a car, cross a road, and perform other daily life activities. Vision is a key part in connecting people with their surroundings. Therefore, an important subject of AI research is the application of machine vision. Machine vision is used in a wide variety of applications, for example, license plate recognition in parking lots, Apple's Face ID, and Tesla's self-driving. But for a machine, a colored image is not more than a three-dimensional matrix of pixels containing red, green, and blue values (Fig. 7.1), how does it perform image identification?

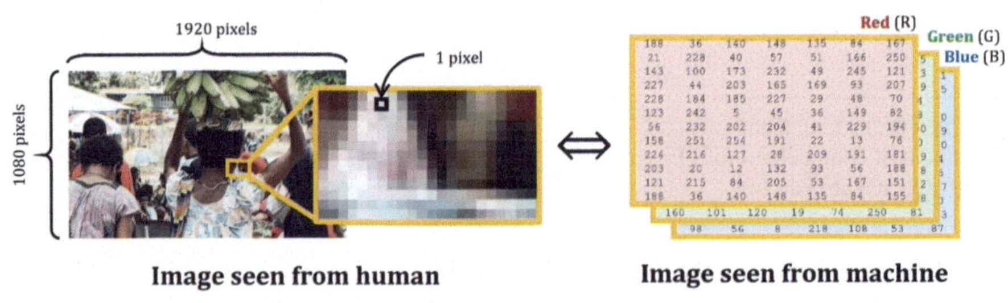

Fig. 7.1 The difference between human vision and machine vision

By visually observing an object, human can identify it by recognizing features such as its outline, color, and size. That is how the traditional methods of image processing, also known as algorithms or rules, were born. These methods design rules for a machine to follow passively and identify features in an image matrix through computation. However, the rules are fixed and inflexible. When the background of an image is more complex, or there is illumination variation, the identification accuracies of the traditional algorithms are greatly reduced. Recognizing such limitations, one of the most famous algorithms since 2010, CNN, was developed. Unlike its predecessors, CNN learns image characteristics by training. In machine vision, CNNs can learn critical information from training images to identify objects, such as lines, outlines, colors, and sizes. CNN, which is formed by a neural network of millions of neurons, mimics the human brain. Each of these neurons stores a small portion of data. Through this characteristic of small data storage, CNN can recognize similar patterns in images, even when the images are transformed. This characteristic enables CNN to achieve high recognition accuracy. The following sections describe CNNs in more detail.

7.3 Neural Network

7.3.1 Neurons, Single-Layer Neural Network, and Multi-Layer Neural Network

Neurons are the most basic components of the human brain and they connect each other through synapses. A human brain has about tens of billions of neurons, with an average of more than a thousand links between each neuron and other neurons, creating a more robust network than the world's most advanced supercomputers. Therefore, scientists used the complex and powerful structure of the brain to develop an AI model, called neural networks, which mimic the human brain's ability to solve complex problems.

Neurons receive signals from other neurons, process them, and then pass them on to other neurons. After processing signals of multi-layered neurons, they form meaningful information or instructions in the brain. The working mechanisms of these brain neurons can be converted into mathematical models. McCulloch and Pitts (1943) developed the most fundamental computational model of a neuron based on logical operators. The method receives multiple inputs, produces a single output, and then evaluates the output against a set threshold, which is built based on "AND," "OR," and "NOT" logics. As shown in Fig. 7.2, an artificial neuron receives input signals $x_1, x_2, \ldots, x_n$ from different neurons, each input is multiplied by a specific weight $w_1, w_2, \ldots, w_n$. These weighted signals are added together, sometimes with a constant deviation value called bias b, to calculate the total signal value. The weights and bias are all parameters trained through learning. When the neuron

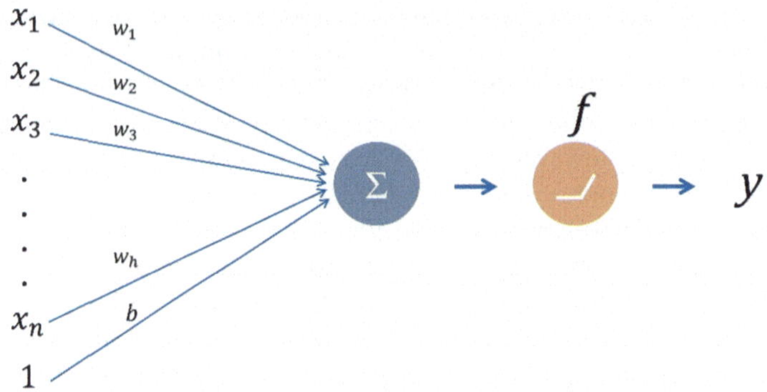

Fig. 7.2 Working principle of an artificial neuron

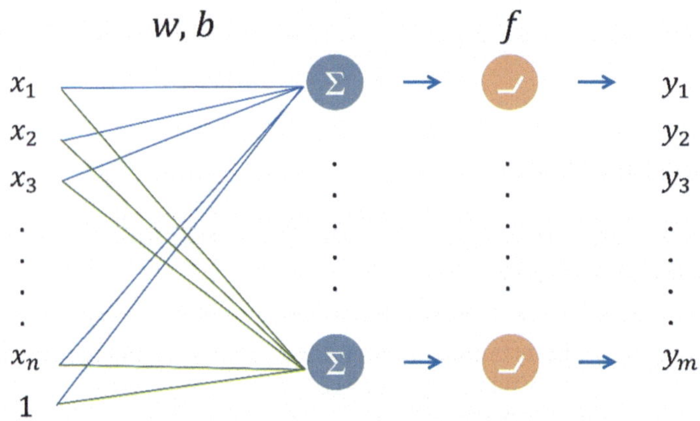

Fig. 7.3 Single-layer neural network

transmits a signal value, it passes through an activation function f, to produce the final result y, which is transmitted to the next neuron.

After knowing how an artificial neuron works, the next step is to learn about the most basic neural network model: a single-layer neural network. As mentioned in the previous paragraph, the signals of multiple neurons are sent to a neuron. If multiple signals are sent to multiple neurons at the same time, then a single-layer neural network is formed, as shown in Fig. 7.3. Neurons in a single-layer neural network can be divided into two layers: the input layer and the output layer. Assuming that the input layer is composed of $n \in N$ neurons, the neuron signals sent to the next layer can be considered as a vector $\boldsymbol{x} = [x_1, \ldots, x_n] \in R^n$. The output layer is made up of $m \in N$ neurons; the output signals can also be thought of as a vector $\boldsymbol{y} = [y_1, \ldots, y_m] \in R^m$. As every output neuron receives signals from every input neuron, these two layers produce $n \times m$ connections, each with a separate weight. The weights between the input and output layers can be represented as a matrix:

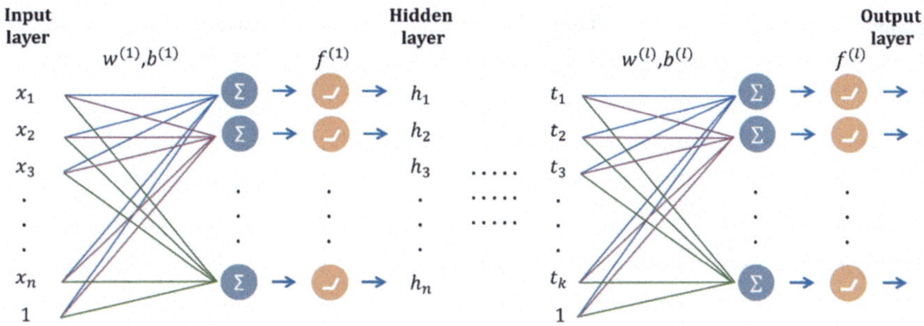

Fig. 7.4 Multi-layer neural network

$$W = \begin{pmatrix} w_{11} & \cdots & w_{1m} \\ \vdots & \ddots & \vdots \\ w_{n1} & \cdots & w_{nm} \end{pmatrix} \in R^{n\times m} \tag{7.1}$$

As a bias is added to each layer, a vector $\boldsymbol{b} = [b_1, \ldots, b_m] \in R^m$ is formed. Thus, the output of a single-layer neural network can be represented as $\boldsymbol{y} = f(\mathbf{W}^{\mathrm{T}}\boldsymbol{x} + \boldsymbol{b})$. This mathematical formula is the most important basic concept of a single-layer neural network.

However, a single-layer neural network is not enough to solve complex cognitive problems. A normal human brain consists of six layers of cerebral cortex, each with its own function. Signals received from the five human senses are transmitted among these six layers, converting simple sensory input into the meaningful interpretation. Therefore, more layers of neurons, called hidden layers, are added between the input and output layers to deal with more complex problems. The mathematical model between two layers is the same as a single-layer neural network. Such a multi-layer neural network architecture forms a simple artificial neural network, as shown in Fig. 7.4. The number of layers is represented as $l \in N$, the output vector of the first layer of neurons is $\boldsymbol{h} = [h_1, \ldots, h_m] \in R^m$, and the output vector of the penultimate layer is $\boldsymbol{t} = [t_1, \ldots, t_k] \in R^k$.

7.3.2 Activation Function

In order to avoid the model having only a linear relationship when passing values between layers, an activation function is added to interpret complex data. There are many types of activation functions, the most commonly used ones are sigmoid, hyperbolic tangent (tanh), and linear rectifier (rectified linear unit, ReLU) functions (Fig. 7.5). ReLU is the most commonly used activation function in recent years. It can improve the prediction accuracy of a neural network model and find the best solutions faster during model training (Glorot et al., 2011).

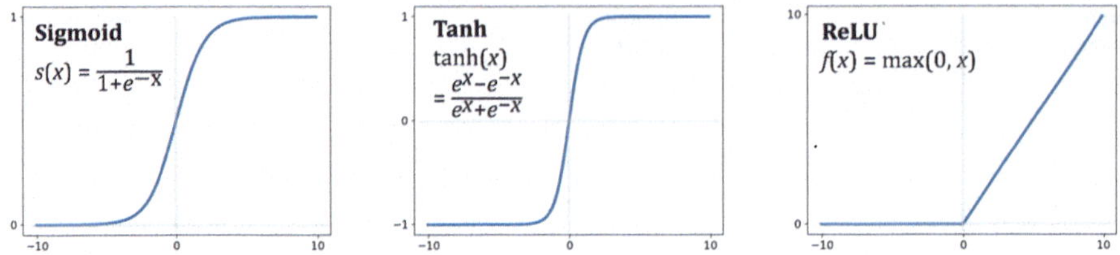

Fig. 7.5 Types of activation functions

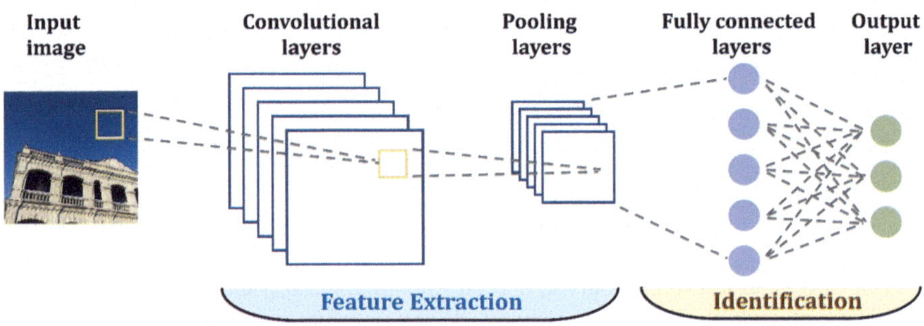

Fig. 7.6 A convolutional neural network

7.3.3 Convolutional Neural Network (CNN)

A CNN uses convolutional layers to learn complex characteristics from input data. Therefore, it is suitable for solving image recognition problems such as faces and objects, as well as analyzing text and sound. Its prediction accuracy is higher than pre-existing methods, which is one of the main reasons AI is popular again. As shown in Fig. 7.6, a basic CNN can be broadly divided into two stages: feature extraction and identification. Feature extraction refers to the extraction of features of an input image through the convolutional layers and preserving those important features through the pooling layers. The features are then identified by a fully connected layer and an output layer to obtain the result.

7.3.4 Convolutional Layer

The convolutional layer mimics human vision by scanning an input image through a sliding window called the kernel. Different characteristics, such as the edges or noises of objects in an image, can be extracted by changing the weights of a kernel. The feature information extracted is called a feature map. For example, in Fig. 7.7, the input data is an image of a plant, a 3 × 3 kernel (green square) is used to scan the entire image and obtain a feature map (the edge of the plant). In this example, the

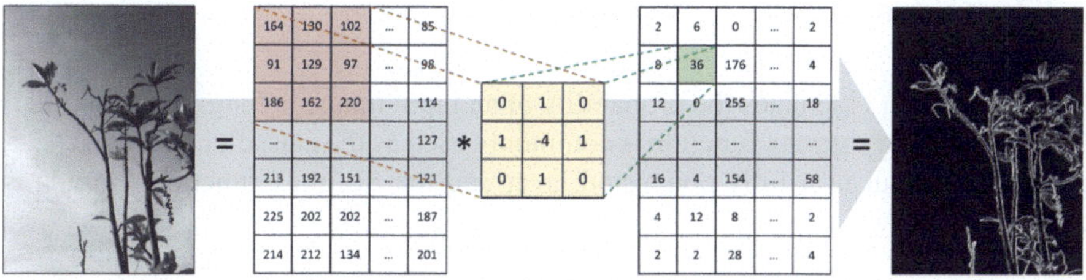

Fig. 7.7 A convolution operation to extract the edges of objects in an image. The 3 × 3 green square is a kernel

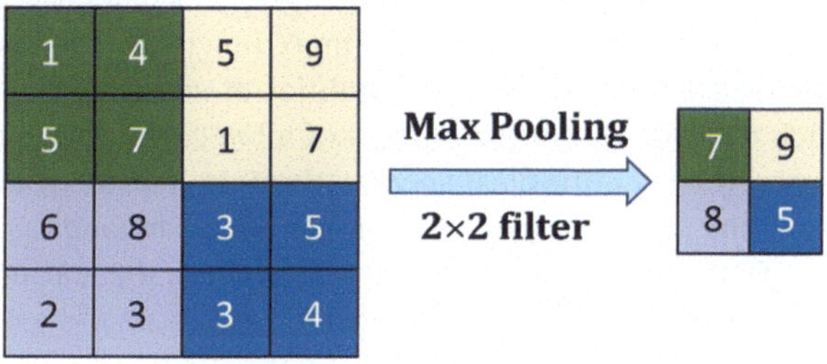

Fig. 7.8 Through max pooling, important features of an input feature map are preserved while reducing data size

kernel is a Laplacian filter that detects the edges of objects in images. Convolutional operations can not only extract features effectively from input images but also use less parameters than fully connected layers. However, in CNNs, the kernel weights can only be obtained through training.

7.3.5 *Pooling Layer*

A pooling layer is often used between convolutional layers to reduce data size while preserving important features. For example, in Fig. 7.8, a 4 × 4 feature map is scanned by a sliding window of 2 × 2 pooling kernel with a stride of 2. Max pooling takes the maximum value in the area of scan. Therefore, the original 4 × 4 feature map becomes a 2 × 2 feature map after the pooling.

7.3.6 *Fully Connected Layer*

Fully connected layer receives information from all neurons of the previous layer. It is generally located before the output layer of a CNN model. If there are m categories of training data (e.g., apple and watermelon categories, then $m = 2$), the output layer will have m neurons. The fully connected layer will connect each pixel of all feature maps in the previous layer to these m neurons.

7.3.7 *Output Layer*

The output layer calculates the probability of an input image belonging to a particular category. The softmax function is commonly used as the classifier. Let $\boldsymbol{x} = [x_1, \ldots, x_m]$ be the output of a fully connected layer and the Softmax function formula is defined as follows:

$$p_i = \text{softmax}\left(x_i\right) = \frac{e^{x_i}}{\sum_{j=0}^{m} e^{x_j}} \tag{7.2}$$

where p_i is the probability of belonging to a category x_i and $\sum_{m}^{j=0} e^{x_j}$ is the sum of all exponential functions. This Softmax function converts the output values of a fully connected layer into the probabilities of an image belonging to various categories, such that the sum of the outputs of all m categories is equal to 1. For example, in a model that identifies leaf species, the output result of the Softmax function is as follows: the probability of an input leaf image belonging to Species A is 0.91, Species B is 0.05, and Species C is 0.04. Therefore, the leaf in the image is predicted to be Species A (the highest probability).

7.4 Training a CNN

7.4.1 *Loss Function*

After defining the structure of a CNN, the next step is to train the kernel weights and biases. A neural network is trained through a loss function, which quantifies the difference between the predicted result $\hat{y}_{ij}$ and the actual result y_{ij}, i.e., the loss. The purpose of the training is to minimize the loss. Optimal weights and biases produce the lowest loss and the best neural network models. There are many kinds of loss functions. Examples of common loss functions are mean squared error, mean absolute error, and cross-entropy loss, as defined in Table 7.1. A loss function is selected according to the type of problem a user wants to solve.

7.4.2 *Optimizer*

The training process of adjusting weights and biases to improve model accuracy is known as parameter optimization. At the start of training, random numbers are used to initialize the weights and biases. After using a loss function to compare the predicted value $\hat{y}_{ij}$ with the actual result y_{ij}, an optimizer iteratively updates the weights and biases to improve the prediction accuracy and minimize the loss. Stochastic

Table 7.1 Commonly used loss functions

Name	Problem type	Formula
Mean squared error	Regression	$\frac{1}{N}\sum_{N}^{i=1}\frac{1}{M}\sum_{M}^{j=1}\left(\hat{y}_{ij}-y_{ij}\right)^2$
Mean absolute error	Regression	$\frac{1}{N}\sum_{N}^{i=1}\frac{1}{M}\sum_{M}^{j=1}\left\lvert\hat{y}_{ij}-y_{ij}\right\rvert$
Cross-entropy loss	Classification	$\frac{1}{N}\sum_{N}^{i=1}\frac{1}{M}\sum_{M}^{j=1}-y_{ij}\log\left(p_{ij}\right)$

gradient descent (Bottou, 2010), momentum (Qian, 1999), AdaGrad (Duchi et al., 2011), and Adam (Kingma & Ba, 2014) are some of the commonly used optimizers.

7.4.3 Hyperparameters

Hyperparameters refer to parameters and settings that can be determined by users to train a model. The hyperparameters need to be set before training a model. If the model outcome is not satisfactory, a user can adjust the hyperparameters and train the model again. Common hyperparameters include layer size, learning rate, batch size, epoch, and regularization. All these hyperparameters are described in the following sub-sections.

7.4.4 Layer Size

Layer size refers to the number of neurons in a single layer. The sizes of the input and output layers must match the input and output data sizes, respectively. For the sizes of the hidden layers, they can be determined by the user. Layer size can affect the performance of a neural network model. If the layer size is too large, it will increase computational time; if the layer size is too small, the neural network model may not have learned sufficiently to identify data features.

7.4.5 Learning Rate

Learning rate sets the pace of weight adjustment. It determines the extent of change an optimizer can apply to the weights based on the loss function result. It needs to be used with an optimizer. As an example, a gradient descent optimizer is formulated as follows:

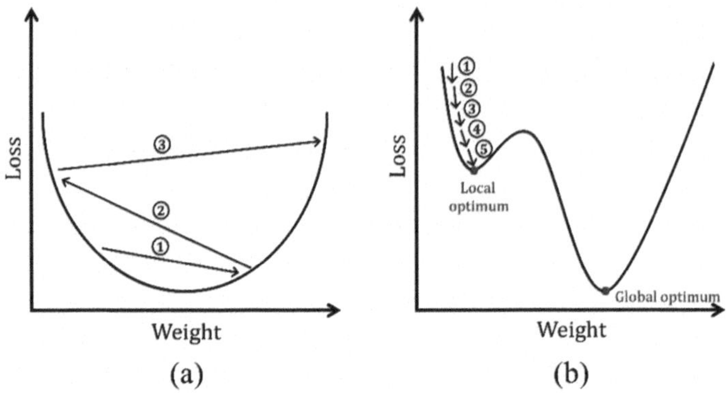

Fig. 7.9 Training problems caused by extreme learning rates. The numbers indicate the weight updates

$$W^{t+1} = W^t - \eta \nabla L \tag{7.3}$$

A new weight $\boldsymbol{W}^{t+1}$ is obtained by using the original weight $\boldsymbol{W}^t$ minus the product of learning rate η and gradient ∇L. The gradient is the partial derivative of the loss function. A model should converge after some time of training. Convergence means the loss function reaches a minimum value. As seen in Formula 7.3, the higher the learning rate, the greater the difference between $\boldsymbol{W}^{t+1}$ and $\boldsymbol{W}^t$. Although a high learning rate may seem desirable to minimize the loss faster and make a model converge sooner, it may prevent the model from achieving optimal solutions. For example, in Fig. 7.9a, the curve represents the loss, i.e., the difference between the predicted result and the actual result, with respect to the weights. The lowest point of the curve is the best solution for the model. The arrows refer to the directions the weights and corresponding losses move after each update. Due to the high learning rate, the change from $\boldsymbol{W}^t$ to $\boldsymbol{W}^{t+1}$ in each update is too great that the model misses the best solution. In contrast, a low learning rate can update a weight slower and take a longer training time. As shown in Fig. 7.9b, while a low learning rate can steadily reduce the loss during training, an overly low learning rate may cause the model to remain in a local optimal solution and prevent it from achieving the global optimal solution.

7.4.6 Batch Size

Batch size refers to the amount of data used for each training session. When training a neural network model, the amount of data that can be used at a time depends on the computer memory size. If the memory size is small, the training needs to be conducted in batches. This situation is prevalent in image recognition model training because image data size is relatively large. Batch size affects not only the training time but also model outcome. If the batch size is too small, it can result in an overly long training time, and the model may not converge. Therefore, hardware capacity should be considered when setting suitable batch sizes.

7.4.7 Epoch

An epoch of training means a model is trained with all training data once. Another term associated with epoch is iteration, the number of batches in a complete training data set. For example, if there are 10,000 training data and the batch size is 500, a model will take 20 iterations to complete an epoch. Often neural network models require multiple epochs of training to find the best solutions. The number of epochs is determined by the model and data complexity.

7.4.8 Regularization

Regularization limits the weights of a model during training, thereby reducing the complexity of the model and avoiding overfitting (Christian & Griffiths, 2016). Overfitting happens when a model has more parameters than the problem complexity requires. As a result, although the model can accurately predict the training data, it cannot accurately predict non-training data, or data it has not seen before. The higher the complexity of a neural network, the stronger its ability to handle complex tasks, but the model may become too rigid, resulting in overfitting. Common regularization techniques are dropout, L1, and L2 penalty functions. Dropout randomly drops out a specific proportion of neurons at each iteration, updating only the weights of the remaining neurons, so that the model avoids overdependence between neurons, becomes less complex and more robust. L1 and L2 penalties are weight decays; a penalty is added to the original loss function L to get a new loss function(L_{REG}). The purpose of the penalty is to punish the weights, such that larger weights generate a greater loss. Consequently, the model tends to use smaller weight values to reduce the complexity of the model (Table 7.2).

7.4.9 Learning Framework

There are many neural network learning frameworks available to build a model. The two most common deep learning frameworks are PyTorch and TensorFlow. PyTorch is an open source Python deep learning framework developed by Facebook's AI Research team. It grew rapidly since its launch in early 2017 and became a leading framework in 2019, due to its ease of use and fewer design packages. TensorFlow is an open source deep learning framework developed by the Google Brain team,

Table 7.2 Common regularization types and their formulas

Regularization type	Formula
L1 penalty	$L_{\mathrm{REG}} = L + \alpha \sum \lvert w_i \rvert$
L2 penalty	$L_{\mathrm{REG}} = L + \alpha \sum (w_i)^2$

which is a part of Google's AI. TensorFlow supports many programming languages. Due to Google's influence in the field of deep learning, TensorFlow has quickly become one of the most popular deep learning frameworks since its launch. As TensorFlow provides the most ubiquitous AI platform and comprehensive support for integration, it is the primary framework used by many companies.

7.4.10 Common CNN Models

CNN applications in machine vision can be divided into three categories: image classification, object detection, and object segmentation. The rest of this section describes each category in detail.

7.4.11 Image Classification

Image classification is the most basic machine vision application (Table 7.3). Its main purpose is to automatically identify and classify images, such as a leaf image classifier (Fig. 7.10) can classify leaf species from leaf photos. LeNet (LeCun et al., 1998) was the most famous CNN model in the early days of CNN. LeNet has three convolutional layers, two pooling layers, and two fully connected layers. There are about 0.06 million parameters, which is considered a relatively simple model for a CNN. AlexNet (Krizhevsky et al., 2012) model uses ReLU activation function to solve the vanishing gradient problem and improve training efficiency. It also uses dropout regularization to reduce the overfitting problem, thereby improving model performance. It has approximately 60 million parameters. ZF-Net model, which also has about 60 million parameters, won the ImageNet Large Scale Visual Recognition Challenge in 2013 (Russakovsky et al., 2013; Zeiler & Fergus, 2014). VGG-16 (Simonyan & Zisserman, 2014) model is improved from AlexNet and it has 13 convolutional layers and three fully connected layers. The convolutional layers can be divided into five sections and each section ends with a pooling layer to reduce the size of the feature maps. Compared with the previous models, VGG-16 is a deep CNN model with 138 million parameters. Since 2014, the structure of CNNs has changed; scholars applied techniques to reduce the number of parameters while improving accuracy. For example, ResNet-50 (He et al., 2016) model uses residual blocks to reduce parameter size and it can effectively preserve image

Table 7.3 Common image classifiers

Model	Year	Parameters	Description	References
LeNet	1998	0.06 M	The early well-known CNN has seven layers	LeCun et al. (1998)
AlexNet	2012	60 M	Applies ReLU activation function and dropout regularization	Krizhevsky et al. (2012)
ZF-Net	2013	60 M	ImageNet Large Scale Visual Recognition Challenge champion in 2013	Zeiler and Fergus (2014)
VGG-16	2014	138 M	A 16-layer network model improved from AlexNet	Simonyan and Zisserman (2014)
ResNet-50	2016	8.6 M	Applies residual blocks to prevent information loss	He et al. (2016)
MobileNet	2017	4.2 M	Uses depthwise separable convolution to effectively reduce the number of parameters	Howard et al. (2017)

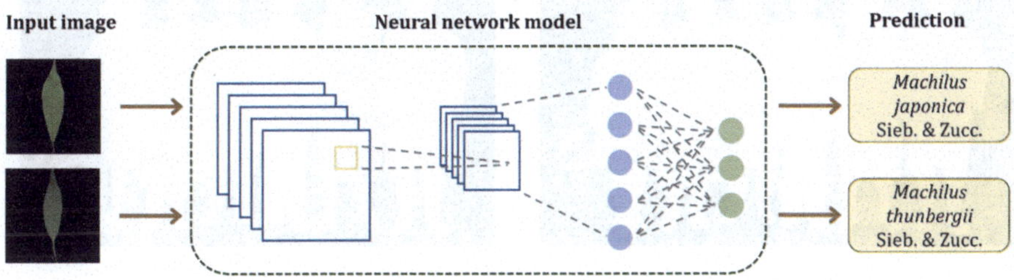

Fig. 7.10 A leaf image classifier

features even after many convolutions. ResNet-50 has about 8.6 million parameters. In another example, MobileNet (Howard et al., 2017) model uses depthwise separable convolution to significantly reduce the number of parameters to approximately 4.2 million while maintaining high accuracy.

7.4.12 *Object Detection*

Detecting objects in an image is a common machine vision application. Object detection includes both locating an object (localization) and determining its category (classification). For example, in Fig. 7.11b, an object detection model automatically locates six individuals in the image and marks them in boxes. Generally, object localization is conducted by generating candidate boxes and then determining whether there are objects in the candidate boxes using a CNN classifier described in the previous paragraph as the backbone. Common object detection models can be broadly divided into two-stage and one-stage network structures. Candidate boxes that do not contain objects are removed to identify candidate boxes that actually contain objects efficiently. A two-stage network structure uses a separate region

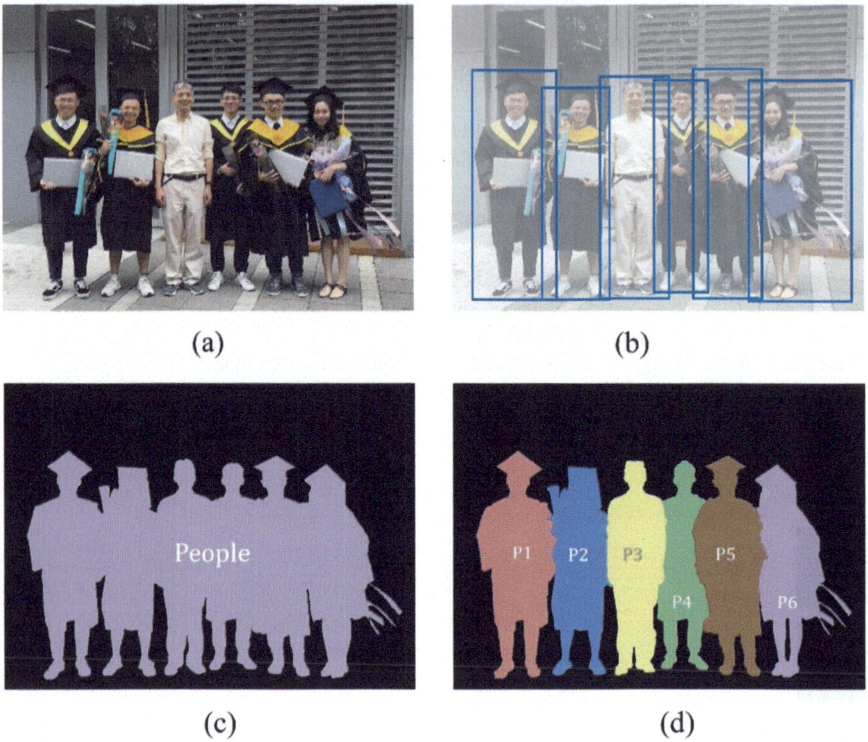

Fig. 7.11 Illustrations of object detection and segmentation: (**a**) input image, (**b**) object detection result, (**c**) semantic segmentation result, and (**d**) instance segmentation result. Semantic segmentation separates objects from the background; instance segmentation separates each object and labels them

proposal network to filter out candidate boxes that do not contain objects. This method often results in higher accuracy than a one-stage network structure. However, a one-stage network structure, which does not have a separate region proposal network, can locate and classify objects simultaneously. Therefore, its computational speed is usually faster than a two-stage network structure and it can achieve real-time detection.

In developing a two-stage network structure, region-based CNN (R-CNN) (Girshick et al., 2014) is the first object detection model. R-CNN uses VGG-16 model as its backbone and performs a selective search to generate candidate boxes, thus increasing the average precision of object detection by more than 50% compared with its predecessors. Fast R-CNN (Girshick, 2015) places region of interest pooling between a convolutional layer and a fully connected layer, so that the size of the candidate boxes agrees with the fully connected layer. This method eliminates a lengthy process in R-CNN, which is cropping and scaling candidate boxes one by one, thus making the detection faster. Faster R-CNN (Ren et al., 2015) introduces region proposal network and anchor. Anchors are pixel positions of a feature

Table 7.4 Common two-stage network models

Model	Year	Description	References
R-CNN	2013	Lays the structural foundation of two-stage network models	Girshick et al. (2014)
Fast R-CNN	2015	Uses region of interest pooling to address inconsistent candidate box size problem	Girshick (2015)
Faster R-CNN	2015	Applies region proposal network to generate anchor boxes and increase computational speed	Ren et al. (2015)

map. Region proposal network produces anchor boxes on a feature map. Each box has a different shape but shares the same anchor point as its center. This method can effectively detect multiple overlapping objects. Anchor boxes replace candidate boxes generated by R-CNN and Fast R-CNN. As anchor boxes are generated on feature maps, the model does not have to go through convolutional layers, significantly improving its computational speed (Table 7.4).

One-stage network structures emerge after 2016. The first model is You Only Look Once v1 (YOLO v1; Redmon et al., 2016). Frame refers to the number of images in a video, a smooth video is shot at a minimum of 24 frames per second. Using VGG-16 model as its backbone, YOLO v1 can process up to 45 frames per second. By contrast, the Faster R-CNN, a two-stage network structure, can detect only five frames per second. Therefore, the processing speed of a one-stage network structure is much faster than that of a two-stage network structure. Single Shot MultiBox Detector (SSD; Liu et al., 2016) is another well-known one-stage network structure. Unlike YOLO v1, which uses only the last convolutional layer to train object localization and classification, SSD simultaneously captures and applies multi-dimensional convolutional layers. As these layers are arranged like a pyramid, SSD is also known as featurized image pyramid. SSD uses the concept of anchors in Faster R-CNN to generate candidate boxes and then applies non-maximum suppression algorithm to eliminate redundant candidate boxes. This method combines the results of object localization and classification while taking into account speed and accuracy. YOLO v2 (Redmon et al., 2016) is built based on YOLO v1. It combines the strengths of SSDs and uses a passthrough layer to rearrange feature maps and improve the detection of small objects. YOLO v3 (Redmon & Farhadi, 2018) is developed based on DarkNet-53 model. It uses a feature pyramid network (Lin et al., 2017) to extract features and applies binary cross-entropy loss function to replace the mean squared error loss function used in softmax activation function, resulting in a significant increase in the average accuracy of the model. YOLO v4 (Bochkovskiy et al., 2020) was proposed in 2020. Its average accuracy exceeds all two-stage network structures and it is faster than the other one-stage network structures. YOLO v4 is built based on YOLO v3. It combines the structures of a variety of CNNs, such as path aggregation network (Liu et al., 2018) improved from feature pyramid network and spatial pyramid pooling (He et al., 2014). Based on a large number of experiments, it is currently the best one-stage object detection model, both in terms of average accuracy and detection speed (Table 7.5).

Table 7.5 Common one-stage network models

Model	Year	Description	References
YOLO v1	2016	Lays the structural foundation of one-stage network models	Redmon et al. (2016)
SSD	2016	Exceeds the accuracy of R-CNN and computational speed of YOLO v1	Liu et al. (2016)
YOLO v2	2016	Applies a passthrough layer to improve detection of small objects	Redmon et al. (2016)
YOLO v3	2018	Uses feature pyramid network and binary cross-entropy to improve performance	Redmon and Farhadi (2018)
YOLO v4	2020	Combines strengths of several CNNs and outperforms predecessors	Bochkovskiy et al. (2020)

7.4.13 Object Segmentation

Object segmentation is another common machine vision application. It can identify object pixels in an image, such as the human pixels identified in Fig. 7.11c, d. Object segmentation models can be divided into two categories: semantic segmentation and instance segmentation. Semantic segmentation separates objects from the background of an image by categories, as shown in Fig. 7.11c. Most of the models in semantic segmentation use a fully convolutional network (FCN; Long et al., 2015) as the backbone. FCN is the first model that applies deep learning to semantic segmentation tasks. It can learn image features and predict object pixels by overlaying multiple convolutional layers to produce an output semantic segmentation map. DeepLab-CRF (Chen et al., 2014) is a semantic segmentation model that combines CNN with a fully connected conditional random field. The fully connected conditional random field is a probability model that can efficiently generate output semantic segmentation maps from feature maps. SegNet (Badrinarayanan et al., 2017) is a model that uses an encoder–decoder structure. The encoder captures the features of an input image. As a CNN contains a large number of neurons, most CNN models reduce feature map sizes through pooling layers to complete calculations in a reasonable time. Therefore, a decoder is applied to enlarge the feature maps (or upsampling), so that the size of the output semantic segmentation map is the same as the input image size. U-Net (Ronneberger et al., 2015) uses the encoder–decoder structure and merges feature maps of the same size in upsampling and downsampling to preserve more feature information and improve prediction accuracy. PSPNet (Zhao et al., 2017) utilizes a pyramid pooling module, which uses pooling operations and different kernel sizes to extract features of multiple feature map sizes for more information. DeepLabv3+ (Chen et al. 2018a, b) replaces the standard convolution in neural networks with dilated convolution. Dilated convolution adds holes in a kernel so that a more extensive range of feature information can be extracted without increasing the number of parameters (Table 7.6).

Instance segmentation differs from semantic segmentation in that the former identifies the boundary between overlapping objects, in addition to the output

Table 7.6 Common semantic segmentation models

Model	Year	Description	References
FCN	2015	The first deep learning semantic segmentation model	Long et al. (2015)
DeepLab-CRF	2016	Applies fully connected conditional random field to predict semantic segmentation	Chen et al. (2021)
SegNet	2017	Uses encoder–decoder architecture	Badrinarayanan et al. (2017)
U-Net	2015	Applies encoder–decoder architecture and combines feature maps in upsampling and downsampling	Ronneberger et al. (2015)
PSPNet	2017	Uses pyramid pooling module	Zhao et al. (2017)
DeepLabv3+	2018	Replaces standard convolution with dilated convolution	Chen et al. (2018a, b)

Table 7.7 Common instance segmentation models

Model	Year	Description	References
Mask R-CNN	2017	Combines Faster R-CNN and FCN	He et al. (2017)
MaskLab	2018	Combines Faster R-CNN, FCN, and instance center direction structure	Chen et al. (2018a, b)
YOLACT	2019	Applies ResNet-101, FCN, and mask coefficients to achieve real-time instance segmentation	Bolya et al. (2019)

segmentation map, as shown in Fig. 7.11d. Mask R-CNN (He et al., 2017) model combines object boundary detection in Faster R-CNN with a FCN to generate object boundaries and semantic segmentation map simultaneously to achieve instance segmentation. MaskLab (Chen et al. 2018a, b) combines object boundary detection in Faster R-CNN, FCN, and instance center direction structure to improve the prediction accuracy of instance segmentation. Instance center direction finds the center pixel of each object and predicts object boundaries. YOLACT (Bolya et al., 2019) is a real-time instance segmentation model that generates a prototype mask through ResNet-101 and a feature pyramid network and predicts instance segmentation using a combination of mask coefficients. ResNet-101 is a model based on ResNet-50 that adds and stacks multiple residual learning modules. A feature pyramid network is composed of kernels of multiple sizes, which are used to extract feature maps of multiple sizes. Prototype mask is a segmentation map formed by stacked output feature maps of feature pyramid networks. The prototype mask is weighted with mask coefficients to produce an instance segmentation map (Table 7.7).

7.5 Applications of AI and Machine Vision in Agriculture

Before the Industrial Revolution, commercial agriculture was a time-consuming, inefficient, and labor-intensive industry. Since the revolution, powered machinery has saved a tremendous workforce and the sector has also officially entered the era of agricultural mechanization. Although machines can replace a massive workforce, many agricultural activities still depend on labor, such as crop health monitoring, crop disease identification, and livestock health management. Leveraging AI and image recognition technologies, which have been booming in recent years, can significantly reduce the labor burden of farmers and increase industry competitiveness. This section introduces the application of deep learning machine vision technologies in recent years of farming, forestry, fishery, and animal husbandry.

7.5.1 Open Field Crops

Tea (Camellia sinensis) is the second most consumed beverage. The global tea market value was approximately $52.1 billion in 2018 and is projected to reach $81.6 billion in 2026 (Kumar & Deshmukh, 2020). Tea diseases and harming insects are significant factors that cause lesions to tea foliages and result in tremendous economic losses. Therefore, quick identification of infection type is essential to stop the disease promptly. In the past, tea diseases were identified by plant pathologists or experienced farmers, who were heavily burdened when diseases broke out. Lee et al. (2020) proposed a CNN detector to help farmers identify tea foliar diseases under practical field conditions. Chen et al. (2021) improved the previously developed detector by replacing the model architecture from Faster R-CNN to cascade R-CNN. Cascade R-CNN extended the two-stage architecture of Faster R-CNN, using multiple Fast R-CNN detectors to acquire bounding box locations more precisely (Cai & Vasconcelos, 2019). A tea disease and harming insect identification system was developed to identify three diseases and six harming insects. Moreover, an instant messaging software or smartphone application was deployed to provide farmers with a real-time and easy-to-use tool. Users can submit a diseased leaf image to the model to instantly diagnose the plant disease type and receive recommended treatments to obtain a suggested treatment and potentially reduce the damage, as shown in Fig. 7.12.

7.5.2 Greenhouse Cultivation

Asparagus (*Asparagus officinalis L.*) is a plant of high economic value and is originated from the temperate zone. The net value of fresh asparagus is over 1.5 billion dollars for only nine million tons per year (Tridge, 2020). In order to introduce

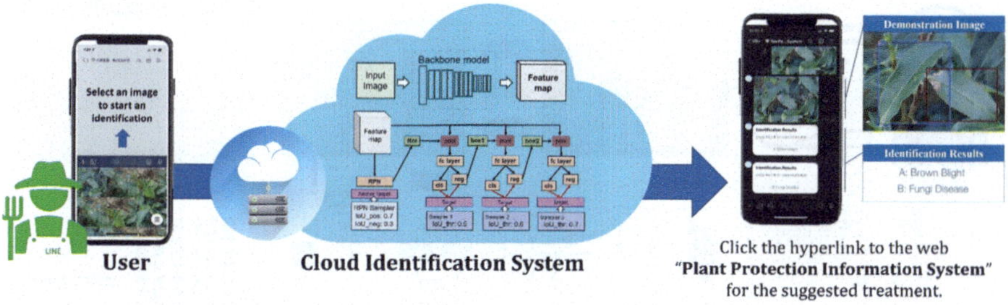

Fig. 7.12 Plant disease identification system. An instant messaging software and disease identification model are connected to provide real-time disease identification results through an application programming interface

asparagus to subtropical countries, the mother stalk method has been adopted for greenhouse asparagus since the 1960s in Taiwan. This method relieves the problems of nutrient deficit and production shortage but escalates the extra labor need for maintenance. Therefore, developing an intelligent monitoring system is beneficial for the ease of field management. Hsiung et al. (2021) established a robot vehicle for field data collection and a Mask R-CNN model to identify the asparagus stalks and spears in the field (Fig. 7.13). For further improving the identification performance, the model was strengthened with Copy-Paste augmentation and semi-supervised learning processes. By copying the existing annotations and learning from the unlabeled images, the improved model had better resilience to field changes. To provide a remote monitoring service, a web application developed with the Django framework was designed to receive the collected images from the robot vehicle, demonstrate the identification results, and present the statistics and historical data.

7.5.3 Forestry

Due to its special geographical environment and climatic factors, Taiwan is rich in forests and possesses forest resources of high economic value. Attracted by the high economic value, illegal logging of precious wood is common and it poses a major threat to forest conservation. Although the government has imposed a ban on logging since the 1990s to protect the forests, there are still many unscrupulous individuals who rely on the inability of the general public and adjudication agencies such as customs to identify wood species, bring economically valuable timber into the black market to make profits. The conventional method of identifying wood species is to cut a wood slice sample, stain it with safranin reagent, and then observe it with an electron microscope. This method is accurate but time-consuming, inefficient, and requires professional training. Wu et al. (2020) proposed applying MobileNetv2 (Sandler et al., 2018) to develop an identification model for wood

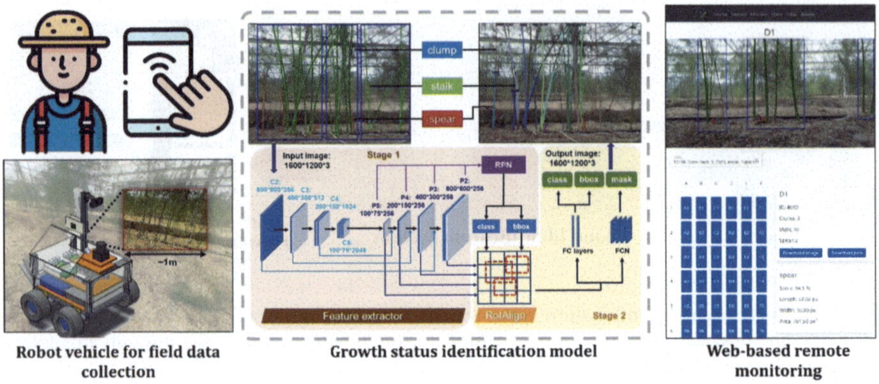

Fig. 7.13 Asparagus monitoring system. The web application collects field image data from a robot vehicle and demonstrated the identification results through a web service

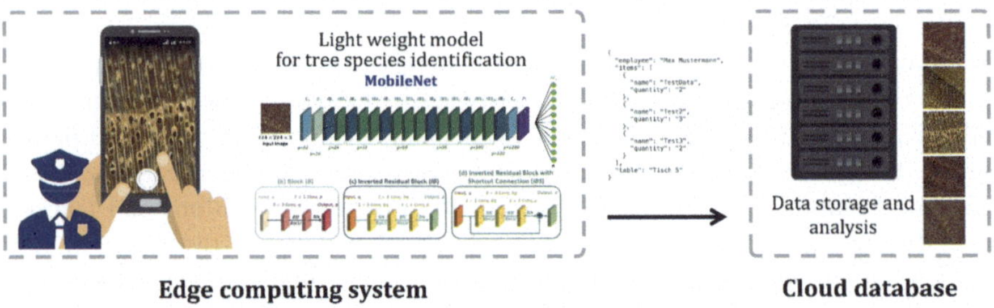

Fig. 7.14 Wood species identification system. The edge computing system built using MobileNetv2 can move part of the computing work from the server to the user's device, such as a mobile phone, and return the processed image of log end sample data to the cloud database for storage and analysis

species. Unlike the standard structure of a convolutional layer, MobileNetv2 uses a special convolutional operation, known as depthwise separable convolution, to reduce the number of parameters and speed up computation. Using a graphical processing unit (GPU) server, it can identify wood species from an image of the log end in less than a second (Fig. 7.14). In addition, MoblieNetv2 can be used on devices with poor computing power, such as mobile phones and single-chip computers, to identify wood species.

7.5.4 Fishery

Fish is an essential resource in the sea, but its population is declining due to overfishing. In order to sustain fishery resources, all countries strictly regulate fish catch. Fish species and fish length are two of the crucial data monitored in fishery

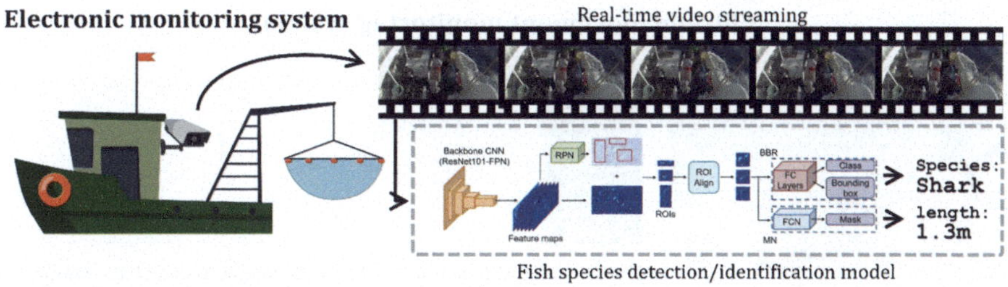

Fig. 7.15 Electronic observer system. Mask R-CNN model can detect target object (i.e., fish) and its position in video frames, then automatically identify the fish species and calculate the fish length

management. The old-fashioned method to gather this data is to have observers identify the species caught and measure the fish length manually. This method does not only cause inconvenience to fishing and port operations; it is also time-consuming, laborious, inaccurate, and subjective. In recent years, many countries have adopted an "electronic observer" system to record fish caught by photographs or videos on fishing vessels, then have the species and fish length manually interpreted. However, this process is still time-consuming. Therefore, Tseng and Kuo (2020) proposed a Mask R-CNN model to interpret fish catch videos automatically. The model detects fish in a video, then segments the fish from its background to measure its length and identify its species, as shown in Fig. 7.15.

7.5.5 Animal Husbandry

Chicken is an important food source. Based on the statistics reported by the Council of Agriculture, Executive Yuan, the annual output value of chicken products in Taiwan was 43.8 billion Taiwan dollars in 2018, accounting for nearly 26% of the total output value of animal husbandry. Chicken farmers in Taiwan monitor chickens by manual inspection. In the case of commercial chicken houses with large numbers of chickens, this approach is labor-intensive, time-consuming, and relies on the owners' experience. Chickens are prone to be inactive due to illness or injury; therefore, chicken activity is an important indicator of health. Missing the best time to administer medicine due to misjudgment of chicken activity results in loss of profits for the farmers. Therefore, Nian and Kuo (2019) proposed installing a camera on the ceiling of a chicken house and using the Faster R-CNN model to detect chicken activity (Fig. 7.16). The model automatically detects chickens in a video frame and then compares the positions of the chickens in the subsequent video frames one by one to obtain the movement path and activity level of individual chickens. This analysis of chicken activity can assist farmers in monitoring the health of chickens and detect unwell chickens.

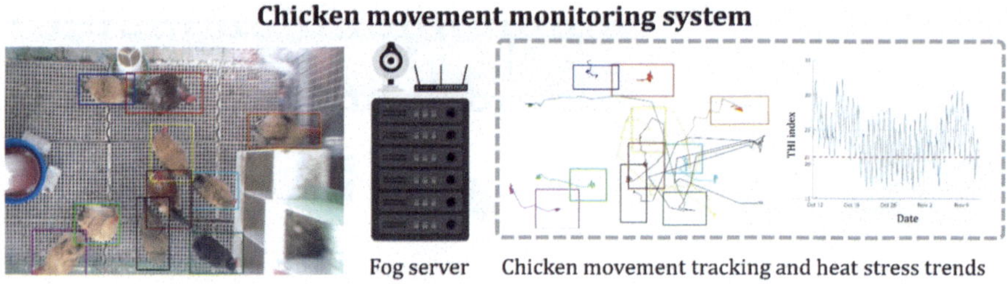

Fig. 7.16 The chicken activity monitoring system uses Faster R-CNN model deployed on a fog server to identify and locate chickens and use this data to determine chicken activity. A fog server can receive and process a variety of sensor data and transfer data between sensors with less latency than cloud servers

7.6 Conclusion

This chapter uses AI in image recognition as an example to explain neural network and CNNs, their principles, training, and frameworks. It also introduces CNN models used in three image recognition applications: image classification, object detection, and object segmentation. Finally, the current applications of image recognition in agriculture are presented: tea disease identification, asparagus growth monitoring, wood and fish species identification, and activity level of individual chickens.

As shown by these successful applications, AI in the past few years has helped outdoor and indoor farming, forestry, fishery and animal husbandry industries reduce the demand for and dependence on various experts and labor-intensive tasks. With the increasing global population, advanced technologies, including narrow AI, are urgently needed to increase food production and reduce agricultural losses caused by natural disasters, diseases, and pests. AI is envisioned to play a pivotal role in assisting the agriculture industry to use existing resources more effectively and efficiently. In addition to solving the acute food shortage, AI can help reduce the damage of the Earth's environmental resources and achieve a sustainable coexistence with other living things on Earth.

Acknowledgements The authors would like to thank Hong-Yang Lin, Hsien-Feng Hsiung, Jui-Yu Lin, Kuan-Ting Yeh, Tsan-Yu Wu, Yi-Hsin Yeh, Yu-Jung Tsai for their assistance in collecting and compiling the chapter content.

References

Badrinarayanan, V., Kendall, A., & Cipolla, R. (2017). Segnet: A deep convolutional encoder-decoder architecture for image segmentation. *IEEE Transactions on Pattern Analysis and Machine Intelligence, 39*(12), 2481–2495.

Bochkovskiy, A., Wang, C. Y., & Liao, H. Y. M. (2020). YOLOv4: Optimal speed and accuracy of object detection. *arXiv preprint* arXiv:2004.10934.

Bolya, D., Zhou, C., Xiao, F., & Lee, Y. J. (2019). Yolact: Real-time instance segmentation. *arXiv preprint* arXiv:1904.02689.
Bottou, L. (2010). 2 On-line learning and stochastic approximations. In D. Saad (Ed.), *On-line learning in neural networks* (pp. 9–42). Cambridge University Press. https://doi.org/10.1017/CBO9780511569920.003
Cai, Z., & Vasconcelos, N. (2019). Cascade R-CNN: High quality object detection and instance segmentation. *IEEE Transactions on Pattern Analysis and Machine Intelligence, 43*(5), 1483–1498.
Chen, L. C., Papandreou, G., Kokkinos, I., Murphy, K., & Yuille, A. L. (2014). Semantic image segmentation with deep convolutional nets and fully connected CRFs. *arXiv preprint* arXiv:1412.7062.
Chen, L. C., Hermans, A., Papandreou, G., Schroff, F., Wang, P., & Adam, H. (2018a). Masklab: Instance segmentation by refining object detection with semantic and direction features. In *Proceedings of the IEEE Conference on Computer Vision and Pattern Recognition*, Salt Lake City (pp. 4013–4022).
Chen, L. C., Zhu, Y., Papandreou, G., Schroff, F., & Adam, H. (2018b). Encoder-decoder with atrous separable convolution for semantic image segmentation. In *Proceedings of the European Conference on Computer Vision (ECCV)*, Munich (pp. 801–818).
Chen, X. M., Lin, C. C., Lin, S. R., & Chen, S. F. (2021). Application of region-based convolution neural network on tea diseases and harming insects identification. In *2021 ASABE Annual International Meeting. Virtual and On-Demand.*
Christian, B., & Griffiths, T. (2016). *Algorithms to live by: The computer science of human decisions*. Henry Holt.
Duchi, J., Hazan, E., & Singer, Y. (2011). Adaptive subgradient methods for online learning and stochastic optimization. *Journal of Machine Learning Research, 12*(61), 2121–2159.
Girshick, R. (2015). Fast R-CNN. *arXiv preprint* arXiv:1504.08083.
Girshick, R., Donahue, J., Darrell, T., & Malik, J. (2014). Rich feature hierarchies for accurate object detection and semantic segmentation. In *Proceedings of the IEEE Conference on Computer Vision and Pattern Recognition*, Columbus (pp. 580–587).
Glorot, X., Bordes, A., & Bengio, Y. (2011). Deep sparse rectifier neural networks. In *Proceedings of the Fourteenth International Conference on Artificial Intelligence and Statistics*, Ft. Lauderdale (pp. 315–323).
He, K., Zhang, X., Ren, S., & Sun, J. (2014). Spatial pyramid pooling in deep convolutional networks for visual recognition. In D. Fleet, T. Pajdla, B. Schiele, & T. Tuytelaars (Eds.), *Computer Vision – ECCV 2014*. Springer. https://doi.org/10.1007/978-3-319-10578-9_23
He, K., Zhang, X., Ren, S., & Sun, J. (2016). Deep residual learning for image recognition. In *Proceedings of the IEEE Conference on Computer Vision and Pattern Recognition*, Las Vegas (pp. 770–778).
He, K., Gkioxari, G., Dollár, P., & Girshick, R. (2017). Mask r-cnn. In *Proceedings of the IEEE International Conference on Computer Vision*, Venice (pp. 2961–2969).
Howard, A. G., Zhu, M., Chen, B., Kalenichenko, D., Wang, W., Weyand, T., Andreetto, M., & Adam, H. (2017). Mobilenets: Efficient convolutional neural networks for mobile vision applications. *arXiv preprint* arXiv:1704.04861.
Hsiung, H. F., Lee, S. H., Wang, J. C., Jiang, J. A., Liu, L. Y., Hsieh, M. H., & Chen, S. F. (2021). Application of mask region-based convolutional neural network on asparagus growth identification. In *2021 ASABE Annual International Meeting. American Society of Agricultural and Biological Engineers. Virtual and On-Demand*, Houston.
Kingma, D. P., & Ba, J. (2014). Adam: A method for stochastic optimization. *arXiv preprint* arXiv:1412.6980.
Krizhevsky, A., Sutskever, I., & Hinton, G. E. (2012). Imagenet classification with deep convolutional neural networks. In *Advances in Neural Information Processing Systems* (pp. 1097–1105).
Kumar, S., & Deshmukh, R. (2020). *Tea market*. Allied Market Eesearch. Retrieved May 10, 2020, from https://www.alliedmarketresearch.com/tea-market

LeCun, Y., Bottou, L., Bengio, Y., & Haffner, P. (1998). Gradient-based learning applied to document recognition. *Proceedings of the IEEE, 86*(11), 2278–2324.

Lee, S., Lin, S., & Chen, S. (2020). Identification of tea foliar diseases and pest damage under practical field conditions using a convolutional neural network. *Plant Pathology, 69*(9), 1731–1739.

Lin, T. Y., Dollár, P., Girshick, R., He, K., Hariharan, B., & Belongie, S. (2017). Feature pyramid networks for object detection. In *Proceedings of the IEEE Conference on Computer Vision and Pattern Recognition* (pp. 2117–2125).

Liu, W., Anguelov, D., Erhan, D., Szegedy, C., Reed, S., Fu, C. Y., & Berg, A. C. (2016). SSD: Single shot multibox detector. In B. Leibe, J. Matas, N. Sebe, & M. Welling (Eds.), *European Conference on Computer Vision. Computer Vision – ECCV 2016*. Springer. https://doi.org/10.1007/978-3-319-46448-0_2

Liu, S., Qi, L., Qin, H., Shi, J., & Jia, J. (2018). Path aggregation network for instance segmentation. In *Proceedings of the IEEE Conference on Computer Vision and Pattern Recognition*, Honolulu (pp 8759–8768).

Long, J., Shelhamer, E., & Darrell, T. (2015). Fully convolutional networks for semantic segmentation. In *Proceedings of the IEEE Conference on Computer Vision and Pattern Recognition* (pp. 3431–3440).

McCulloch, W. S., & Pitts, W. (1943). A logical calculus of the ideas immanent in nervous activity. *The Bulletin of Mathematical Biophysics, 5*, 115–133.

Nian, S. Y., & Kuo, Y. F. (2019). Monitoring chicken activity in commercial chicken houses using deep convolutional neural network. In *2019 Conference on Bio-Mechatronics and Agricultural Machinery Engineering*, Taichung, Taiwan.

Qian, N. (1999). On the momentum term in gradient descent learning algorithms. *Neural Networks, 12*(1), 145–151.

Redmon, J., & Farhadi, A. (2018). Yolov3: An incremental improvement. *arXiv preprint* arXiv:1804.02767.

Redmon, J., Divvala, S., Girshick, R., & Farhadi, A. (2016). You only look once: Unified, real-time object detection. In *Proceedings of the IEEE Conference on Computer Vision and Pattern Recognition* (pp. 779–788).

Ren, S., He, K., Girshick, R., & Sun, J. (2015). Faster R-CNN: Towards real-time object detection with region proposal networks. In *Advances in neural information processing systems* (pp. 91–99). Curran Associates.

Ronneberger, O., Fischer, P., & Brox, T. (2015). U-net: Convolutional networks for biomedical image segmentation. In *International Conference on Medical Image Computing and Computer-Assisted Intervention* (pp. 234–241). Springer.

Russakovsky, O., Deng, J., Krause, J., Berg, A., & Li, F. (2013). *ILSVRC-2013*. Retrieved from http://www.imagenet.org/challenges/LSVRC

Sandler, M., Howard, A., Zhu, M., Zhmoginov, A., & Chen, L. C. (2018). Mobilenetv2: Inverted residuals and linear bottlenecks. In *Proceedings of the IEEE Conference on Computer Vision and Pattern Recognition*, Salt Lake City (pp. 4510–4520).

Senior, A. W., Evans, R., Jumper, J., Kirkpatrick, J., Sifre, L., Green, T., Qin, C., Žídek, A., Nelson, A. W., & Bridgland, A. (2020). Improved protein structure prediction using potentials from deep learning. *Nature, 577*(7792), 706–710.

Silver, D., Huang, A., Maddison, C. J., Guez, A., Sifre, L., Van Den Driessche, G., Schrittwieser, J., Antonoglou, I., Panneershelvam, V., & Lanctot, M. (2016). Mastering the game of Go with deep neural networks and tree search. *Nature, 529*(7587), 484–489.

Simonyan, K., & Zisserman, A. (2014). Very deep convolutional networks for large-scale image recognition. *arXiv preprint* arXiv:1409.1556.

Tridge. (2020). *Fresh Asparagus - Global production and top producing countries*. Retrieved September 10, 2021, from https://www.tridge.com/intelligences/asparagus/production

Tseng, C. H., & Kuo, Y. F. (2020). Detecting and counting harvested fish and identifying fish types in electronic monitoring system videos using deep convolutional neural networks. *ICES Journal of Marine Science, 77*(4), 1367–1378.

Wallach, I., Dzamba, M., & Heifets, A. (2015). AtomNet: A deep convolutional neural network for bioactivity prediction in structure-based drug discovery. *arXiv preprint* arXiv:1510.02855.

Wu, S. Y., Kuo, Y. F., & Lee, K. M. (2020). A preliminary study of convolutional neural network in wood species identification. In *2020 The Chinese Forest Products Association Symposium*, Taichung, Taiwan.

Zeiler, M. D., & Fergus, R. (2014). Visualizing and understanding convolutional networks. In D. Fleet, T. Pajdla, B. Schiele, & T. Tuytelaars (Eds.), *Computer Vision – ECCV 2014*. Springer. https://doi.org/10.1007/978-3-319-10590-1_53

Zhao, H., Shi, J., Qi, X., Wang, X., & Jia, J. (2017). Pyramid scene parsing network. In *Proceedings of the IEEE Conference on Computer Vision and Pattern Recognition*, Honolulu (pp. 2881–2890).

Chapter 8
Smart Farming Management

Du Chen and Xindong Ni

8.1 Introduction of Smart Farming

Lack of resources per capita, labor shortage, severe environmental situation are always scientific problems in the development of agricultural modernization. Smart farming is the inevitable trend of agricultural modernization, together with a higher level of intensity, precision, and coordination to achieve precise and intelligent agricultural production, which aims to solve the difficult problems on agricultural modernization construction fundamentally. Smart farming is a modern agricultural form supported by Internet of Things, big data, artificial intelligence, agricultural robot, and other technologies, which is a higher stage of agricultural development following traditional agricultural, mechanized and automatic agriculture.

8.1.1 From Traditional Agriculture to Intelligent Agriculture

The development of human society is a process in which laborers exert their intelligence and talents and constantly create new labor tools to understand nature, adapt to nature, and transform nature. Also, the evolution of labor tools reflects the evolution from traditional agriculture to intelligent agriculture.

Traditional agriculture is the product of the initial agricultural society, which is dominated by human and animal power. The main tools of human labor are used to exploit land resources and various primary one-handed tools and livestock forces, which can alleviate human physical labor to a certain extent, but do not

D. Chen (✉) · X. Ni
College of Engineering, China Agricultural University, Beijing, China
e-mail: tchendu@cau.edu.cn

S. Ma et al. (eds.), *Sensing, Data Managing, and Control Technologies for Agricultural Systems*, Agriculture Automation and Control,
https://doi.org/10.1007/978-3-031-03834-1_8

fundamentally liberate human production activities from heavy physical labor and focus on human labor. Although agricultural production has been formed, the production scale is small, and the production technology and management level are relatively backward, the ability to resist natural disasters is weak as well. Additionally, The immature agricultural system restricted the development of agricultural productivity but laid the foundation for agricultural industrialization. In the era of traditional agriculture, the most obvious label is the summary of farming technology.

In the first Industrial Revolution, the invention and use of steam engine marked the revolutionary development of production tools in human society, in which new tools gained power through energy conversion, and machines replaced manual labor and tools, opening the curtain of industrial society. During the course of more than 300 years of industrial society, tools characterized by energy conversion have achieved two historic leaps of mechanization and electrification, which have had a profound impact on the production and life of human society. With the development of the industrial revolution, agricultural machinery tools have begun to emerge and are widely used in agricultural production. Mechanized agriculture, which is based on mechanized production, replaced the traditional production tools of human and animal power and changed the farming mode of facing loess back to the sky. The backward and inefficient traditional agriculture has been transformed into advanced and efficient mechanized operation, which has greatly improved the productivity and productivity level.

With the fusion application of computer technology, electronic information and communication, and other modern information technology and automation equipment in the agricultural field, mechanized agriculture will transform information agriculture, with the application of modern information technology and local agricultural operation automation and intelligence as the main characteristics. On the basis of strengthening sensor network, computer network, data communication network and other basic information infrastructure construction, set up information internet sharing platform to realize information exchange and knowledge sharing, making modern information technology and intelligent agricultural equipment widely used in each link of agriculture tillage implementation, such as plowing, planting, managing, harvesting. Improving resource utilization, automation, land yield, and labor productivity. At present, we are in the stage of mechanization agriculture to information agriculture transformation. With the help of technologies such as Internet of Things, big data, cloud computing, and artificial intelligence, we will complete the transformation in the near future.

8.1.2 The Inevitable Trend of Intelligent Agriculture

As the highest stage of modern agriculture, smart farming is formed on the basis of the integration of network information resources, sensor technology, Internet of Things, big data, cloud computing, artificial intelligence, and robot technology, realizing unified management of agricultural production under the agricultural standardization system. At the same time, information technology is applied comprehensively and systematically to each link of agricultural production and finally realizes unmanned agriculture. In the face of the development trend of modern agriculture and the increasingly huge food demand, improving the agricultural productivity, resource utilization rate, and land yield rate is still the ultimate goal of agricultural development. In addition, on the basis of the integration of soft and hard technologies, breakthrough in basic theory, key technological innovations, accurate perception, intelligent decision-making, and accurate control can be realized, and finally computers can replace human brains, and machines replace manpower, which truly liberates mankind from complex agricultural labor.

8.2 Smart Farming Solutions Based on Internet of Things

With the progress of science and technology and the industrial upgrading, Internet of Things comes into being at the historic moment. The combination of information technology such as the Internet of Things with agriculture has resulted in smart farms. Smart farm based on Internet of Things is an important development direction of agricultural modernization, which will redefine the farm management, integration of information technology, and scientific and technological means to solve the problems of the different levels of the agricultural production to reduce the waste of resources and improve productivity.

8.2.1 The Rise and Application of the Internet of Things

Internet of things is a network of interconnected devices, in which devices can communicate with each other and generate relevant data in the operating environment. The key and core of the Internet of Things is the information exchange and data transmission among the elements of the Internet of Things, in which the equipment network of data sharing is jointly constructed between the network and equipment through wireless transmission, cloud computing, big data, and artificial intelligence.

The concept of the Internet of Things was formally established at the World Summit on the Information Society in Tunis in 2005. The theme of the conference is "ITU Internet Reports 2005: The Internet of Things," which pointed out that we are standing on the edge of a new era of information, and the goal of information and communication technology (ICT) has evolved from facilitating communication between people to connecting people and things to things. Thus, the era of ubiquitous Internet of Things communications is upon us. What is more, the Internet of Things gives us a new dimension in the world of information and communication technology, connecting anyone, even extending to connecting anything at any time, any place. Eventually the Internet of Things came into being.

The Internet of Things has been widely used in various fields of production and life, whose application is constantly promoting the change of production and lifestyle. Flexible manufacturing, intelligent management, remote monitoring, and other fields are iteratively updated based on the Internet of Things technology. The American Grid applies the Internet of Things in power generation, transmission, transformation, distribution, and electricity consumption, using strong bidirectional communication, sensor networks, and distributed computers to improve power exchange and work efficiency and improve system operation stability [research on the architecture and key technology of Internet of Things (IoT) applied on smart grid]. In urban construction and management, applications with the Internet of Things as the core, such as traffic control, traffic flow dredging, and intelligent parking enable people, vehicles, and roads to cooperate closely and improve the transportation environment and increase the utilization rate of road resources. The application of the Internet of Things in environmental monitoring mainly monitors the urban environment and ecological environment intelligently and carries out comprehensive perception through biology, optics, chemistry, infrared, telemetry, and other sensor technologies. Based on monitoring data and data processing technologies such as environmental analysis, decision support and cloud computing and so on, the best solution can be obtained to achieve effective control.

The Internet of Things is a promising technology, which can provide efficient and reliable solutions for informatization and modernization transformation in many fields. At present, many intelligent agricultural solutions have been developed based on the application of Internet of Things technology in the agricultural field. Due to the complexity and challenges of agriculture itself, the combination of agriculture and Internet of Things is bound to bring a huge revolution to agriculture.

8.2.2 Key Technologies of Agricultural Internet of Things

Agricultural Internet of things refers to equipment through agricultural information awareness, in accordance with the contract agreement, the agricultural system production tools, such as plants, animals, environment elements, and production tools in the physical components and all kinds of virtual objects connected to the Internet in order to exchange information and communicate, realizing intelligent identification, positioning, tracking and monitoring and management of agricultural objects and processes. The human–machine–thing integration and interconnection of the agricultural Internet of Things can help humans cognize, manage, and control various elements and processes in agriculture in a more refined and dynamic way and greatly improve human cognitive ability of agricultural plants and animals, the regulatory capacity of agricultural complex systems, and the processing ability of agricultural emergencies.

Intelligent Agriculture System Based on Internet of Things

The agricultural Internet of Things is mainly composed of four parts: the perception layer, the transmission layer, the processing layer, and the application layer. The perception layer realizes the acquisition and collection of information about agricultural objects and object behaviors, such as environment, crops, plants, and animals, through the sensor network composed of various detection and monitoring devices and sensor nodes; The multi-source data obtained from the perception layer enters the transmission layer and is transmitted to the processing layer through wired or wireless. The functions of the processing layer include data storage, data processing, data visualization, etc. When data enters the processing layer, which is classified, stored, and processed in parallel with the support of cloud computing and big data technologies, so that information can be extracted in a short time; In the application layer, decision support, modeling, and prediction can be made according to the data results obtained from the processing layer to achieve the management and control of agricultural production process.

Wireless Sensor Networks

In the agricultural Internet of Things system, the wireless sensor network deployment based on sensor technology is the source and basis of data, and the interconnection of things in the Internet of Things system is also realized through interconnected data. As a key technology of the sensing layer, wireless sensor network is composed of a large number of sensor nodes deployed in the monitoring areas. It is a multi-hop ad hoc network system established by wireless communication, which can cooperate with the sensing acquisition through a large number of integrated sensors deployed in key locations, together with perception, acquisition, monitoring, and transmission of the sensing object information within the coverage of the sensor network, transmitted through wireless communication and sent to the user in the form of self-organizing multi-hop network to achieve data acquisition, target tracking, monitoring, and early warning.

Big Data

Technological progress enables each sensor node at the perception layer to have a certain storage and computing capacity and to acquire perception data at a near real-time speed and to generate large-scale data at the same time. If heterogeneous multi-source data cannot be effectively integrated, the operating efficiency of the Internet of Things system will be greatly reduced. The emergence of big data greatly relieves the pressure of data processing of the Internet of Things and plays an important role in the Internet of Things system. Big data is the fuel of the Internet of Things and artificial intelligence driven by big data becomes the brain. The processing layer represented by big data promotes the efficient operation of the Internet of Things. With the support of big data, data is not only the number it represents but also the relationship and derivative interaction between a large number of data, which makes data become information.

Cloud Computing

Cloud computing provides a powerful boost to the development of big data. Big data relies on computing power storage and other services of cloud computing to mine data value and cloud computing supports big data as computing resources. As the technical basis of the application layer, cloud computing provides decision support for the operation of the agricultural Internet of Things in various scenarios. Cloud computing is a new computing mode that connects various computing storage and software resources through different network protocols to provide services to users according to specific requirements.

8.2.3 Internet of Things Intelligent Agriculture Platform

In the process of the widespread application of Internet of Things technology, system planning and framework design are prone to produce different results due to different starting points. Meanwhile, with the continuous development of application scenarios and user requirements, all kinds of new technology and new ideas will gradually merge into the field of Internet of things system and the architecture of the Internet of Things will also determine the technical requirements, operating mechanism, and development direction of the Internet of Things.

System Structure

The perception layer of agricultural Internet of Things has strong heterogeneity. In order to realize interconnection and interaction between multi-source heterogeneous information, the open, multi-level, and scalable network system of the four-level Internet of Things platform can support huge data flow.

1. Perception layer

   Agricultural platforms are developed to solve specific agricultural problems. Because of the complexity of agricultural problems, the basic sensor network construction needs to cover enough comprehensive monitoring data, using RFID remote sensing technology and all kinds of sensor nodes to obtain information in monitoring area, including environmental data, soil information, weather conditions, crop characterization, pests, and diseases that may affect agricultural production processes. Due to the complex and changeable use environment, the sensor developed for agricultural field needs high performance requirements: First of all, it should have stability to ensure that it can work normally in high temperature and high humidity environment and be able to cope with environmental changes; The second is applicability. The sensor needs to open protocols and interfaces that match the agricultural platform and the performance needs to adapt to the system; Finally, economy. The implementation of agricultural Internet of Things requires the use of a large number of sensor nodes. If the cost is high, it will not be conducive to the promotion and application of agricultural Internet of Things technology.

2. Transport layer

   As the link between the perception layer and the processing layer, the transmission layer is mainly responsible for transmitting the monitoring data obtained by the perception layer to the processing layer quickly, accurately, and safely without barriers. It solves the problem of long-distance signal transmission within the monitoring range covered by the sensing layer, responding to different information processing needs. The transmission layer CAN be wired through CAN bus or RS485 bus protocol and monitoring signals are obtained through Bluetooth, Wi-Fi, ZigBee, and other wireless sensor networks and then transmitted to the remote server through Ethernet, Wi-Fi, GPRS, and other forms so as to realize the series connection between the perception layer and the processing layer.

3. Processing layer

   The sensor network built in the monitoring area will transmit a large number of heterogeneous multi-source data, which leads to extremely complex data types. In order to process big data, it is necessary to preprocess the data obtained in the acquisition stage first, extract the characteristic information and relationship association of multi-source data, and store the data in a unified data structure after data integration and collection. At the same time, for big data, not all information is valuable information, and interference items and error items need to be removed and filtered.

   Most of the previous modeling analysis serves a single target and the work efficiency is low. In order to provide data support for intelligent decision-making at the application layer, cloud computing technology integrated analysis resources to increase computing capacity and expand storage space, providing efficient and fast computing power for the Internet of Things. Cloud computing technology uses virtual means to achieve unified scheduling and management of all kinds of resources, greatly improving the efficiency of resource use; In the calculation process can be joined or quit at any time, and computational processing power is flexible. In addition, cloud computing changes the previous local storage or local server storage mode, and data information stored in the cloud can obtain more storage space, and data security will be greatly improved as well.

4. Application layer

   The data is processed through a cloud-based processing layer to produce solutions and processing results for different agricultural problems, and it can predict the future state based on current and past monitoring data, providing users with enough time to respond.

Network Topology and Protocols

Due to commercial technology maturity and other reasons, perceptive devices deployed in the monitoring area have significant differences in function, interfaces, and data transmission protocols. Wireless sensor network, bus technology, internet, and other communication networks have different data structures and transmission modes. How to effectively realize the interconnection between different networks and different devices has become the core problem of the implementation of agricultural Internet of Things.

As the data transmission link of the Internet of Things, the transport layer needs a variety of data interfaces to connect multiple heterogeneous data sources when receiving data. The hardware of the transport layer mainly includes gateway interface driver and embedded node. Besides, input interface includes RS232, RS485, Wi-Fi, etc.; output interface includes Wi-Fi, RJ45, GPRS, and other modes, which can support users to select the transmission mode according to the actual conditions of application scenarios. In the process of data transmission, the perception layer data is not simply sent to the processing layer, according to different communication protocols, the original signal is converted into data in a unified standard format for data fusion, data encapsulation, and other processing and then sent to the cloud or server. In this way, data redundancy caused by the underlying multi-source heterogeneous perception network can be avoided.

8.3 Smart Farming Management Applications

With the mature development of Internet of Things technology, information communication technology and automatic control technology are introduced into farm management, and farm intellectualization is an inevitable trend. By accurately monitoring the farm environment and the growth process of farm animals and plants, high-yield, high-quality, and efficient farms can be achieved. In a smart farm environment, edge computing manages big data generated by Internet of Things devices at the edge of the network, enabling shorter response time, quality of service, and more secure services. The Internet of Things and distributed ledger technology enable resource monitoring and traceability of agricultural products, allowing farmers to source their products and assure consumers of their quality. This section will analyze intelligent farm management from three aspects, including the application of intelligent farm crop management, animal husbandry management, and agricultural product supply traceability.

8.3.1 Application in Crop Management

Smart farms use technical resources to help at all stages of production, and their applications in crop management are mainly reflected in vehicles and machinery control, crop monitoring, disease prevention, soil management, irrigation control, and so on.

In agricultural technology, most automated machinery is operated in a fixed manner and requires manual operation or regular supervision to avoid mechanical failures or errors. Horng et al. (2019) developed a remote crop harvesting system. The system determines crop maturity through target detection by training neural network model and then uses robot arm to harvest mature crops and uses deep learning to perform intelligent image recognition on collected image, proposing the target detection model of MobileNet SSD (Single Shot multibox Detector), with an average accuracy of 84%. A prediction model of arm motion using a four-layer hidden layer sensor model is designed, and its average picking accuracy reaches 89%.

Environmental factors such as light, CO_2, temperature, humidity, water, and nutrients are key factors in producing high-quality crops. Harun et al. (2019) proposed an improved Internet of Things monitoring system for the growth optimization of Chinese cabbage and developed an intelligent embedded system to realize the real-time acquisition of plant experimental environmental parameters and automatic control and operation of LED. Meanwhile, the Internet of Things technology was used as a remote monitoring system through the spectrum of light-emitting diode (LED), photoperiod and light intensity are used to control indoor climatic conditions to improve plant yield and shorten turnaround time.

Plant disease and nutrient deficiency are one of the sources of economic losses in the field of agriculture. In some cases, the color and characteristics of some plant diseases and nutrient deficiency are similar, such as early downy mildew and nitrogen deficiency. Apk and Sps (2019) proposed an intelligent agricultural decision support system based on the Internet of Things for identifying plant disease hazards and nutrient imbalance. In this system, a multi-stage parameter optimization ELM classifier featuring selection algorithm (IGA-ELM) based on an improved genetic algorithm is adopted, which is applied to benchmark high dimensional biomedical data set and real-time application (plant disease data set). By reducing the features by 58.50% and 72.73%, respectively, the classification accuracy was improved by 9.52% and 5.71%.

Soil suitability analysis is a necessary prerequisite for crop planting to help achieve maximum yield, and soil testing is widely used by agricultural experts and farmers to determine soil characteristics required for agricultural production. Vincent et al. (2019) proposed an expert system combining sensor network with artificial intelligence systems such as neural network multi-layer Perceptron for agricultural land suitability evaluation. The system helps farmers assess the agricultural lands, with four suitability-decision levels being "more-suitable," "suitable," "sub-suitable," and "non-suitable." The results show this multi-category classification system with four-hidden layers is valid for evaluating the farmlands. Besides, such well-trained model will be used for evaluating future assessment and classifying the land after every cultivation.

Internet of Things solutions for irrigation control have also been developed for use in a variety of agricultural environments. Angelopoulos et al. (2020) proposed a decentralized intelligent irrigation method for strawberry greenhouse and realized a comprehensive intelligent Irrigation system in a real strawberry greenhouse environment in Greece. They designed and implemented and verified the intelligent Irrigation solution for greenhouse strawberry and developed Smart Irrigation Network system to maintain soil moisture between 50% and 55% and the system is significantly superior to traditional strawberry irrigation methods in terms of soil moisture change and water consumption.

Considering the problem of water utilization in water-scarce areas and vision-free monitoring for people far from farmland, Nawandar and Satpute (2019) proposed a crop monitoring and automatic irrigation system that uses Unified Sensor Poles to obtain crops, planting date, and soil data, using these data calculation of evapotranspiration and irrigation schemes by using neural network to make decisions, and MQTT and HTTP are systematically used for data transfer. A sample crop test-bed was selected to demonstrate the results, and the system achieved an overall water saving effect of about 30% and 67%, respectively, compared with conventional drip irrigation and conventional irrigation.

Castaeda-Miranda and Castao-Meneses (2020) proposed an intelligent anti-freezing irrigation management system. Considering the relative humidity, air temperature, solar radiation, wind speed and other environmental factors, artificial neural network is used to optimize the prediction of greenhouse internal temperature and predict the internal temperature of greenhouse by artificial neural network, and the start of water pump is controlled by fuzzy expert system. Besides, real-time information interconnection is realized by mobile phone system (GSM/GPRS) and Internet (TCP/IP) service to acquire and monitor, and the temperature of greenhouse and farmland can be predicted through fuzzy control and neural network. The existence of frost is confirmed by fuzzy reasoning and defuzzy method and used to start the anti-freeze water distribution system.

Goap et al. (2018) came up with a kind of intelligent irrigation architecture based on Internet of Things, which provides a closed loop control of water supply in order to realize the fully autonomous irrigation Schemes. A hybrid machining learning method was developed to predict the soil moisture content, with using in-field sensor nodes and weather forecast data. The method also estimated the difference and changes of soil moisture that is caused by different weather conditions. Overall, this method realized the data collection, transmission, and processing of the weather forecast information and farmland physical parameters, such as soil moisture, air temperature, air relative humidity, soil temperature and radiation, etc. The system was developed and deployed on a pilot scale, in which sensor node data is collected wirelessly through the cloud and network services, and the Web-based information visualization and decision support system provides real-time information observation based on sensor data and weather forecast data analysis. Additionally, the hybrid machine learning algorithm based on Support Vector Regression (SVR) + K-means has high accuracy and minimum Mean Squared Error(MSE), which has been applied to irrigation planning module.

Munir et al. (2019) have come up with a Smart Watering System for small and medium-sized gardens and fields aided by an Android app system, which uses an accessible and affordable set of sensors to capture real-time data. By using blockchain and fuzzy logic to determine Watering schedules, and using the Android Smart Watering System (prototype) application, multiple users and devices can participate in plant monitoring and remote interaction. The experimental results show that the accuracy of the intelligent water system reaches 95.83%, and the system can effectively and safely handle the watering process of plants as well.

Keswani et al. (2018) developed a low-cost and high-precision distributed wireless sensor network environment for the Internet of Things, which is used to accurately monitor soil and environmental parameters. The study in-depth compared different optimization methods, i.e. stochastic gradient descent that support adaptive learning rates, feed-forward neural network with gradient descent algorithm for pattern classification, and the best practice was performed to predict the soil moisture per hour together with interpolation method for producing the soil moisture content distribution map. Combined with the interpolation method, the soil water content distribution map was generated, and the valve control command was processed by the weather condition modeling system based on fuzzy logic. The control command was operated according to different weather conditions, which met the unified irrigation requirements under almost all weather conditions.

8.3.2 *Application in Animal Management*

The Internet of Things is about connecting people, processes, data, and things and is changing the way we monitor and interact with things. The positive combination of information and communication technologies with sophisticated data analysis methods will also transform traditional livestock farming. Improving farming practices by gaining actionable insights in order to increase efficiency, yield, and productivity, and help farmers manage their farms well. The intelligent perception and analysis of animal individual information and behavior is the core of animal husbandry.

Pigs are particularly vulnerable to heat stress, which triggers behavioral and physiological responses that can negatively impact productivity in the summer. The remote monitoring system can monitor the movement behavior of pigs and the environment of piggery. Zeng et al. (2021) proposed a ZigBee-based three-layer wireless sensor network system, which is used for real-time monitoring of four environmental parameters, including temperature, relative humidity, carbon dioxide concentration, and ammonia concentration in pregnant female pigpen. The indoor environmental monitoring system is composed of three layers: livestock environmental sensing, wireless transmission service, and multi-client application. The transmission performance of data packets collected by sensor nodes through the system is evaluated and the null records and abnormal observation data are preprocessed. The results showed that the average outlier rate of CO_2 was 6.5%, and the other sensors all worked well. And the real-time monitoring and timely intervention of the microclimate of pregnant female piggery were realized through the system, thus providing intelligent decision for automatic and accurate management of livestock.

Some Internet of Things applications have been used for intelligent wastewater monitoring and control system. Chung (2020) established a farm-scale demonstration base with self-designed intelligent piggery sewage treatment facilities and remote monitoring function through the application of the Internet of things, improving the efficiency of pig farm wastewater treatment. The intelligent pig farm wastewater treatment system was applied to a pig farm with 1000 pigs located in I-Lan County, Taiwan. The experimental results showed that the overall removal rates of biochemical oxygen demand, chemical oxygen demand, and suspended solids were 94%, 86–87%, and 96%, respectively.

Livestock monitoring is an important aspect of livestock management, and advanced technology has made it possible to automatically track and monitor cattle. Ilyas and Ahmad (2020) have proposed a smart solution for livestock tracking and geo-fencing using the most advanced Internet of Things technology. This study created a geo-safe zone for cattle based on Internet of Things and GPRS, remotely monitoring and controlling the cattle with dedicated Internet of Things sensors. Intelligent systems collect data about the location, health and physical and emotional states of livestock, and the proposed livestock management system reduces the time and energy complexity of systems and integrated modules, reducing the cost of farming, and realizing remote monitoring.

Behavior may be the most critical of the different parameters monitored for the animals themselves, providing important information about the health or reproductive status of the animals and helping farmers understand how the animals are performing in the environment. Debauche et al. (2019) proposed a Lambda cloud architecture, which provides help in storage, real-time processing and large-scale data storage and analysis, and innovatively coupled to a scientific sharing platform for archiving and processing high-frequency data, allowing scientists to collect, store, process, and share information to integrate Internet of Things applications for livestock monitoring. An iphone with IMU (Inertial Measurement Unit) was worn on the livestock to measure the cattle's behavior. Obtained signals were transmitted to the gateway through UDP protocol. Data redundancy was eliminated through the combination of various related variables for edge computing, processing, modeling and decision-making.

Data-driven approaches are transforming many industries, including dairy farming, and offering an opportunity for predictive control and prevention of certain undesirable events. Data-driven decision-making methods and measures help improve the productivity of dairy farms. Taneja et al. (2019) offered an end-to-end Internet of Things application that uses fog and cloud analysis to make data-driven decisions for intelligent dairy farming. The system uses behavioral analysis to generate early alerts about animal health to help farmers monitor their livestock. Moreover, the system uses fog assist and cloud support to analyze data generated by wearable devices on cows' feet to detect abnormalities in animal behavior related to diseases such as lameness, improving productivity, and milk production by detecting potential diseases early. The system was tested on a real-world intelligent dairy farm of 150 dairy cows in Ireland and the results showed that the system could send out lameness detection alerts 3 days before manual observation.

Dairy production faces many challenges, including improving resource efficiency, being more environmentally friendly, being able to provide detailed information to consumers, and ensuring the safety and quality of the final product. Through the new global edge computing architecture, Alonso et al. (2020) proposed an application platform of artificial intelligence and blockchain technology for edge computing of the Internet of Things in the intelligent farm environment, which is used to monitor the status of cows and feed grains in real time. Edge computing reduces the use of cloud computing, storage and network resources, reduces the response time of services, and improves the quality of service and the security of applications. Distributed Ledger Technology provides security, data integrity and traceability. The platform was deployed and tested in a real scenario on a dairy farm, verifying the effectiveness of edge computing in reducing data traffic and improving the reliability of communication between the edge layer of the Internet of Things and the cloud.

8.3.3 Application in Agricultural Product Supply Traceability

Blockchain technology will provide real-time tracking, certification, protection, and monitoring functions for the food supply chain process, and the agricultural supply chain traceability system based on the Internet of Things can ensure the food safety and quality of every production link. Most of the blockchain conversation revolves around commodity transactions and the tracking of agricultural inputs and production outputs, particularly with regard to traceability of food safety. Through blockchain, agricultural data from sources such as soil sensors, weather satellites, drones, and distributed storage of agricultural equipment enables us to build trust and ensure sustainable agricultural development. Besides, by pooling and analyzing this data, it improves decision-making and automation at both the individual farm level and the community level.

Agricultural data from joint yield monitoring is mixed, considering the validity of the data, especially as other people may have influenced data quality at various steps along the data path. Lei et al. (2020) proposed an agricultural data storage method based on distributed ledger technology DLT to ensure the integrity of agricultural data. The distributed ledger keeps a complete record of all activities that take place on a fish farm and is linked to data recorded by farm sensors so that the data is indivisible. In addition, smart contracts are being used to automate agricultural data processing, including filtering outliers before generating records to the ledger, the data stored in the blockchain, and the intelligent contract can trigger and perform a particular operation, define access control rules, for specific participants to provide access to network resources or execute permissions within the business network, using authorized blockchain network as the foundation facility, which enhances transaction security while maintaining data transparency. A proof of concept was built on top of the designed architecture to test all system functionality by using Hyperledger Fabric and the fish farm system.

The safety of agricultural products has always been a concern for people. Considering the uncertainty of the environment and the multi-factor nature of the agricultural product Traceability System (BAPTS) based on blockchain, Yang et al. (2020) proposed a multi-criteria decision-making (MCDM) framework based on Q-RoFWPMM operator for evaluating and selecting a blockchain based agricultural product traceability system (BAPTS) design. This method combines muir-head mean (MM) and power average (PA) operators into Q-ROFS to generate two aggregation operators (Q-ROFPMM and Q-ROFWPMM), which provides an effective method for information aggregation evaluation. The selection of agricultural products traceability system based on blockchain (BAPTS) is verified by an example. Finally, the sensitivity and comparative analysis results are given to verify the validity and accuracy of the proposed method.

In the supply chain of agricultural products, traceability system is an effective means to ensure the quality and safety of agricultural products. Li et al. (2021) built a simplified agricultural products traceability system (STSAP) based on Android platform. Agricultural Products Traceability System (STSAP) consists of six functional modules: account management, site management, planting material management, agricultural activity management, processing management, and traceability document management. STSAP is used to document the supply chain of agricultural production, including planting, harvesting, transportation, and sales. When the customer uses WeChat to scan the QR code, STSAP automatically invokes API to read traceability files, helping consumers track production process and transaction records.

Liao and Xu (2019) proposed a blockchain traceability system based on intelligent agriculture that integrates wireless sensor network. The system realizes the agricultural products traceability system based on Ethereum, collecting front-end data and storing it in the blockchain system, and utilizes the blockchain itself which has the characteristics of decentralization, tamper-proof, security and encryption, combined with back-end database management and traceable QR code, to provide consumers with safe and reliable QR code, reliable and true agricultural product traceability information.

8.4 Prospects and Challenges of Agricultural Internet of Things Technology

The agricultural Internet of Things covers agricultural information acquisition, sensor instrument technology, data transmission, network communication technology, data fusion, intelligent decision, expert system and automation control technology, etc., which is an interactive platform of things to things and people, as well as an open platform. With the continuous development of science and technology, more and more advanced science and technology will be integrated into the agricultural Internet of Things to promote the development of agricultural Internet of Things, and the development of agricultural Internet of Things will become the basis and power to promote the development of modern and sustainable agriculture.

8.4.1 Development Prospects of Agricultural Internet of Things

Agricultural Internet of Things technology can cover the entire agricultural production process and play an important role in agricultural production management, becoming a powerful boost for intelligent agriculture. It will also get better development opportunities in the following aspects.

1. As the technical support of agricultural information perception, agricultural sensor technology plays a crucial role in agricultural Internet of Things. The development of sensing equipment for agricultural production will be an essential link in agricultural Internet of Things industry. With the development of detection methods and the increase of coverage, more and more attention will be paid to the research and development of micro-sensor nodes for agricultural animals and plants, soil and environment with low-cost, stable performance, low power consumption, small size and high detection accuracy, to provide intelligent agricultural management more perfect, more accurate monitoring data for the internet to provide more comprehensive data to support agriculture.
2. At present, agricultural machinery equipment is developing toward large-scale, accurate, and intelligent direction while meeting the needs of different agricultural operations. As an important tool in the process of agricultural operations, agricultural machinery and equipment will gradually be integrated into the agricultural Internet of Things, becoming the intelligent execution terminal of agricultural Internet of things, and play a crucial role in agricultural production operations. At the same time, agricultural equipment in the process of operation is increasingly dependent on crop information and environmental perception, agricultural data transmission, operation planning and intelligent decision-making, and agricultural Internet of Things technology can play an important role in these aspects. Therefore, the combination of agricultural machinery

equipment and Internet of Things technology is bound to become an important development direction of Internet of Things technology.

3. As a technology that can be monitored from agricultural production, agricultural processing, circulation to consumption, agricultural Internet of Things will increasingly play an important role in the field of food safety. Agricultural Internet of Things sensing information can be used as important information for the full traceability of agricultural production process, and sales of agricultural products in circulation link through QR code, electronic tags, and other technology to obtain the accurate origin so that consumers can not only trace all the information of the origin and circulation of the agricultural product through the whole process, but also query the information of fertilizer and pesticide dosage in production through the combination of the traceability system and the Internet of Things system to record and track the whole process information of agricultural products logistics. Through the combination of food safety traceability system and agricultural Internet of Things system, consumers' rights and interests will be more effectively protected and food safety will be guaranteed.

8.4.2 The Challenge of Intelligent Agricultural Management Application Development

1. The research and development of agricultural sensors is slow and the current products are not practical and not easy to use and manage; Conversely, in farmland environment, problems such as power supply damage and maintenance of sensing equipment have seriously hindered the implementation of agricultural Internet of Things. With the promotion of agricultural Internet of Things, agricultural sensors are bound to develop toward the direction of small size, low power consumption, high detection accuracy, stable, and reliable work. According to different agricultural Internet of Things application scenarios, it is urgent to accelerate the development of sensing technologies and products for the detection of animal, crop, soil, environment, and other parameters.
2. Nowadays, the data transmission reliability of sensor network of agricultural Internet of Things is poor and data collection is unstable. The characteristics of farmland environment and the demand of sensor technology put forward higher requirements for the data transmission of agricultural Internet of Things. In addition, low power sensor network transmission anti-interference, self-organizing network, and other technology development will provide technical basis for data transmission reliability. As 5G technology matures, difficulties in back-end data processing caused by slow network transmission and unstable communication will be alleviated, and there is no time to accelerate the iterative optimization of transmission layer technology.
3. The goal of the application and promotion of agricultural Internet of Things technology is to realize the on-demand control and fine management of agricultural production, which must rely on agricultural knowledge model to

support the processing layer to make correct decisions. At present, a large amount of agricultural data can be obtained through sensor network, but due to the immature algorithm model, most of which has not been fully mined and utilized. For the moment, decision processing is mainly based on timing control and single variable control. However, due to the lack of agricultural intelligent decision algorithm and model, it is difficult to realize on-demand control and multi-variable control, resulting in low intelligence degree of Internet of Things platform.

4. At present, a relatively complete agricultural Internet of Things technical standard system has not been established. Due to the lack of application standards and specifications, the standardized use of The Internet of Things technology in the agricultural field is restricted, and the standardization degree of agricultural sensors is not enough, and the reliability is difficult to guarantee, so it cannot achieve extensive integrated application; In the deployment process of sensor networks, there is no unified guidance and specifications, most of which are adjusted according to actual requirements and use custom transport protocols, which are quite arbitrary; The fusion mining and application layer development of perception data have no standards to follow and cannot be shared, which is not conducive to the promotion and implementation of agricultural Internet of Things.

Agricultural Internet of Things technology is an important support of intelligent agriculture. To promote its application, it is necessary to further improve its universality, reliability, and intelligence level. The agricultural Internet of Things platform should also gradually form a unified platform in accordance with the standardization, personalized, cloud computing requirements to realize system integration of agricultural Internet of Things, big data, artificial intelligence, and new information technology.

References

Alonso, R. S., Sitton-Candanedo, I., Garcia, O., Prieto, J., & Rodriguez-Gonzalez, S. (2020). An intelligent edge-IoT platform for monitoring livestock and crops in a dairy farming scenario. *Ad Hoc Networks, 98*, 102047.1–102047.23.

Angelopoulos, C. M., Filios, G., Nikoletseas, S., & Raptis, T. P. (2020). Keeping data at the edge of smart irrigation networks: A case study in strawberry greenhouses. *Computer Networks, 167*, 107039.1–107039.10.

Apk, A., & Sps, B. (2019). IoT based smart farming: Feature subset selection for optimized high-dimensional data using improved GA based approach for ELM. *Computers and Electronics in Agriculture, 161*, 225–232.

Castaeda-Miranda, A., & Castao-Meneses, V. M. (2020). Internet of things for smart farming and frost intelligent control in greenhouses. *Computers and Electronics in Agriculture, 176*, 105614.

Chung, H. C. (2020). Establishing a smart farm-scale piggery wastewater treatment system with the internet of things (IoT) applications. *Water, 12*.

Debauche, O., Mahmoudi, S., Andriamandroso, A. L. H., et al. (2019). Cloud services integration for farm animals' behavior studies based on smartphones as activity sensors. *Journal of Ambient Intelligence and Humanized Computing, 10*, 4651.

Goap, A., Sharma, D., Shukla, A. K., & Krishna, C. R. (2018). An IoT based smart irrigation management system using machine learning and open source technologies. *Computers and Electronics in Agriculture, 155*, 41–49.

Harun, A. N., Mohamed, N., Ahmad, R., Rahim, A., & Ani, N. N. (2019). Improved Internet of Things (IoT) monitoring system for growth optimization of Brassica chinensis. *Computers and Electronics in Agriculture, 164*, 104836.

Horng, G. J., Liu, M. X., & Chen, C. C. (2019). The smart image recognition mechanism for crop harvesting system in intelligent agriculture. *IEEE Sensors Journal, 20*(5), 2766–2781.

Ilyas, Q. M., & Ahmad, M. (2020). Smart farming: An enhanced pursuit of sustainable remote livestock tracking and geofencing using IoT and GPRS. *Wireless Communications and Mobile Computing, 2020*(6), 1–12.

Keswani, B., Mohapatra, A. G., Mohanty, A., Khanna, A., Rodrigues, J. J. P. C., Gupta, D., et al. (2018). Adapting weather conditions based IoT enabled smart irrigation technique in precision agriculture mechanisms. *Neural Computing and Applications, 3*, 277.

Lei, H., Ullah, I., & Kim, D. H. (2020). A secure fish farm platform based on blockchain for agriculture data integrity. *Computers and Electronics in Agriculture, 170*, 105251.

Li, G. Q., Chen, D. D., Zhang, J. T., Hu, F., & Zheng, G. Q. (2021). Construction of simplified traceability system of agricultural products based on android. In *IOP Conference Series: Earth and Environmental Science*.

Liao, Y., & Xu, K. (2019). Traceability system of agricultural product based on block-chain and application in tea quality safety management. *Journal of Physics Conference Series, 1288*, 012062.

Munir, M. S., Bajwa, I. S., & Cheema, S. M. (2019). An intelligent and secure smart watering system using fuzzy logic and blockchain. *Computers and Electrical Engineering, 77*, 109.

Nawandar, N. K., & Satpute, V. R. (2019). IoT based low cost and intelligent module for smart irrigation system. *Computers and Electronics in Agriculture, 162*, 979–990.

Taneja, M., Jalodia, N., Malone, P., Byabazaire, J., & Olariu, C. (2019). Connected cows: Utilizing fog and cloud analytics toward data-driven decisions for smart dairy farming. *IEEE Internet of Things Magazine, 2*, 32.

Vincent, D. R., Deepa, N., Elavarasan, D., Srinivasan, K., Chauhdary, S. H., & Iwendi, C. (2019). Sensors driven AI-based agriculture recommendation model for assessing land suitability. *Sensors, 19*(17), 3667.

Yang, Z., Li, X., & He, P. (2020). A decision algorithm for selecting the design scheme for blockchain-based agricultural product traceability system in q-rung orthopair fuzzy environment. *Journal of Cleaner Production, 290*(1), 125191.

Zeng, Z., Zeng, F., Han, X., Elkhouchlaa, H., & Lü, E. (2021). Real-time monitoring of environmental parameters in a commercial gestating sow house using a Zigbee-based wireless sensor network. *Applied Sciences, 11*(3), 972.

Chapter 9
Emerging Automated Technologies on Tractors

Jianzhu Zhao and Enrong Mao

9.1 CANBUS Technology of Agricultural Machinery

9.1.1 Overview of Agricultural Machinery CAN

With the rapid development of electronic technology, there are more and more electronic control equipment for agricultural vehicles and machines. The traditional point-to-point connection between electronic control units has the disadvantages of many harnesses and complex wiring, resulting in low network reliability and difficult sharing of data resources, which makes the coordinated control between machines and tools difficult to be realized. The serial BUS connecting the electronic control unit greatly simplifies the vehicle wiring system and has low cost. At the same time, it realizes the high data sharing between the electronic control units and improves the system reliability and fault diagnosis level. The application of serial BUS has become the inevitable trend of the development of electronic control network technology for agricultural vehicles and machines.

CAN (controller area network), namely controller area network, is a serial communication network designed by Bosch Company of Germany in 1983, which can effectively support distributed control and real-time control. It is used to solve the data exchange between many control and test instruments in modern vehicles. Due to the advantages of CAN and its special design, CAN specification was formulated as an international standard by International standard Organization (ISO) in 1993.

J. Zhao (✉)
China Agricultural University, Beijing, China
e-mail: zhjzh@cau.edu.cn

E. Mao
College of Engineering, China Agricultural University, Beijing, China
e-mail: gxy15@cau.edu.cn

S. Ma et al. (eds.), *Sensing, Data Managing, and Control Technologies for Agricultural Systems*, Agriculture Automation and Control,
https://doi.org/10.1007/978-3-031-03834-1_9

German LAV (German Association of Agricultural Machinery and Tractors) established a committee in 1988 to formulate a new agricultural machinery BUS standard LBS (DIN9684 standard) based on CAN BUS (CAN1. 0). In 1991, ISO started the standardization of agricultural machinery BUS, and TC23/SC19 (the 19th Sub Technical Committee of the 23rd Technical Committee, which works on the electronization of agricultural machinery) formulated a new international standard based on the standards of five parts in DIN9684. SC19 working group 1 (WG1) is especially responsible for the formulation of International Standards for Agricultural Machinery BUS. By 1992, TC23/SC19/WG1 decided to adopt CAN2.0B as the basis of the standard and named the standard ISO 11783.

9.1.2 Content and Structure of ISO 11783 Standard

As shown in Table 9.1, ISO 11783 standard is divided into 14 parts.

ISO 11783 can be divided into application layer, network layer, data link layer, and physical layer. The physical layer and data link layer of CAN2.0B are used as the underlying protocol, then the communication and addressing mechanism is established on the data link layer of CAN2.0B. The concept of "address" is used in ISO 11783 to prevent multiple controllers from using the same CAN flag. The relationship between ISO 11783 and CAN is shown in Fig. 9.1. The physical layer and data link layer of CAN are encapsulated through ISO 11783.

Table 9.1 Content of ISO 11783 standard

Parts	Name
Part 1	General standard for mobile data communication
Part 2	Physical layer
Part 3	Data layer
Part 4	Network layer
Part 5	Network management
Part 6	Virtual terminal
Part 7	Implement messages application layer
Part 8	Power train messages
Part 9	Tractor ECU
Part 10	Task controller and management information system data interchange
Part 11	Mobile data element dictionary
Part 12	Diagnostics services
Part 13	File server
Part 14	Sequence control

Fig. 9.1 Structure of ISO 11783 standard

application layer of ISO 11783
network layer of ISO 11783
data link layer of ISO 11783 CAN DLL
physical layer of ISO 11783 CAN PHL

Fig. 9.2 ISOBUS certification mark

9.1.3 ISOBUS (Deng, 2004)

The technical development of ISOBUS began in 1991 and ISO11783 was introduced as the tractor standard in 2001. ISOBUS defines the communication protocol, interface, and operation guide between tractors, machines and tools, and mobile equipment, solves the problems of universality and compatibility of communication data between tractors and machines and tools, as well as the information interaction between mobile equipment and on-board system. It has become a general standard followed by all agricultural machinery manufacturers, which greatly promotes the development of precision agriculture. Enterprises adopting ISOBUS need to be certified by AEF (Agricultural Electronics) committee. The certified products can automatically identify each other and stick special ISOBUS marks (as shown in Fig. 9.2). By 2019, more than 190 enterprises have passed the certification. The high-power tractors of John, New Holland, Case IH, CLAAS, and other companies have been equipped with ISOBUS interface as standard, which can match with various machines and tools using ISOBUS and realize interactive control. ISOBUS is used as the schematic diagram of electronic communication between machines and tools, tractors and computers, as shown in Fig. 9.3.

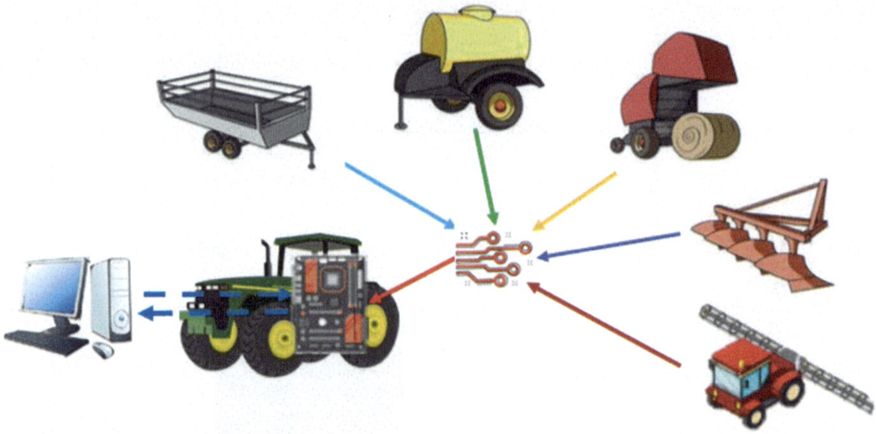

Fig. 9.3 Schematic diagram when ISOBUS is used for electronic communication between implements, tractors, and computers (Chen et al., 2017)

9.1.4 Application of CANBUS Technology in Agricultural Vehicles

CANBUS technology has been applied to some tractors manufactured by John Deere, kessnew Netherlands, Klass, and other companies. On related tractors, the electronic combination instrument, control console unit, chassis control unit, and traction control unit are integrated through CAN network, which realizes the integration of monitoring and control of subsystems such as engine, transmission system, and hydraulic system and greatly improves the convenience and reliability of use and maintenance. For example, the information of the central display screen comes from each subsystem and is shared with each controller. Because there are few can transmission lines, the control console can be moved, and most control devices such as shift lever, front-wheel drive switch, differential lock, power take-off shaft, and three-point suspension can be concentrated on it. The power shift transmission realizes the separation of the cab control system and the transmission component control of the transmission due to CAN, effectively improves the reliability, and can share the necessary information collected by other sensors. The control of front-wheel drive, differential lock and power take-off shaft is based on the information collected by relevant sensors throughout the vehicle. For example, the control of front-wheel drive can determine the control state according to the sensor signals such as brake pedal, parking brake, and driver output speed or the front-wheel drive switch signal on the console.

Tractor BUS network structure is shown in Fig. 9.4 based on ISO 11783 protocol. There was no master controller for the network and it consists of two

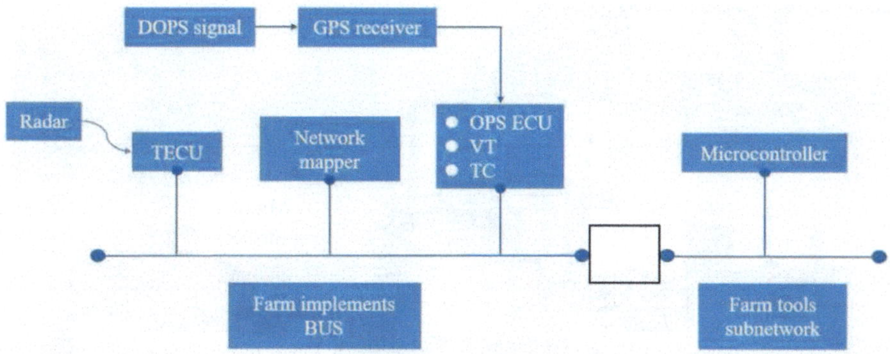

Fig. 9.4 Tractor BUS network structure based on ISO 11783 protocol

communication BUS, namely the tractor BUS and tools of BUS. There were the tools of network BUS from tools in figure. The BUS is through the tractor ECU and implements ECU linked to bridge the ECU.

Virtual terminal (VT) is connected to the tools on the BUS. Task controller is an ECU, it is usually located in the instructions of the tractor, which is used to provide some tools work, such as providing fine agricultural operation of prescription instructions. Gateway management computer contains an interface, compatible with management computer and data exchanged was allowed between the controller and management computer. In the task and task controller, interface between controller and tools and management is standardized communication between application software in the computer; however, computer and task management interface between controller not standardized.

The network information can be communicated and shared between components. For example, for the task controller and communication between GPS ECU, when the navigation information was defined, mission controllers can receive location information. Similarly, when engine torque information was defined, the engine torque of engine ECU can provide current torque curve for transmission. A lot of information in different repetition rates is transmitted, individual information 100 times per second the repetition rate of transmission in the network, and this kind of information of BUS capacity by about 5%. Planning and management information are needed to avoid excessive use of BUS capacity. In addition, the CAN BUS signal can be filtered by tractor ECU to avoid BUS overloading which affects entire BUS congestion.

Figure 9.5 shows the variable seeding system based on ISO 11783 CAN BUS technology for Valtra company. The system contains five ECU, which is used for the control of variable rate fertilization and management, data collection, and ISO 11783 communication between devices in a network. Four ECUs are installed on the tractor, GPS ECU, mission controllers (TC), virtual terminal (VT), and tractor ECU (TECU) respectively. ECU of differential GPS vehicle GPS signal processing.

Fig. 9.5 Variable seeding system based on ISO 11783 CAN BUS technology

VT for homework monitors. TC to manage. Homework prescription diagram, data storage and send through the ISO 11783 network expected input to control the amount of fertilizer application. TECU vehicle speed information for processing radar sensor. In order to analyze network communication program, add a Vector in a network company of PCMCIA CAN BUS network sniffer, through the ISO 11783 net data information exchange to monitor and analyze its application effect. Fifth ECU located fertilization machine, to a microcontroller, through the ISO 11783 network receive instruction from a TECU, according to the prescription chart control components to realize variable assignments.

9.2 Telematics

9.2.1 Telematics Overview

Telematics is a compound word of telecommunications (Telecommunications) and information science (Informatics) for long-distance communication. The system is based on wireless network communication technology, with cars as the main body, using wireless communication technology to connect with external networks to realize an interactive system of one-way or two-way transmission between people and external information resources. Through this system, the interaction with the vehicle can be realized at a long distance. The on-board Telematics module (T-BOX) is the core component of the system. It will be loaded on the vehicle to realize the connection between the vehicle and the network (also known as the cloud).

9.2.2 Accurate Management of Agricultural Machinery Operations Based on the On-Board Telematics System

Many types of agricultural machinery are operated at the same time and it is difficult to achieve efficient and accurate management by only relying on manual labor. All agricultural machinery on the farm is managed by the on-board Telematics system, as shown in Fig. 9.6 The Telematics system provides services through the computer system built into the agricultural machinery, wireless communication technology, satellite navigation system, and the Internet. It is to connect agricultural machinery equipped with computers, satellite navigation, wireless communication, and other equipment to the Internet through wireless networks to provide information and management services.

The on-board Telematics system is used to record the status of agricultural machinery in real time through the on-board computer, such as location, speed, working status, energy consumption, fuel tank oil volume, harvester storage bin status, etc., and farm maps, road locations, driving routes, and other agricultural machinery locations are viewed through wireless connection to the Internet. It can understand the location and status of each agricultural machinery in real time, make reasonable planning and arrangements for agricultural machinery operations, and

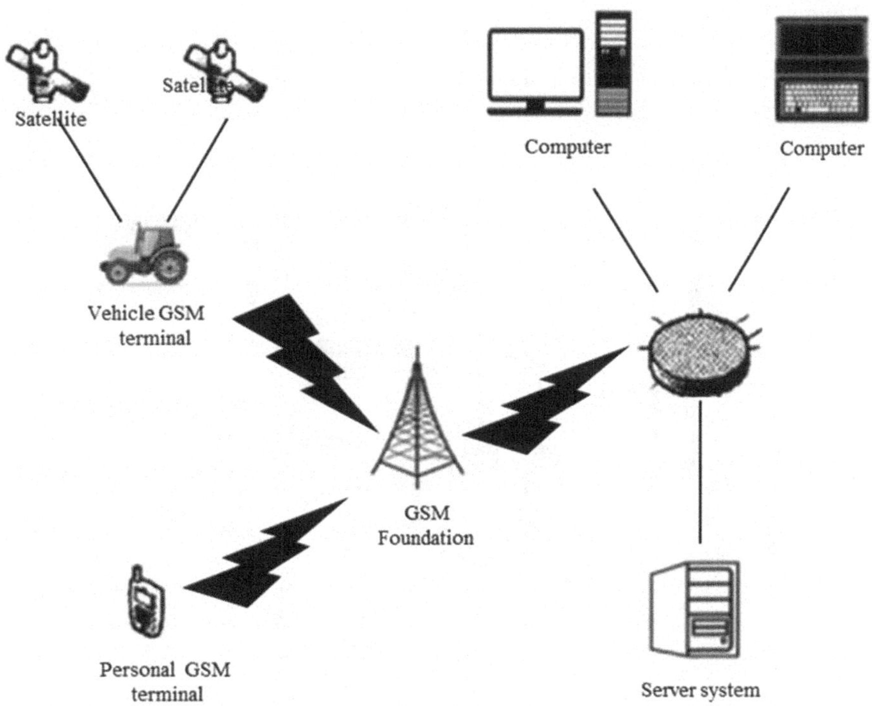

Fig. 9.6 Schematic diagram of Telematics system

realize dynamic, collaborative, efficient, and accurate management of agricultural machinery operations.

9.3 Hitch Electro-Hydraulic Control

9.3.1 Overview of Electro-Hydraulic Control Suspension

The hydraulic lifting system is an important working device of the tractor. The agricultural tractor is used as a farmland power machine and its power is ultimately transmitted to the farm machinery through the tractor's working device. The hydraulic lifter is used as the implementation part of the machine control and its function directly affects farmland operations. Therefore, the performance of the hydraulic lifting part of the tractor is an important index to judge the quality of the tractor.

Hydraulic suspension device uses hydraulic pressure as power to suspend, manipulate, and control the work of agricultural machinery. There are many types of hydraulic suspension devices and their functions and performances are different, but their main functions are: using hydraulic as the power to lift agricultural tools, manipulating the lifting of agricultural tools, controlling the depth of farming of agricultural tools, or maintaining the height of agricultural tools from the ground and hydraulic output.

In recent years, electro-hydraulic control suspension technology has been widely used in tractors and its system is shown in Fig. 9.7, mainly includes controller, control panel, solenoid valve, hydraulic cylinder, hydraulic pump, CAN BUS, speed

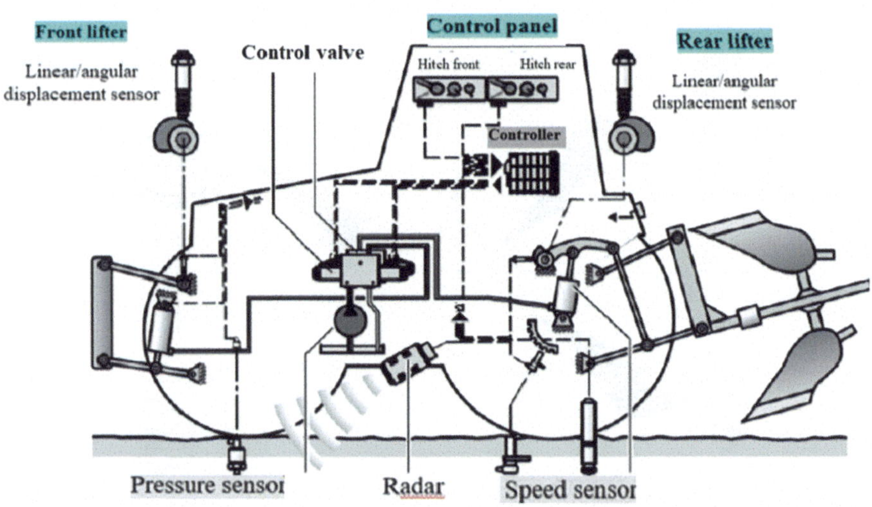

Fig. 9.7 Schematic diagram of tractor electronic hydraulic suspension system

sensor, displacement/angle sensor, tension sensor, radar, and so on. The application of CAN BUS makes it possible to share data among the electronic control units of tractor, which is helpful to realize the cooperative work among the control systems. The tilling depth setting knob and the working mode selection switch are arranged on the control panel. The driver only needs to select the corresponding working mode and set the target tilling depth according to the working demand and the working condition of the tractor, the utility model can realize the automatic control of the tillage depth of farm tools, and the operation is simple and fast.

9.3.2 Control of Tillage Depth with Suspended Farm Tools

When the hanging unit is working in the field, it should ensure the uniform tillage depth, stable engine load, and good tractor traction adhesion.

However, in the actural tillage operations, corresponding measures must be taken according to various of field conditions such as soil specific resistance, topography, dry land and paddy field, etc. Otherwise, the excessive change of tillage depth will not only affect the quality of tillage but also make the engine load fluctuate too much, which will affect the productivity and economy of the unit.

Similarly, the loading condition of the driving wheel affects the tractor's traction and attachment performance, as well as the productivity and economy of the unit. Therefore, different tillage depth control methods must be adopted to meet the operational requirements. The basic control methods are floating control (that is, height regulation), force control (that is, force regulation), position control (that is, position regulation), and force-position comprehensive control (that is, force-position comprehensive regulation).

In the electro-hydraulic suspension system, the driver selects the adjusting mode on the control panel according to the work demand, sets up the control signal, the controller (ECU) receives the input command of the control panel, reads out the driver's operation information. At the same time, the information of the suspension system is collected by the force sensor and the angle sensor, and the working position and state of the suspension system are determined, comparing the actual tillage depth with the target set by the driver, the system control quantity can be obtained, and the electromagnetic valve can be output in the way of control signal until the tillage depth reaches the designated position, thus achieving the control of the target tillage depth, the working position and the opening size of the solenoid valve determine the expansion and movement speed of the hydraulic cylinder.

Electrically controlled hydraulic suspension adopts electronic control technology. On the control panel, there are control switches and adjusting knobs. The driver inputs control information through the control panel. After the controller receives the driver's operation information, compared with the sensor's feedback signal, when the error exceeds the set value, the single-chip microcomputer sends out commands to control the electromagnetic valve according to the control strategy and pushes the hanging mechanism through the hydraulic cylinder to complete the control of the farm implements. Using sensors to measure the feedback signal of the suspension system is simple, convenient, and highly accurate. Compared with mechanical hydraulic suspension, electronically controlled hydraulic system has unparalleled advantages of mechanical hydraulic suspension system, as follows:

1. Use electrical signals to transmit information, so that the control panel and sensing elements are not affected by the layout of the tractor cab and other parts and can be flexibly arranged according to needs.
2. The electronic control system is convenient to process the feedback signals of different control parameters and it is easier to realize the integrated control of the suspension system.
3. Using microcomputer control, high-end control algorithms such as fuzzy control and sliding film control can be applied to the control of the suspension system, which is beneficial to improve the overall control level and performance of the suspension system.
4. Using advanced electronic information and communication technology, real-time monitoring of various working conditions of the tractor can be carried out, which is convenient to realize the energy management and fault diagnosis of the whole tractor.
5. The electronically controlled hydraulic suspension system has high control accuracy, which promotes the development of precision agriculture and is conducive to the research of unmanned driving.

With the advancement of agricultural modernization, the traditional mechanical hydraulic suspension system no longer meets the needs of the development of the times and is gradually being replaced by electronically controlled hydraulic suspensions. Electronically controlled hydraulic suspensions can complete more complex control of agricultural tools and have higher control accuracy and it is a necessary system for the development of agricultural machinery automation and intelligence.

9.4 Smart Tractor Visual Navigation

9.4.1 Visual Navigation Overview

Visual navigation uses the camera to collect, filter, and calculate images of the surrounding environment, complete its position determination and target-oriented path planning, and make navigation decisions. In recent years, visual navigation extracts environmental information such as images and radio signals for analysis, combines image data and digital terrain data to complete positioning and path planning, and is often used to assist inertial navigation. In recent years, the focus of visual navigation research is mainly reflected in algorithm optimization. Algorithm optimization can improve the accuracy of visual navigation, increase the success rate of navigation in invisible environments, and enhance the autonomy of visual navigation (Yang et al., 2021).

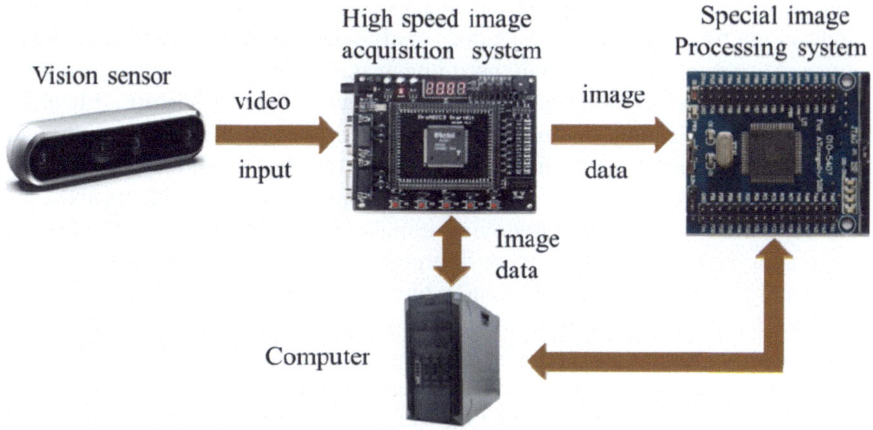

Fig. 9.8 Basic building blocks of visual navigation system

Vision is the use of computers to realize human visual functions—the perception, recognition, and understanding of the three-dimensional scene of the objective world. With the deepening of visual research and the development of semiconductor and computer technology, visual information is increasingly being applied to the practice of navigation. Among them, visual navigation is a navigation method that uses visible light and invisible light imaging technology. It has the advantages of good concealment, strong autonomy, fast and accurate measurement, and cheap and reliable. In the past 30 years, with the continuous emergence of new concepts, new methods, and new theories, visual navigation has been widely used in unmanned aerial vehicles, agricultural mobile machinery, and indoor and outdoor robots.

The basic composition of the visual navigation system: the visual navigation system is generally computer-centric, mainly composed of modules such as visual sensors, high-speed image acquisition systems, and dedicated image processing systems, as shown in Fig. 9.8.

The visual sensor acquires the feature image of the surface of the measured object and converts it into a digital signal by the high-speed visual image acquisition system. The rapid digital image processing and can be conducted by high-speed machine vision software, which can also be used to extract the coordinates of image feature information. It is realized by the computer Fast calculation of parameters, such as space geometric parameters and position and posture of the measured object.

9.4.2 *Status of Visual Navigation System*

For visual navigation systems, visual sensors provide original and direct visual information, generally called visual images. The processing of visual images and the extraction of feature information are the prerequisites and foundations for the application of visual navigation systems.

1. Visual image preprocessing: Visual images are often polluted by strong noise and smooth filtering is needed to reduce or eliminate the influence of such strong noise. At present, the commonly used image smoothing filtering includes mean filtering, median filtering, Gaussian transform, and wavelet transform. Mean filtering is easy to design and performs well when the signal spectrum and noise spectrum have significantly different characteristics, but it will make the edge of the image blurred. Median filtering can overcome the above problems and keep the edge undisturbed while removing impulse noise. However, in the face of large area noise pollution, median filtering is not as good as mean filtering in smoothing noise. Wavelet transform, which is called "digital microscope," can perform local analysis in time-frequency domain at the same time, which has become an important development direction of denoising.
2. Visual image feature extraction: Visual image feature extraction is an important method for image recognition and classification and also the basis for understanding, processing, and decision-making of image information. In a visual image, information with distinctive features, such as edge, corner, circle or ellipse center, and image shape features are usually extracted as visual image feature information. In visual image, image edge is the main feature information of visual image.
3. Visual positioning methods: At present, there are mainly two visual navigation and positioning methods, which are visual odometer and ground-based punctuation matching.

   (a) Visual odometer: it is a navigation and positioning method that uses monocular or binocular camera to obtain image sequence and then estimates carrier motion information by feature extraction, matching and tracking.
   (b) Ground punctuation matching navigation: Ground punctuation matching navigation takes some special scenes in the environment as landmarks in advance. On the premise of knowing the coordinates and shapes of these ground punctuation marks, the robot determines its position by detecting ground punctuation marks. According to the difference of local punctuation, it can be divided into artificial punctuation matching navigation and natural punctuation matching navigation (Guan & Wang, 2014).

9.4.3 Application of Visual Navigation in Agricultural Machinery

Machine vision has the characteristics of low cost and rich information and is suitable for irregular plots or signal blocking environments. When using visual navigation, a visual sensor is usually installed above the agricultural machinery cab to collect image information in front of the agricultural machinery and finally extract the navigation baseline through preprocessing and crop line detection.

Table 9.2 Characteristics of crop line inspection methods

Detection method	Advantage	Shortcoming
Vertical projection	Simple calculation and good anti-noise effect	More affected by weeds
Hough transform	Strong anti-interference ability	High complexity and time-consuming
Linear regression	Simple calculation	Affected by noise
Stereo vision	Depth information is less affected by light	Large amount of calculation, only suitable for higher crops

1. Image preprocessing: Factors such as weather, weeds, shadows, and non-target areas in a farmland environment can interfere with crop detection and it is difficult to obtain ideal results in direct detection. The color distinction between the target area and the non-target area can be increased through special band vision sensors or grayscale feature factors; part of the shadow interference can be eliminated by converting the RGB color model to HSV, HSI, YCbCr and other color models; by setting the image in a reasonable manner the region of interest to be processed can reduce the interference of non-target crop rows and at the same time reduce the amount of calculation.
2. Crop line inspection: At present, a large number of researches on crop row extraction methods have been carried out at home and abroad, mainly including vertical projection, Hough transform, linear regression, stereo vision, etc. The characteristics are shown in Table 9.2.

At present, visual navigation technology has been applied to automatic pesticide application, automatic weeding, and automatic harvesting. However, due to the influence of the farmland environment on the stability of image collection, there are still problems such as blurred images and missing information. The robustness of visual navigation technology needs to be further improved (Zhang et al., 2020).

9.5 Continuously Variable Transmission

9.5.1 Overview of Continuously Variable Transmission

Continuously variable transmission (CVT) is an ideal transmission form for vehicles. It can continuously change the transmission ratio according to the road conditions and the working state of the engine, so that the engine always works near the best working point or best working line, which means it is improving the fuel economy of the whole machine and reducing noise. At the same time, there is no shifting step, which reduces the impact of shifting and improves the driving comfort of the whole machine. Nowadays the CVT is divided into 3 types: mechanical CVT (M-CVT), electric CVT (E-CVT), and fluid CVT (F-CVT).

9.5.2 The Form of Continuously Variable Transmission

The metal belt type continuously variable transmission is a representative form of M-CVT. Due to its structure limitation, the power that can be transmitted and the speed range are limited. At present, it is still difficult to obtain application in high-power tractors.

The electric continuously variable transmission system consists of an electric generator, a control system, and a traction motor. The electric drive adopts electric wheel drive technology, and the power source and the drive motor are connected by a flexible cable, which not only frees up the constraints of the transmission system in the design space, makes the layout of the whole vehicle very flexible, but more conducive to the reasonable distribution of axial load. It has the advantages of large transmission power range, easy control, high transmission efficiency, but because of its own high quality and high cost and usually integrates motors, brakes, and other parts into electric wheels, it is comparable to ordinary wheels of the same specification. The specific mass increases more, the unsprung mass of the vehicle increases, and the ride comfort and ride comfort decrease. Therefore, this type of transmission system is only used on mining dump trucks, large scrapers, and wheel loaders.

Fluid continuously variable transmission also can be divided into three types: hydraulic mechanical, hydrostatic, and hydromechanical. Due to low transmission efficiency and power limitation of hydraulic components, hydraulic machinery and hydrostatic transmission are not widely used in tractors. But a few small and medium power models sometimes use hydrostatic transmission such as Versatile200 Series of New Holland company, L Series of Kubota company of Japan (power between 8.6 and 14.9 kW), 20 Series and 77 Series tractors of Fiat company of Italy, etc.

The hydraulic-mechanical continuously variable transmission (HM-CVT) takes into account the high efficiency of mechanical transmission and the smooth and low impact characteristics of hydrostatic transmission. Conversely, it only needs to use common mechanical transmission mechanisms and ordinary hydraulic components to achieve high efficiency and high-power output. As shown in Fig. 9.9, HM-CVT is usually composed of mechanical transmission mechanism, pump-motor hydraulic stepless transmission system, power split and confluence planetary gear mechanism, electronic control device and drive system. The power output from the engine is divided into mechanical and hydraulic two-way transmission to the drive axle. The mechanical transmission is usually transmitted by planetary gears and multi-stage gears (John Deere's, Auto Power) or by multi-stage planetary gears (ZF's S-Matic, Eccom), while the hydraulic transmission adopts variable pump and variable motor speed control. At the output end of the hydraulic transmission, the hydraulic energy and mechanical energy are recombined and input into the drive axle.

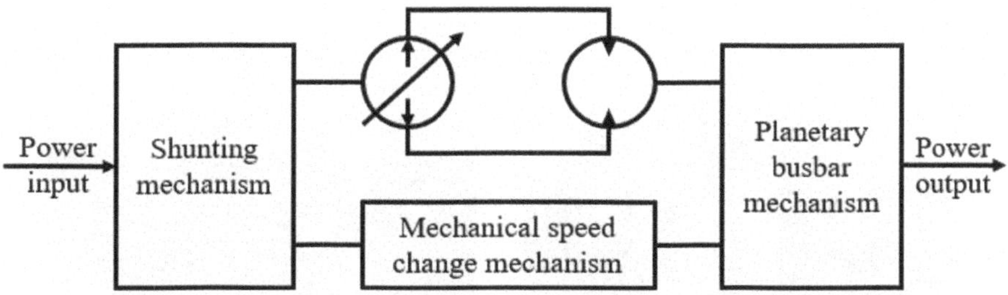

Fig. 9.9 Schematic diagram of hydromechanical continuously variable transmission

9.6 Driverless Tractor

9.6.1 Overview of Driverless Tractor

The development of precision agriculture has promoted the continuous change of agricultural equipment and technology and the emergence of driverless tractor and driverless agricultural machinery has brought new content to the development of precision agriculture. Especially with the increasing shortage of agricultural labor force and the higher requirements for the operation efficiency of agricultural machinery, the emergence of driverless tractors and driverless agricultural machinery marks an exciting great technological progress.

At present, driverless agricultural equipment is still in the concept machine stage. The main development units are agricultural equipment manufacturing enterprises. Typical driverless tractors without cab include Magnum of Case corp of the United States and "super tractor 1" of China Yituo Group Co., Ltd. (No.1 Tractor Manufacturing Factory). The Case corp of the USA has launched the driverless concept tractor magnum (Fig. 9.10a), which combines the latest breakthroughs in positioning, remote control, data sharing, and agronomic management. The operation process of the tractor starts from the input of the field boundary and the controller will automatically plan the driving path according to the boundary and the width of the machines and tools. The most efficient coordination route is planned when using multiple interconnected machines. The operator can use desktop computers, tablet computers, and other terminals to monitor the operation of the tractor. The camera installed on the tractor body shows the operation status and working environment of the tractor to the operator for reference in real time. The operator can browse the engine speed, fuel level, implement settings, and other tractor parameters and modify them manually. To ensure the safety of unmanned driving, sensors such as radar, laser ranging sensor, and camera are installed on the tractor to detect obstacles. The tractor can use big data information such as real-time meteorological information for independent decision-making. When the weather conditions become so bad that it cannot continue to operate, the tractor will automatically stop operation and automatically resume operation when the conditions improve. China Yituo Group Co., Ltd. exhibited Chinese first truly driverless tractor "super tractor 1"

(a) (b)

Fig. 9.10 Driverless tractor. (**a**) Case corp magnum unmanned tractor. (**b**) Super tractor 1 of China Yituo Group Co., Ltd.

(Fig. 9.10b) at the 2016 China International Agricultural Machinery Exhibition. The "super tractor 1" was provided by Beijing Sylincom Technology Co., Ltd. with the whole machine system scheme, satellite communication, and other core technologies, and the driverless technology was provided by the Institute of microelectronics of the Chinese Academy of Sciences, sensor provided by Hefei Institute of materials, Chinese Academy of Sciences, and China Yituo Group Co., Ltd. provides technology and application scenario data such as tractor frame and transmission system. "Super tractor 1" was composed of five core systems: driverless system, power battery system, intelligent control system, central motor and drive system, and intelligent network system. It has the functions of vehicle status monitoring, fault diagnosis and treatment, machine control, and energy management and realizes the intelligent identification and control functions such as constant tillage depth and constant traction. "Super tractor 1" can realize the functions of obstacle detection and obstacle avoidance, path tracking and agricultural tool operation through path planning technology and driverless technology (Liu et al., 2020).

Unmanned tractors will promote the intensive cultivation of agriculture to a new level:

1. Greatly improve agricultural production efficiency. This is because the application of automatic control technology will greatly increase the operating speed of unmanned tractors and agricultural machinery. Conversely, the control system of unmanned tractors will not feel the same as human fatigue and can achieve continuous operation 24 h a day.
2. Greatly improve the accuracy of agricultural operations. Unmanned tractors using global positioning system technology can control the accuracy of agricultural operations within 2-3 cm, while the best accuracy of manual driving can only reach an error of 10 cm. The routes traveled during farming, pesticide spraying, and harvesting will not overlap, saving fuel, reducing the use of pesticides and fertilizers, not only reducing agricultural production costs, but also reducing environmental pollution.
3. Further increase crop yields. The heavier the tractor is, the tighter the soil will be and it will be difficult for the roots of crops to grow. Therefore, people tend to

use more powerful tractors to make the pear deeper and the weight of the high-power tractor will form a deeper compacted soil layer. Lighter driverless tractors break this vicious circle and can increase crop yields by 10%.

In addition, because an unmanned tractor does not have a driver, there is no need to install a cab and related operating mechanisms, but many sensors need to be installed, which will change the structure, layout, and design ideas of traditional tractors.

9.6.2 System Structure and Work of Unmanned Tractor

Unmanned tractors are also called robotic tractors. In order to realize unmanned driving, the tractor needs to have an automatic driving function, have the ability to recognize itself and the surrounding environment, and be able to judge the recognition results to drive and operate correctly. Figure 9.11 shows the structure block diagram of a remotely controlled unmanned tractor system, which mainly includes two parts: the farm monitoring center and the unmanned tractor. Remote information communication is carried out between the monitoring center and the tractor via the wireless local area network and corresponding communication equipment.

The unmanned tractor adopts global positioning system technology and installs a GPS receiver on the tractor to determine the tractor's ground position. Through the real-time GPS signal and the field digital map stored by the tractor's on-board controller, the automatic navigation of the tractor can be realized, that is, the on-board controller controls the steering of the tractor's electro-hydraulic steering mechanism and the turning angle of the steering wheel is fed back by the turning angle sensor. The image returned by the machine vision sensor (camera) on the tractor is displayed on the monitor of the farm monitoring center and the operator can navigate and correct the tractor from a distance by watching the screen. Inertial navigation device is a supplement and correction to GPS navigation to achieve high-precision combined navigation.

The monitor in the farm monitoring center continuously receives and displays vehicle location, vehicle speed, images, and other information. Based on this information, the operator can remotely monitor the vehicle.

9.6.3 Key Technologies of Unmanned Tractors

1. Autonomous navigation: Autonomous navigation is an important function of driverless tractor. The tractor should be able to determine its own walking direction based on environmental knowledge and the target position or sequence of positions, so as to reach the target position as efficiently and reliably as possible. There are many ways to solve navigation problems, such as map navigation,

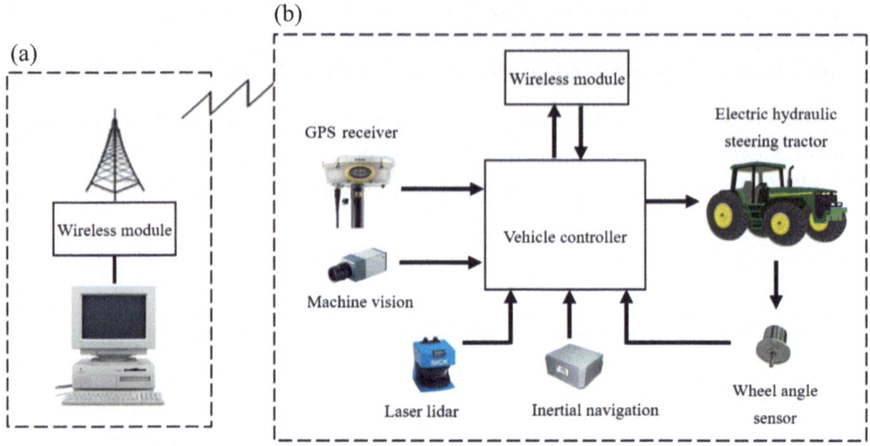

Fig. 9.11 System block diagram of unmanned tractor. (**a**) Farm monitoring. (**b**) Tractor

beacon navigation, satellite navigation, visual navigation, and other sensor navigation. At present, unmanned tractors mainly rely on GPS navigation, combined with machine vision and inertial navigation. Although there are multiple navigation methods, these methods have a high degree of similarity, and they are all based on the basic navigation architecture shown in Fig. 9.12.

2. Control technology: Control technology is the core of unmanned tractors, mainly including direction control and speed control. Unmanned driving is actually the use of electronic technology to control tractors for human-like driving. As the tractor model changes with time, its system parameters will change, the model parameters are extremely complex, and the model equations are nonlinear. Research on vehicle control focuses on improving the adaptability and anti-interference of the control algorithm. At present, the control algorithm of unmanned tractors is being researched from traditional PID control to adaptive control and intelligent control.
3. Information fusion technology: The positioning and navigation of unmanned tractors should have higher intelligence. It is necessary to fuse various sensor information or some prior knowledge to realize a full understanding of environmental information and facilitate unmanned tractors to make correct decisions. Information fusion can improve the reliability and resolution of the system, increase the measurement space dimension, and broaden the range of activities, thereby improving the adaptability and robustness of the system under complex conditions. In order to improve the performance of the system, it is necessary to continuously improve and perfect the information fusion algorithm in combination with new theories and it is also necessary to strengthen the research on the evaluation of the effect of information fusion.
4. Human–machine collaboration and multi-machine collaboration: Human–machine collaboration is an effective way to solve the contradiction between the

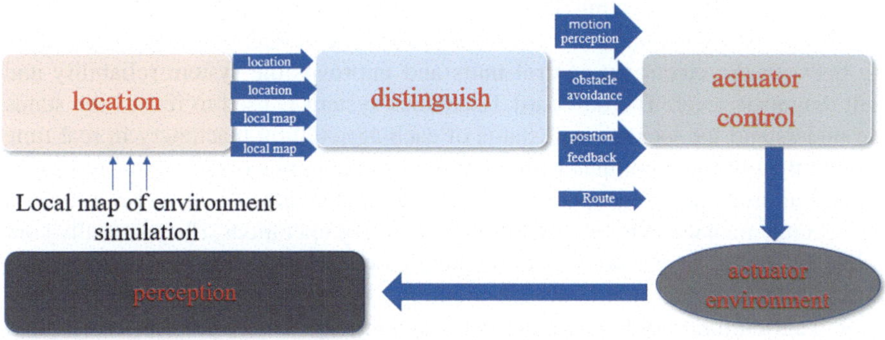

Fig. 9.12 Basic navigation architecture

intelligent development level of unmanned tractors and complex tasks. Human participation can give full play to human experience, initiative, and ability to respond to unexpected events, enhance the tractor's ability to deal with emergencies and imprecise events, and enhance the robustness of the system.

With the further development of unmanned tractor technology, multiple unmanned tractors or agricultural machinery will be required to coordinate workbenches in the field, thereby significantly improving the intelligence and work efficiency of tractor operations, but there are problems with formation travel, collision prevention, path planning, etc. In addition, although the research on unmanned tractors for multi-machine cooperative operation is challenging, it should be emphasized.

5. Safety: The safety of unmanned tractors mainly refers to the safety of people, the environment, and the safety of the machine itself. The unmanned tractor itself should have safety judgment and processing capabilities, and when this capability fails, the operator can perform emergency processing outside. Of course, obstacle detection and obstacle avoidance capabilities of unmanned tractors have always been one of the researches focuses and difficulties. Most research is based on obstacle detection based on machine vision and lidar. In addition, other sensors such as proximity switches are installed on the tractor to carry out research on the tractor's collision protection function.

9.7 Summary

As the core power equipment of agricultural machinery, tractor plays a more and more important role in agricultural production. Intelligent technology will be necessary for realizing the modernization, automation, intelligent perception and decision-making of agricultural production, efficient operation, automatic driving and remote control of intelligent tractors. The topics of CANBUS/ISOBUS, Telematics, hitch electro-hydraulic control, visual guidance, continuously variable speed driving and automatic driving were discussed in this chapter.

The serial BUS connecting the electronic control unit greatly simplifies the vehicle wiring system and has low cost. At the same time, it realizes the high data sharing between the electronic control units and improves the system reliability and fault diagnosis level. The on-board Telematics system is used to record the status and understand the location and status of each agricultural machinery in real time through the on-board computer, then make reasonable planning and arrangements for agricultural machinery operations, and realize dynamic, collaborative, efficient, and accurate management of agricultural machinery operations. Electronically controlled hydraulic suspensions can complete more complex control of agricultural tools and have higher control accuracy and is a necessary system for the development of agricultural machinery automation and intelligence. At present, visual navigation technology has been applied to automatic pesticide application, automatic weeding, and automatic harvesting. However, due to the influence of the farmland environment on the stability of image collection, there are still problems such as blurred images and missing information. The robustness of visual navigation technology needs to be further improved. CVT could continuously change the transmission ratio according to the road conditions and the working state of the engine, so that the engine always works near the best working point or best working line, which means it improves the fuel economy of the whole machine and reducing noise. At the same time, there is no shifting step, which reduces the impact of shifting and improves the driving comfort of the whole machine. Driverless tractor could improve agricultural production efficiency and the accuracy of agricultural operations greatly and further increase crop yields. In addition, because an unmanned tractor does not have a driver, there is no need to install a cab and related operating mechanisms, but many sensors need to be installed, which will change the structure, layout and design ideas of traditional tractors.

References

Chen, Z. Q., Wu, Y., Chen, S. X., & Bian, Q. L. (2017). The practice and enlightenment of precision agriculture in German——Taking gut Derenburg farm for example. *Chinese Journal of Agricultural Resources and Regional Planning, 38*(05), 222–229.

Deng, S. (2004). Looking forward to connected car 2.0. Retrieved from http://www.elecfans.com/qichedianzi/20161116449110.html

Guan, X. J., & Wang, X. L. (2014). Review of vision-based navigation technique. *Aero Weaponry, 05*, 3–14.

Liu, C. L., Lin, H. Z., Li, Y. M., Gong, L., & Miao, Z. H. (2020). Analysis on status and development trend of intelligent control technology for agricultural equipment. *Transactions of the CSAM, 51*(01), 1–18.

Yang, W. Y., Li, D. B., Sui, Y., & Shen, Y. P. (2021). Overview of the development of satellite independent navigation technology abroad in 2020. *Aerodynamic Missile Journal, 1*, 25–30.

Zhang, M., Ji, Y. H., Li, S. C., Cao, R. Y., Xu, H. Z., & Zhang, Z. Q. (2020). Research progress of agricultural machinery navigation technology. *Transactions of the CSAM, 51*(04), 1–18.

Chapter 10
Applied Time-Frequency Control in Agricultural Machines

Zhenghe Song, C. Steve Suh, and Xiuheng Wu

10.1 Features of Automated Agricultural Machines

Application of automation and technology of intelligence is seeing better performance, improved safety, operation economy, and environmental friendliness in agricultural machines, attracting attentions from both academia and industry alike as a result. Detailed analysis for farming machines from the perspective of automated control is necessary for designing a suitable controller to enhance their performance.

The working environment farming equipment is subjected to both open and complex, meaning that the automatic actuator often suffers from an uncertain large load and significant noise jamming. To handle fluctuant heavy load, a large stiffness system such as an electric-hydraulic system or electric motor with a large-ratio gear reducer is usually employed in these machines, often running in a low frequency range. Due to the application of a hydraulic system or large-ratio gear reducer, nonlinear characteristics inherent of kinetic compressed oil or backlash between gears are often introduced into the actuator system, resulting in containing high frequency information in the feedback signal of the system. This information can mingle with noise jamming. It is therefore a challenge to recognize the nonlinear feedback information and make full use of it to stabilize the system and improve dynamic performance.

Z. Song · X. Wu (✉)
College of Engineering, China Agricultural University, Beijing, China
e-mail: wxh599@cau.edu.cn

C. S. Suh
Department of Mechanical Engineering, Texas A&M University, College Station, TX, USA

S. Ma et al. (eds.), *Sensing, Data Managing, and Control Technologies for Agricultural Systems*, Agriculture Automation and Control,
https://doi.org/10.1007/978-3-031-03834-1_10

Additionally, to be energy efficient, some actuators in farming machines such as an active hydro-pneumatic suspension need to work only in a narrow-frequency range for the reason that if working in the frequency domain the actuator will consume massive amount of energy, thus a condition not acceptable to the user. It would require that controllers designed for these systems are expected to enhance the dynamic response in their working frequency range. However, "waterbed" effect exists in the controller design process pervasively, which would lead to the decreasing of the system dynamic performance in other frequency range under the same controller. Properly balancing these requirements is a critical issue while also a difficult task to achieve for conventional controller designs.

While the majority automatic systems in agricultural machines respond in finite frequency range, it is required that they display different dynamic performances in different frequency ranges. In addressing this need, a novel time-frequency control approach is developed, one that features real-time analysis for feedback signals both in the time and frequency domain to determine the system's running status.

10.2 State-of-Affair of Time-Frequency Control

From the viewpoint of simultaneous time-frequency control, traditional varying-parameter controller implies the thought of time-frequency control. Before the application of computers and advanced computing software, it is hard to depict the system response under different inputs immediately. Analysis performed to the frequency domain not only can characterize system response, it can also measure stability margin. However, nearly all controllers operate in the time domain. An optimal controller with invariable parameters based on analysis in the frequency domain cannot perform well in many working conditions. So, a feasible thought is to change the controller's parameters in response to the variation of the working condition. Because the response of a controlled system has different frequency performance under different working conditions, the controller's parameters can change with the feedback signal in which frequency information is contained. In this way, these types of controllers realize the function preliminarily that proper schemes are switched in time under the time-frequency analysis of system output.

Development of control theory makes many new controllers with varying parameters been invented. With the development of fuzzy mathematic strategy and system identification, fuzzy control and adaptive control are researched and applied. Taking fuzzy PID controller as an example, the error and derivative of error are chosen as the input to the controller, based on the prior knowledge, different coefficients are selected under different inputs, and that the derivative of error includes the current frequency information of the running system. In adaptive control system, coefficients of the controller are often adjusted based on the derivative of output-to-input, and it also includes the frequency information. Overall, through the quantified frequency index concluded by the input and output of the system, the varying rules of controller parameter are designed, this process accords with the connotation of time-frequency control approach. Thus, the main purpose of the time-frequency control is to improve the system performance.

The widespread use of microchips for digital computation accelerates the application of time-frequency analysis tools. It is convenient for discrete signal processing, which supplies a straightforward method for real-time time-frequency analysis, enabling time-frequency control to be physically realized with efficiency. Several studies (Tan, 2016; Lin et al., 2005, 2006) developed a series of adaptive wavelet neural network controllers, which combined wavelet analysis with neural networks. These controllers take full advantage of the two tools, with discrete wavelet analysis being used to extract information in the frequency domain in real-time and neural network to tackle nonlinearity. It was shown that neural network can identify unknown systems faster and get more precise results when it is incorporated with discrete wavelet transform to better guide the controller design. For the same purposes, other studies (Lin & Li, 2012; Zekri et al., 2008; Hung et al., 2015) have developed a kind of controller which combines discrete wavelet transform with fuzzy control and neural network for nonlinear system control. Essential information in the feedback signal can be resolved by discrete wavelet transform, which are then used to guide the varying rules of controller parameters, resulting in control performance being improved without significantly increasing the computing load. Therefore, with the development of discrete wavelet-based approach, exploring discrete wavelet analysis with established control methods has received extensive interest in recent years.

Compared with the varying-parameter controllers, the output of the controller based on discrete wavelet analysis is continuous, which avoids the impact generated by switching the controller when parameters are changed. Because discrete wavelet analysis is equivalent to real-time decoupling in the frequency domain for a signal, in theory, it is more effective with synthesizing different frequency components for quantifying characters than addressing them based on prior knowledge. Based on this, Parvez and Gao (2002, 2005) applied the discrete wavelet transform (DWT) to traditional PID control and named it as multiresolution PID because of the multiresolution feature in the time and frequency domain. In their studies, DWT was used to decompose an error signal. Several signal components were obtained whose frequencies are from low to high which together transpire the system dynamics and external disturbance with clarity. Then different sub-controllers were designed for each signal component based on its physical contents. At last, these sub-controllers were synthesized into one output. This scheme was later employed by Tsotoulidis et al. (2013) and Khan and Rahman (2008) to control brushless direct current motors and obtain good performance. At the same time, Sun et al. (2000) proposed an adaptive wavelet PID controller. In combination with the previous research, a book on wavelet PID was published (Tolentino et al., 2012). All these works promoted the use of discrete wavelet analysis in controller design.

Although algorithms that imply the idea of time-frequency control have been developed much earlier, the concept of time-frequency control was proposed formally by Suh and his colleagues (Suh et al., 2002; Dassanayake & Suh, 2007a, b;

Liu & Suh, 2012, 2013; Wang & Suh, 2017; Zhang & Suh, 2022; Yang & Suh, 2021). It is common knowledge that variation in parameters in a nonlinear system can lead to the deterioration of its dynamic response from stability to bifurcation or even chaos. It was found that the response of a nonlinear system is characterized by varying spectrum that is also broad in bandwidth. Suh et al. showed that high-speed cutting operation, albeit being excited by a single harmonic, demonstrated specific characteristics in the simultaneous time-frequency domain that are commonly shared by all nonlinear systems. To be specific, in the frequency domain, there are many frequency components in the response of a nonlinear system at any given moment. In other words, the system response is broadband. The stronger the nonlinearity of the system, the more frequency components, or the wider of the response bandwidth. The frequency components of the nonlinear system response also vary in time. The stronger the nonlinearity is, the more drastic the temporal variation will be. Therefore, nonlinear system responses must be characterized in the simultaneous time-frequency domain to determine system nonlinearity effectively. With the insight and knowledge, Suh et al. (2002) formulated a wavelet-based time-frequency control theory that exerts proper control as soon as the spectral components of the system response are identified. The philosophy of the control theory puts emphases on designing control laws that are carried out simultaneously in the time and frequency domain.

The main purpose of the time-frequency controller is to analyze the current state of the system to determine the nonlinear characteristics of the system to control it. Considering that wavelet time-frequency analysis is widely used for real-time applications, Suh et al., presented a time-frequency control architecture features filter-x least mean square (FXLMS) algorithm and discrete wavelet transform (DWT) in leteratures mentioned before. The design was shown to work particularly well with systems of high nonlinearity. In addition, the controller demonstrated a better computational efficiency because FXLMS and DWT are implemented together as a filter. As an adaptive controller demonstrating good performance in the time domain, the FXLMS component in the controller design plays the critical role of manipulating the wavelet coefficients, which carry both time and frequency information. The time-frequency control technique emphasizes that the control algorithm must be implemented in the time domain and frequency domain at the same time. Control errors are small in the time domain and the oscillation broadband spectrum can be suppressed in the frequency domain—the two indicators showing the stability of the system is under control. Lately, a time-frequency PID control approach was developed by our team, which fuses the new idea with the traditional controller (to be elaborated in the following sections). It can not only improve the dynamic performance but also suppress disturbance through designing different parameters in different sub-frequency domains as decomposed by using the discrete wavelet transform (Wu et al., 2017).

In summary, time-frequency control is a broad control concept feasible for the control design of farming equipment subject to a wide range of farming operation needs. Due to the different frequency domain characteristics of different types of systems, the designed time-frequency controllers are quite different. Therefore, in the design of a time-frequency control algorithm, it is often necessary to have a more comprehensive analysis of the time-frequency characteristics of the controlled system, and it is necessary to make a clear design purpose of the controller, whether it is to stabilize the system or to improve control performance. Exploring these properties, time-frequency controller with excellent performance can be designed to mitigate any operation condition.

10.3 Applications of Time-Frequency Controller for Automated Farming Equipment

This section introduces the principle of two time-frequency control configurations and the corresponding applications of the two controller designs are described in detail.

10.3.1 Time-Frequency FxLMS Control of Electric Motors for Support-Cutting of Sugar Cane

Background

Sugarcane cutters play an important role in cane harvest. Their performance has a strong impact on stubble cutting quality that in turn affects the yield of the crop in the following year. Base cutting is critical to the performance of the harvester as well as the application of the equipment in the cane-planting area. Harvesters with high efficiency, a low rate of broken and low uprooting incidents are sought in recent years. Factors influencing the sugarcane harvest effect from different aspects are studied and analyzed by many researchers. Two kinds of main reasons affecting base cutting performance are concluded. One reason is the sugarcane growing pattern or stalk pattern, which means different situations of sugarcane growing, likes cane stool density, the angle between cane stalk and vertical direction, or between adjacent two cane stalks. The other is the cutting type, which means a different shape of the blade and a different mounting angle of the blade. Although analytical and physical studies focusing on optimizing the structural and kinetic parameters of base cutters are abundant, problems remain including high rates of broken stubble, heavy losses, and uprooting. Till now, most research aimed at the traditional disc-type base cutters which are dominant in cane harvesters around the world.

Traditional disc-type base cutters are designed in a free-cutting mode, so that the cutting blades chops sugarcane with high angular inertia. A blade has good cutting performance when the chopping angle is perpendicular to the cane stalk. As the angle between the chopping blade and the cane stalk decreases, the risk of the blade splitting the stalk becomes greater. In daily cutting operation, due to the large variety of stalk patterns, it is practically impossible to always maintain a perfect chopping angle that is normal to the cane stalk. During cutting, blade impact increases stalk deflection to render the angle less than 90°; as a result, the stubble could be split, broken, and even uprooted. It is intuitive to replace free-cutting (chopping) with support-cutting (shearing) because two-sided support prevents the cane stalk from moving freely during cutting, thus effectively reducing cane deflection and cane damage. In addition, support-cutting can achieve higher cutting efficiency at a lower cutting speed. Cutting quality improves because support-cutting is relatively insensitive to cane growth patterns. Support-cutting for other crops just as wheat or rice is studied extensively as it has advantages of lower energy consumption, but fewer researchers focus on support-cutting of sugarcane harvest at present. A kind of support-cutting device for sugarcane comprised of a set of stationary blades and a set of kinetic blades is researched, the cutting dynamic process is simulated using finite element software named ANSYS/LS-DYNA to determine that how is cutting performance likes cutting force and broken rate influenced by different blade edge angle.

Furthermore, because of the unpredicted sugarcane growing pattern, the stationary blade of the supporting cutting device just mentioned before does not align fully with each row of sugarcane in field, so it is restricted in practice for these devices. The base cutter with two sets of moving blades in opposite directions can overcome the shortage of stationary blades because it provides the supporting action in dynamic process. In order to study the cutting performance of this kind of cutter, a test platform for sugarcane support-cutting is developed, in which, a time-frequency FXLMS controller is developed for electric driving motor to obtain satisfied cutting performance (Wu et al., 2019).

System Description of the Support-Cutting Device

Shearing a cane by support-cutting is similar to clipping a stick using a pair of scissors where two sets of discs with cutting blades spin in opposite directions to shear a cane stalk. In this study, two cutting discs were driven separately by a variable-frequency electric motor to achieve coaxial contra-rotation. A stable system being controlled is required to keep the speeds of the two motors in a proper range. These considerations led to the test platform design shown in Figs. 10.1 and 10.2, which features a mechanical sub-system and a control unit. The mechanical sub-system has a cutting device and a feeding device as its components. The control unit includes a master computer, a controller, two frequency three-phase inverters, and two torque/rotating speed transducers (Fig. 10.1). Table 10.1 lists the specifications of the test platform.

The cutting device consists of two variable-frequency electric motors and two cutting discs. The two cutting discs are driven by the two motors and rotating in opposite directions. Two neighboring upper and lower blades provide support for one another to enable support-cutting. Theoretically, the vertical clearance between the neighboring upper and lower blades should be no more than 0.5 mm to ensure the cutting quality. However, in practice, the clearance is set at 2 mm to prevent

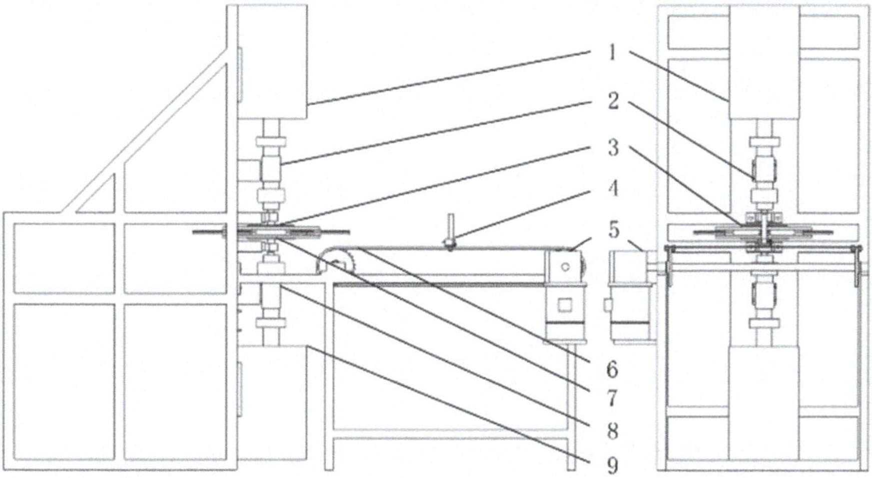

Fig. 10.1 Schematic diagram of the mechanical system. 1-Upper cutting electric motor. 2-Upper torque/rotating speed transducer. 3-Upper cutting disc. 4-Sugarcane clamp. 5-Feeding electric motor. 6-Conveying chain. 7-Lower cutting disc. 8-Lower torque/rotating speed transducer. 9-Lower cutting electric motor

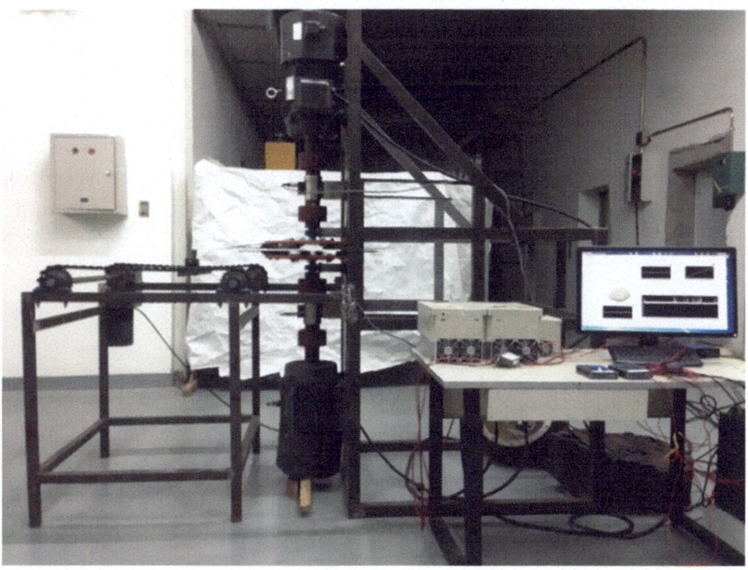

Fig. 10.2 The photo of the cutting test platform

Table 10.1 The cutting test platform specifications

Parameters/unit	Value
Power of the cutting motor *P*/kW	7.5
Power of the feeding motor *P*/kW	0.75
Rotating speed of the cutters *n*/(r/min)	0~900
Feeding speed *v*/(m/s)	0~1.5
Cutting clearance *x*/mm	2
Tilt angle of the cutting *d*/°	0~20
Length of the root stubble *L*/mm	30

Table 10.2 Main parameters of the cutter disc

Parameters/unit	Value
Diameter of the cutter disc/*mm*	455
Length of the blade/*mm*	120
Width of the blade/*mm*	50
Thickness of the blade/*mm*	5

crashing of the upper and lower blades in testing. The clearance of the test platform is adjustable by changing the thickness of the cushion block between the blade and the disc. The specifications of the cutter discs are found in Table 10.2.

An integrated transducer (TQ-660, Shitongkechuang Technology, Beijing, China) was used to acquire feedback signals of the torque and speed. An NI myRIO-1900 digital controller was used to implement the control algorithm. Real-time dynamic data were viewed in LabVIEW. The controller and the variable-frequency electric motors were linked using a SANCH S1100 series frequency three-phase inverter (Sanch Electric Corporation) for motor speed control. The feeding motor was driven by a frequency converter to allow for different feeding speeds.

Design of Time-Frequency FXLMS Controller

As a utility tool for time-frequency analysis, DWT is widely used for its superior ability in feature extraction of short data sequence. There is a fast pyramid algorithm, developed by Mallat, that can simplify the complex process of decomposition and reconstruction into inner product of vectors or matrix multiplication, allowing the processing of a succession of discrete value of the signal to be implemented in real-time.

Figure 10.3 illustrates a two-level DWT processing of a signal $\boldsymbol{x}(k)$ using the subband coding scheme, which includes wavelet decomposition and reconstruction where $\hat{g}$ (k) and $\hat{h}$ (k) are high-pass and low-pass decomposition filters, respectively, whereas $g(k)$ and $h(k)$ are high-pass and low-pass reconstruction filters,

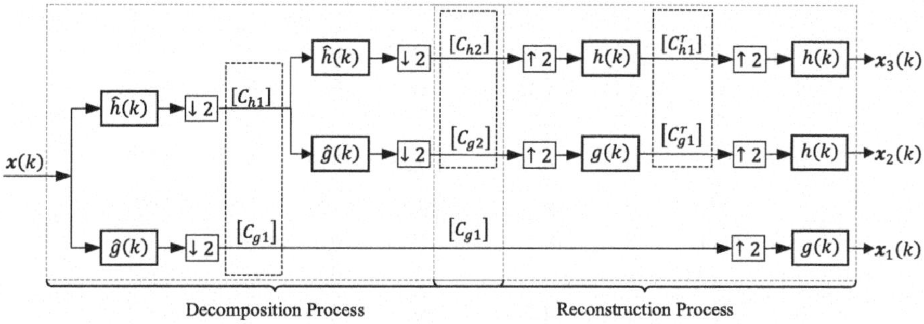

Fig. 10.3 DWT process of a signal

respectively. The latter pair forms a quadrature conjugate mirror filter pair with the decomposition filters. Once the convolution between $\boldsymbol{x}(k)$ and decomposition filters is done, and the down-sampling are finished, the first-level detail coefficients C_{g1} and trend coefficients C_{h1} are obtained. They contain the high frequency and low frequency information of the original signal $\boldsymbol{x}(k)$. In the same token the second-level detail coefficients C_{g2} and trend coefficients C_{h2} can be acquired using C_{h1}. Many groups of coefficients representing the frequency information from low to high can be generated by repeating the procedures. By engaging the groups of coefficients with the filters $g(k)$ and $h(k)$ in the reconstruction process, we can obtain the signal components $\boldsymbol{x}_1(k)$, $\boldsymbol{x}_2(k)$ and $\boldsymbol{x}_3(k)$ as shown.

Assuming the wavelet filter has four coefficients, the first-level decomposition process convolutes the input $\boldsymbol{x}(k)$ with a high-pass filter $\hat{g}(k)$ and a low-pass filter $\hat{h}(k)$ can be carried out by multiplying the signal with a linear transformation matrix T_d, as

$$\boldsymbol{X}_1[N] = \boldsymbol{T}_d[N]\boldsymbol{X}[N] \tag{10.1}$$

where the $\boldsymbol{X}_1[N]$, $\boldsymbol{T}_d[N]$, $\boldsymbol{X}[N]$ are defined below

$$\boldsymbol{X}_1[N] = \begin{bmatrix} \boldsymbol{A}_1 \\ \boldsymbol{D}_1 \end{bmatrix} = \begin{bmatrix} a(1) \; a(2) \; \cdots \; a(N/2) \; d(1) \; d(2) \; \cdots \; d(N/2) \end{bmatrix} \tag{10.2}$$

$$T_d[N]=\begin{bmatrix} h_d(4) & h_d(3) & h_d(2) & h_d(1) & \cdots & 0 & 0 & 0 & 0 \\ 0 & 0 & h_d(4) & h_d(3) & \cdots & 0 & 0 & 0 & 0 \\ \vdots & \vdots & \vdots & \vdots & \ddots & \vdots & \vdots & \vdots & \vdots \\ 0 & 0 & 0 & 0 & \cdots & h_d(2) & h_d(1) & 0 & 0 \\ 0 & 0 & 0 & 0 & \cdots & h_d(4) & h_d(3) & h_d(2) & h_d(1) \\ h_d(2) & h_d(1) & 0 & 0 & \cdots & 0 & 0 & h_d(4) & h_d(3) \\ g_d(4) & g_d(3) & g_d(2) & g_d(1) & \cdots & 0 & 0 & 0 & 0 \\ 0 & 0 & g_d(4) & g_d(3) & \cdots & 0 & 0 & 0 & 0 \\ \vdots & \vdots & \vdots & \vdots & \ddots & \vdots & \vdots & \vdots & \vdots \\ 0 & 0 & 0 & 0 & \cdots & g_d(2) & g_d(1) & 0 & 0 \\ 0 & 0 & 0 & 0 & \cdots & g_d(4) & g_d(3) & g_d(2) & g_d(1) \\ g_d(2) & g_d(1) & 0 & 0 & \cdots & 0 & 0 & g_d(4) & g_d(3) \end{bmatrix} \tag{10.3}$$

$$\boldsymbol{X}[N]=\begin{bmatrix} x(1) & x(2) & \cdots & x(N) \end{bmatrix}^T \tag{10.4}$$

The second-level decomposition process can be expressed as

$$\boldsymbol{X}_2[N]=\begin{bmatrix} \boldsymbol{A}_2[N/4] \\ \boldsymbol{D}_2[N/4] \\ \boldsymbol{D}_1[N/2] \end{bmatrix}=\begin{bmatrix} \boldsymbol{T}_d[N/2] & 0 \\ 0 & \boldsymbol{I}[N/2] \end{bmatrix}\boldsymbol{X}_1[N]=\begin{bmatrix} \boldsymbol{T}_d[N/2] & 0 \\ 0 & \boldsymbol{I}[N/2] \end{bmatrix}\boldsymbol{T}_d[N]\boldsymbol{X}[N] \tag{10.5}$$

Correspondingly, the reconstruction process has

$$\boldsymbol{X}_0[N]=\boldsymbol{T}_r[N]\begin{bmatrix} \boldsymbol{T}_r[N/2] & 0 \\ 0 & \boldsymbol{I}[N/2] \end{bmatrix}\begin{bmatrix} \boldsymbol{A}_2[N/4] \\ \boldsymbol{D}_2[N/4] \\ \boldsymbol{D}_1[N/2] \end{bmatrix} \tag{10.6}$$

where the reconstruction matrix T_r has the relationship as below

$$\boldsymbol{T}_r[N]\boldsymbol{T}_d[N]=\boldsymbol{I}[N],\ \boldsymbol{T}_r[N]=\boldsymbol{T}_d^{-1}[N]=\boldsymbol{T}_d^T[N] \tag{10.7}$$

Define that

$$\boldsymbol{T}_r^A[N]=\boldsymbol{T}_d^T[N]\begin{bmatrix} \boldsymbol{T}_d^T[N/2] & 0 \\ 0 & \boldsymbol{I}[N/2] \end{bmatrix} \tag{10.8}$$

Thus, each component of the signal can be computed as

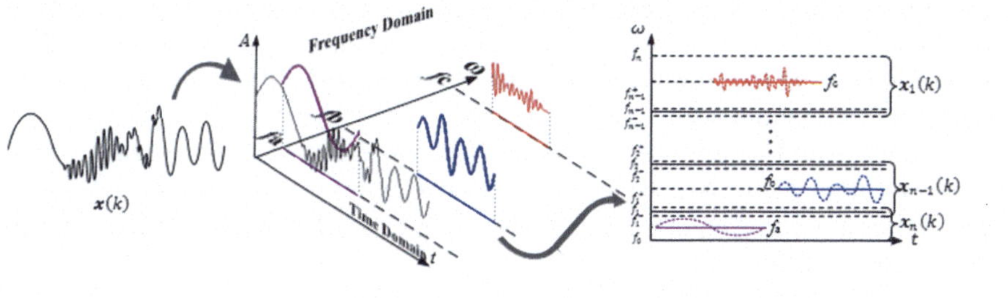

(a) original signal **(b)** DWT **(c)** TF performance of the signal

Fig. 10.4 Diagram of the decomposing process of a signal

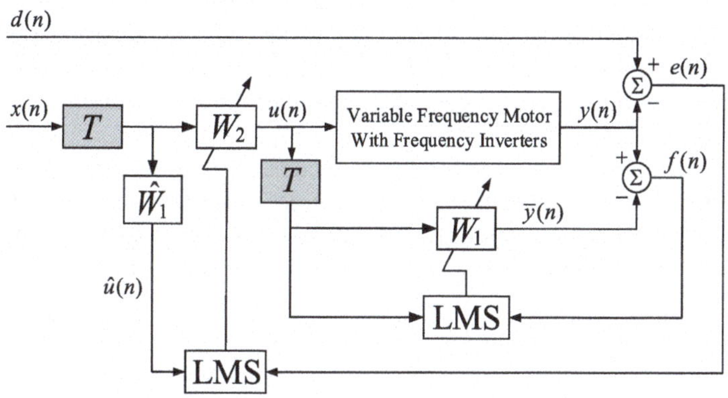

Fig. 10.5 Block diagram of the time-frequency FXLMS controller

$$\begin{cases} \boldsymbol{X}^3[N] = \boldsymbol{T}_r^A[N]_{(1:N,1:N/4)}\boldsymbol{A}_2 \\ \boldsymbol{X}^2[N] = \boldsymbol{T}_r^A[N]_{(1:N,N/4+1:N/2)}\boldsymbol{D}_2 \\ \boldsymbol{X}^1[N] = \boldsymbol{T}_r^A[N]_{(1:N,N/2+1:N)}\boldsymbol{D}_1 \end{cases} \tag{10.9}$$

where $\boldsymbol{T}_r^A[N]_{(r_1:r_2,c_1:c_2)}$ represents the new matrix created by taking the data from row r_1 to row r_2 and column c_1 to column c_2 of the reconstruction matrix $\boldsymbol{T}_r$.

According to the Shannon Sampling Theorem, if the sampling frequency is set to be $2f_n$, then the highest frequency of the actual signal one can resolve is f_n. By assuming that the highest frequency of the signal $\boldsymbol{x}(k)$ is f_n, the DWT process can decompose the signal into several frequency subsets from f_0 to f_n, where f_0 represents the lowest frequency of the signal whose general value is zero. Figure 10.4 illustrates the vivid process of how the complex nonlinear signal $\boldsymbol{x}(k)$ is expressed (decomposed) as simple sub-signals for easier handling. Following the two-level decomposition in Fig. 10.3, the value n in Fig. 10.4 is 3. Although aliasing would emerge in adjacent frequency subsets, it would not affect the result of dividing the frequency domain and designing the controller set later.

The configuration of the time-frequency controller shown in Fig. 10.5 featured FXLMS and DWT as 2 components. The identification filter $\boldsymbol{W}_1$ was used to identify the dynamics of the system and the filter $\boldsymbol{W}_2$ controlled the system adaptively by updating its own parameters.

The block $\boldsymbol{T}$ in Fig. 10.5 represents an $N \times N$ wavelet transform matrix which embodies a filter bank. According to the Mallat pyramidal algorithm, the square matrix is composed of high and low filter coefficients. The configuration has two error sequences as follows:

$$e(n) = d(n) - \bar{y}(n) \tag{10.10}$$

$$f(n) = y(n) - \bar{y}(n) \tag{10.11}$$

where $e(n)$ is the output error between the desire output and system actual output, which is used to update the control filter $\boldsymbol{W}_2$, while $f(n)$ is the identification error for adjusting the weight of $\boldsymbol{W}_1$. $d(n)$ is the desired stable response or reference output rotating speed at time n, $y(n)$ is the actual output of the system at time n, and $\bar{y}(n)$ is the estimated response.

The reference input vector at time n, defined as $\boldsymbol{X}(n)$ in Eq. (10.12), includes the reference input $x(n)$ that can be set to a nonzero constant if there is not a specific reference input sequence. Similarly, Eq. (10.13) gives the controller output vector, defined as $\boldsymbol{U}(n)$ at time n, and includes the controller output $u(n)$ that can be computed using Eq. (10.14)

$$\boldsymbol{X}(n) = \left[x(n), x(n-1), \ldots, x(n-N+1)\right]^T \tag{10.12}$$

$$\boldsymbol{U}(n) = \left[u(n), u(n-1), \ldots, u(n-N+1)\right]^T \tag{10.13}$$

$$u(n) = \boldsymbol{W}_2^T(n)\boldsymbol{T}\boldsymbol{X}(n) \tag{10.14}$$

where $\boldsymbol{W}_2$ is an adaptive FIR filter of length N

$$\boldsymbol{W}_2(n) = \left[w_{2,0}(n), w_{2,1}(n), \ldots, w_{2,N-1}(n)\right]^T \tag{10.15}$$

It is evident that $u(n)$ can be changed by adjusting $\boldsymbol{W}_2$. The least mean square (LMS) algorithm of the error $e(n)$ and the steepest descent method were employed to update the weights of $\boldsymbol{W}_2$

$$W_2(n+1) = W_2(n) + \mu_2 \boldsymbol{T}\hat{\boldsymbol{U}}(n)e(n) \tag{10.16}$$

in which $\hat{\boldsymbol{U}}(n)$ is

$$\hat{U}(n) = \left[u(n), u(n-1), \ldots, u(n-N+1)\right]^T \tag{10.17}$$

$$\hat{u}(n) = \boldsymbol{W}_1^T(n)\boldsymbol{T}\boldsymbol{X}(n) \tag{10.18}$$

and $\boldsymbol{W}_1$ is the FIR identification filter of length N used to calculate the estimated response $\bar{y}(n)$

$$\boldsymbol{W}_1(n) = \left[w_{1,0}(n), w_{1,1}(n), \ldots, w_{1,N-1}(n)\right]^T \tag{10.19}$$

$$\bar{y}(n) = \boldsymbol{W}_1^T(n)\boldsymbol{T}\boldsymbol{U}(n) \tag{10.20}$$

Similarly, the LMS algorithm of the identification error $f(n)$ and the steepest descent method were employed to update $\boldsymbol{W}_1$

$$\boldsymbol{W}_1(n+1) = \boldsymbol{W}_1(n) + \mu_1 \boldsymbol{T}\boldsymbol{U}(n) f(n) \tag{10.21}$$

Parameters μ_1 and μ_2 are the step sizes that control the incremental estimation of $\boldsymbol{W}_1$ and $\boldsymbol{W}_2$ at each iteration step. They influence the performance of the controller to a large extent. Overall, the time-frequency controller uses a control method where the feedback signal is not directly used in the calculation of the controller output. Instead, it is sent to update the coefficients of the controller to change the output. As a result, the control error was kept within a range by minimizing the mean square root of the error.

Considerations were given to the selection of a wavelet function that provided the best control performance. Following along with the numerical experiments, a db3 mother wavelet was used to create the wavelet matrix using six high-pass and six low-pass filter coefficients. In theory, the larger the N of the control filter or the identification filters is, the more parameters can be involved to control the system, thus a better controller. However, computational capacity is the limiting factor during a microprocessor selection. Once the request of real-time control and the high response frequency of the motor are synthesized, the length of the filter N is set to 64 and the sampling frequency and the computing frequency are set at 500 Hz. As described above, step sizes μ_1 and μ_2 are the key parameters for controller design. A set of step-size rules is given in the equations below.

$$\mu_1 = 1 \times 10^{-7} \tag{10.22}$$

$$\mu_2 = \begin{cases} 1 \times 10^{-6}, & (e(n) > 600) \vee (t < 20\text{s}) \\ 1 \times 10^{-7}, & \text{the else} \end{cases} \tag{10.23}$$

Equation (10.23) shows that a variable step size is used to update the control filter $\boldsymbol{W}_2$. When the error is larger than a threshold value of 600, a larger step size is required to minimize the error. In the first 20 s a larger step size is required to update $\boldsymbol{W}_2$ quickly because its initial value is zero. In other situations, a small step size is used for the system to meet the control objective with high precision when the error is throbbing near zero.

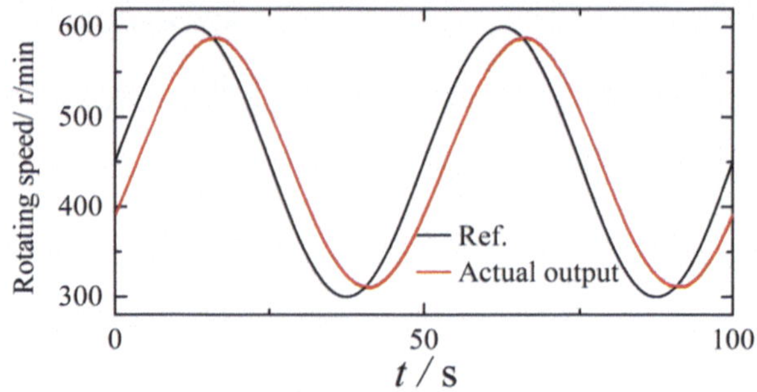

Fig. 10.6 Results of sinusoidal tracking of the disc rotating speed

Results and Discussions

Experiments including a series of idle tests and sugarcane cutting tests were performed with the experiment equipment shown in Fig. 10.2. Idle tests were to verify the feasibility of the controller and the cutting tests to prove the effectiveness of support-cutting.

Figure 10.6 is the response of sinusoidal tracking of the disc rotating speed. Allowing for the phase lag between the actual output and the target, the system response is a perfect sine wave, indicating that cutter speed is controlled smoothly by the time-frequency controller. The lock-step responses of the upper and lower discs were considered next. Figure 10.7a shows step responses of the speeds of the upper and lower discs. Figure 10.7b plots the absolute speed difference of the 2 discs. The zoom-in figure in Fig. 10.7b provides the detail of the speed differences starting at $t = 25$ s. The peaks in Fig. 10.7b correspond to the moment each step begins, which shows the two discs speeds are different in spite of the fact that they are under the same reference input. This is due to the inherent difference of the 2 frequency three-phase inverters. However, when the responses reach steady-state, the difference decreased rapidly to be less than 2.5 rpm. In the entire process, the overshoot of the step response was within 1%.

Cutting tests were conducted to study the load impact on the speeds of the two electric motors with and without control. The corresponding test results are shown in Fig. 10.8 in which Fig. 10.8a, b are the control-free and controlled speed variations, respectively. Figure 10.8c illustrates the absolute speed difference of the two motors with control and without control. When a cutting blade touches a cane stalk, the speed of the cutter disc drops sharply, thus the dips in Fig. 10.8a, b. The speed drops 5 rpm without control in Fig. 10.8a and 2 rpm with control in Fig. 10.8. It is evident that speed control was improved with the online time-frequency controller.

When the two motors were properly controlled, the speed difference between the upper and lower discs was reduced more than under the condition without control subject to cutting resistance. The speed difference between the two motors is plotted in red in Fig. 10.8c where the maximum value was registered at 5 rpm. The black solid line in Fig. 10.8c represents the speed difference with control where the maximum value is less than 2.5 rpm. It is evident that saltation of speed was significantly reduced.

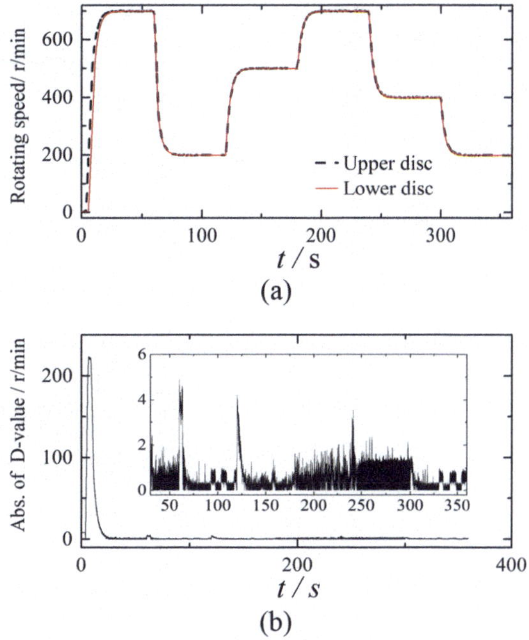

Fig. 10.7 Results of synchronization step response. (**a**) Step responses of upper and lower cutting discs. (**b**) Absolute value of the difference of rotating speeds between upper and lower discs

Instantaneous frequency (IF) is another indicator for evaluating the performance of the time-frequency controller. Intrinsic mode functions (IMFs) can be acquired by employing empirical mode decomposition (IMFs) to decompose the upper disc speed signal to obtain the corresponding instantaneous frequency of the IMFs. The algorithm is also called the Hilbert–Huang Transform (HHT). Generally, a narrow-frequency bandwidth and low-IF oscillation amplitude are indications of stable cutting. Figure 10.9a, b below give the IF of rotational speed without and with control. Figure 10.9c, d compare the IFs of the second and third IMFs without control and under control.

It can be seen from Fig. 10.9a, b that the IF of the speed signal without control oscillates more violently than its controlled counterpart where the first IF (in black) in Fig. 10.9a switched more frequently between 0 Hz and 30 Hz. Similar observations can be made in Fig. 10.9c, d where the maximum amplitude of the second IF with control is approximately 15 Hz, while the one without control is 25 Hz. The time-frequency responses of the controlled speed were consistently better than those without control.

The essential characteristics of the cutting process can be understood from the frequency-domain perspective. The cutter disc was not running as stably as expected. The speed varied constantly even in the non-cutting state. Minor amplitude variation was not a reliable indication of cutting stability. Variations with high frequency are harmful to sugarcane stalks in the cutting process because changing each time causes a tiny impact to the microscopic tissue of the stalk and is the initial fracture

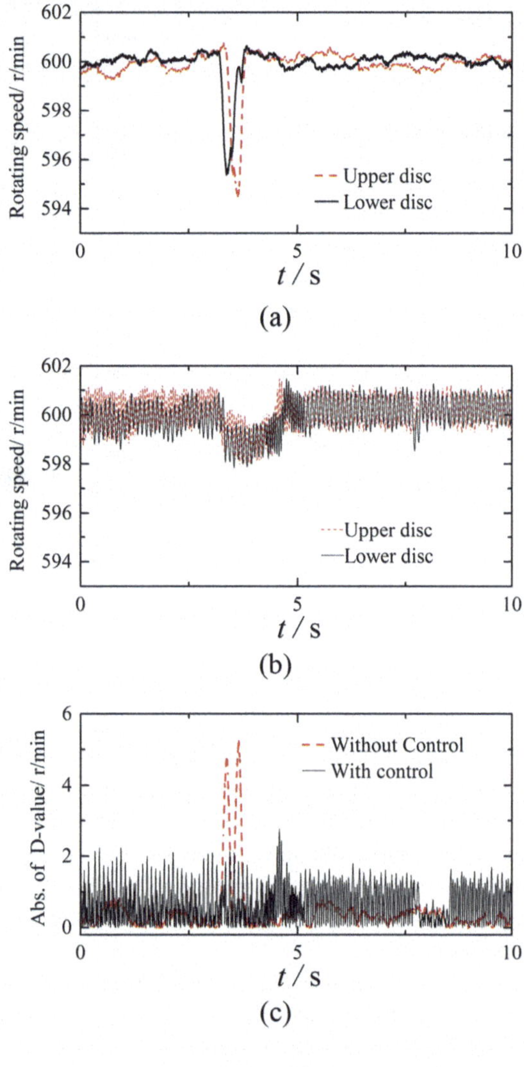

Fig. 10.8 Comparison of the difference rotating speed without and with control. (**a**) Rotating speeds without control. (**b**) Rotating speeds with control. (**c**) Absolute value of the rotating speed difference between two motors with and without control

that ultimately becomes zig-zag splitting. Thus, ensuring stable rotating speed of the cutter disc under time-frequency control is meaningful to the sugarcane base cutter if decreased broken rate is desired.

A series of indoor cutting tests were performed to compare the cutting qualities of free-cutting (on 30 stalks) and support-cutting (on 30 stalks). For support-cutting, the speeds of the upper and lower discs were both set at 200 rpm. For free-cutting, the upper disc was powered off and the speed of the lower disc set at 200 rpm. Cane stalk cutting quality corresponding to support-cutting and free-cutting is shown in Figs. 10.10 and 10.11. Figures 10.10a and 10.11a show cutting damages highlighted in red. Figure 10.10b and 10.11b give the details of stalk splits seen in Fig. 10.10a and 10.11a.

The damage rate, which is calculated by dividing the number of total stalks (30) by the number of damaged stalks, is used to appraise cutting quality. Comparing Fig. 10.10 with Fig. 11.11, support-cutting reduced the stalk damage rate from 26.67% to 6.67%.

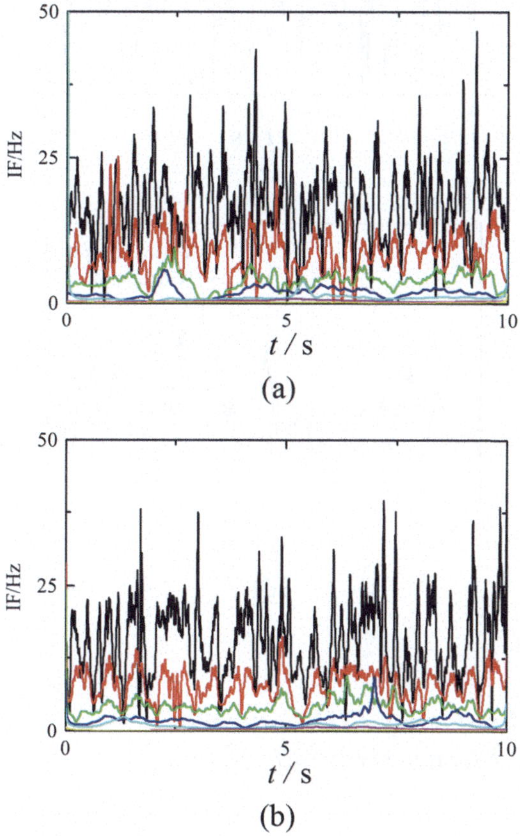

Fig. 10.9 Comparison of instantaneous frequency for rotating speed without and with control. (**a**) Instantaneous frequency of rotating speeds without control. (**b**) Instantaneous frequency of rotating speeds with control. (**c**) IF comparison of the second IMFs of rotating speeds. (**d**) IF comparison of the third IMFs of rotating speeds

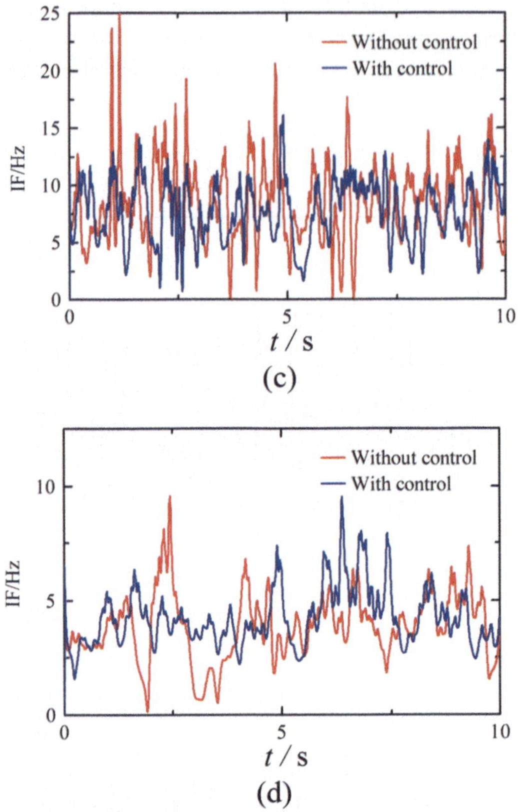

Fig. 10.9 (continued)

10.3.2 *Time-Frequency PID Controller for the Electro-Hydraulic System*

Background

Nonlinear control methods including sliding mode (or variable structure) control, back-stepping control, adaptive control, and fuzzy control have seen increasing applications in electro-hydraulic systems whose dynamic behaviors are highly nonlinear due to complex structures of many components, time-varying parameters, and uncertainties. Electro-hydraulic systems found in farming equipment and machinery with characteristically large power are often employed to handle system responses that are of low frequency. Given the low operation speed and thus weak nonlinearity, PID controllers are in general effective in mitigating the impact of nonlinearity on stability and performance. PIDs are commonly employed to work with other control concepts and implemented as digital signal processors for the real-time control of many an industrial application including electro-hydraulic systems.

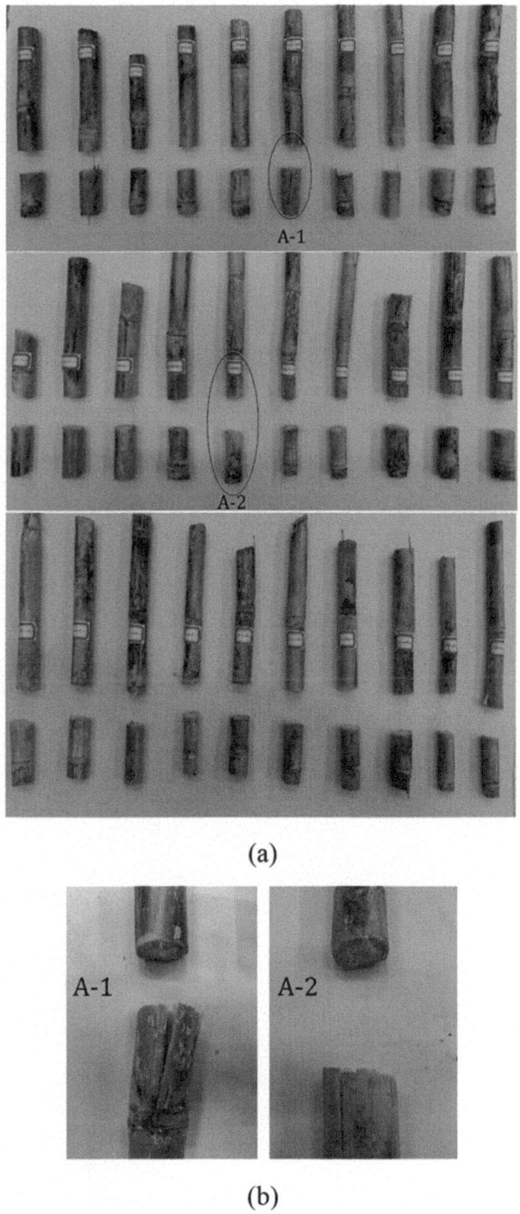

Fig. 10.10 Performance of the support-cutting. (**a**) Results of all cutting tests. (**b**) Detail of split roots

Theoretically, PID parameters can be optimally tuned to improve dynamic performance, reduce steady-state error, and overcome system oscillation. However, doing so will inevitably lead to chatter in daily applications due to the nonlinearities of omnipresent high frequency, external load disturbance, and noise jamming. That is, adjusting parameters however slightly can induce high frequency responses that

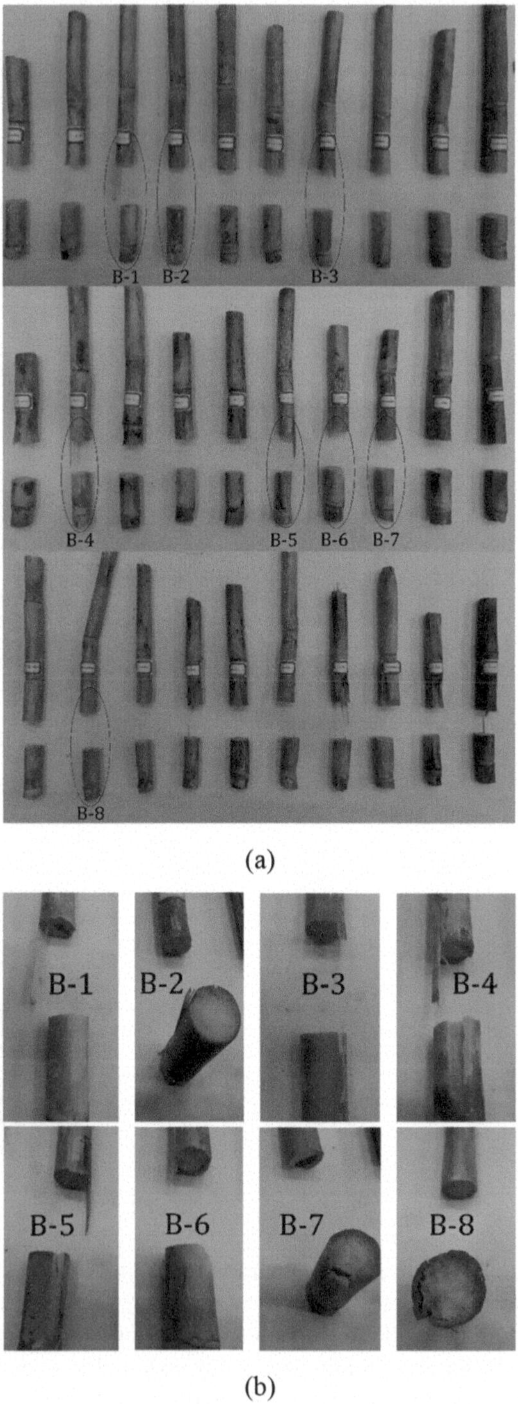

Fig. 10.11 Performance of traditional free-cutting. (**a**) Results of all cutting tests. (**b**) Detail of split roots

are nonlinear and inadvertently magnify the external load fluctuation and noise disturbance, rendering low energy efficiency and risking system breakdown. Conversely, conservative controller design aiming to weaken the control action in exchange for system stability sacrifices system performance at the expense of precious resources. This is a dilemma demanding a solution to an improved controller design that is effective in mitigating disturbance, negating high frequency nonlinear oscillation, and in the meantime enhancing system performance.

The sensitivities of the proportional, differential, and integral control actions of the PID method are different from each other in different frequencies. Strengthening the differential action in low frequency part of the error signal is effective in reducing overshoots, suppressing oscillations, stabilizing the system stable, and improving system response and dynamic performance. However, it can also negatively amplify the high frequency part of the error signal that is indicative of nonlinear high frequency chattering and noise interference. Similarly, the integral control can eliminate the steady-state error and improve the precision of the control system. But enhancing the integral action to the low frequency can cause the phenomenon called wind-up or integral saturation. Because the oscillation of the low frequency error signal can last a long period of time, integral saturation can force the solenoid valve spool that is already reaching the end of the valve to jerk and impact. Conversely, the high frequency error is insensitive to larger integral coefficient because of the rapid fluctuation involved. Therefore, it is essential that a PID controller is designed by exploring the parameters for targeting different ranges of frequency response.

An improved electro-hydraulic system controller design is reported herein with exploring all the prominent TF features. A nonlinear dynamic system model of a hydraulic cylinder is developed, which is controlled by a solenoid proportional valve. The model is used for identifying the optimal control parameters and performing numerical experiments that consider various nonlinearities including friction, dead zone, leakage, and the variable effective bulk modulus. The working principle of the wavelet-based TFC and the design of a TFPID controller incorporating PID control are presented. At last, the hardware-in-the-loop experiments employed based on the mathematical model. Comparisons are also made with the traditional PID controller to validate the controller design.

System Description of Electro-Hydraulic System

Consider the hydraulic cylinder controlled by a solenoid proportion valve of a universal configuration such as the one shown in Fig. 10.12. Comprised of a double-ended hydraulic cylinder, a 4/3-way solenoid proportion valve, and an unknown load, the system is commonly applicable to either force or position control. A comprehensive nonlinear model of the system is formulated.

According to Newton's second Law, the dynamic equation of the hydraulic actuator can be described as

$$M\ddot{x} + B_e\dot{x} + K_e x = (p_1 - p_2)A - F_L - F_F \tag{10.24}$$

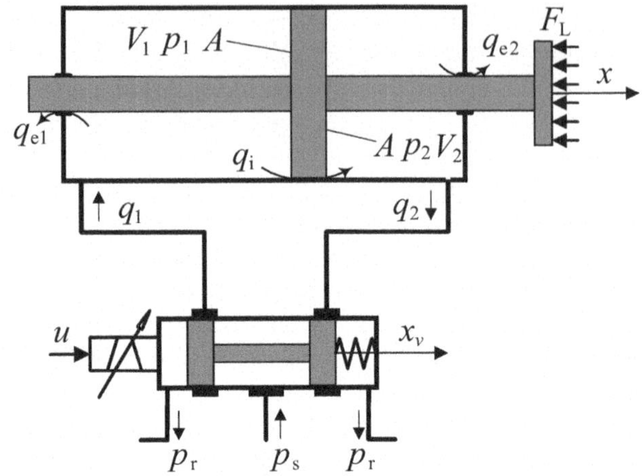

Fig. 10.12 Schematic of the hydraulic cylinder controlled by a solenoid proportion valve

where x is the displacement of the piston, m is the total mass of the piston along with all the linked objects, B_e and K_e are equivalent damping and the spring stiffness coefficient, respectively, p_1 and p_2 are pressure of the left and right chamber, respectively, A is active area of the piston, and F_L is the unknown force of nonlinearity due to the external disturbance. F_F is the friction force on the piston, which is one of the primary nonlinear factors affecting the dynamics of the actuator piston. The particular friction is influenced by many factors. In general, it is considered to be a function of the position and velocity of the piston. Many empirical models have been established and applied to specific hydraulic actuators.

A particular friction equation of motion of the actuator piston following from the Lund–Grenoble model is adopted. It is formulated based on sound hypotheses and is rigorously derived on the basis of many kinds of friction can be considered. Most importantly, it agrees well with empirical data in most situation with adjusting parameters in the formulation properly. The friction model which is a set of coupled equations is concisely given below in Eq. (10.25).

$$\begin{aligned} \frac{dz}{dt} &= \dot{x} - \frac{\sigma_0 |\dot{x}|}{g(\dot{x})} z \\ g(\dot{x}) &= \alpha_0 + \alpha_1 e^{-(\dot{x}/v_{sk})^2} \\ F_F &= \sigma_0 z + \sigma_1 \frac{dz}{dt} + \alpha_2 \dot{x} \end{aligned} \quad . \tag{10.25}$$

where z is an intermediate variable, $g(\dot{x})$ a function describing the steady-state friction characteristics at a constant velocity, v_{sk} the Stribeck velocity defined as the most unstable velocity on the Stribeck curve, α_0 the Coulomb friction, α_1 the Stribeck friction, α_2 the viscous friction parameter, σ_0 the spring constant, and σ_1 the damping coefficient.

For a double-ended hydraulic cylinder experiencing leakage, the following pressure continuity equations featuring the effective bulk modulus β_e for the cylinder chambers can be derived:

$$\begin{aligned} \dot{p}_1 &= \frac{\beta_e}{V_1 + Ax}\left(q_1 - A\dot{x} - q_i - q_{e1}\right) \\ \dot{p}_2 &= \frac{\beta_e}{V_2 - Ax}\left(-q_2 + A\dot{x} + q_i - q_{e2}\right) \end{aligned} \tag{10.26}$$

where V_1 and V_2 are the original control volumes of the left chamber and right chamber, respectively, including the volume of the servo valve, pipeline, and cylinder chambers. q_1 is the supplied flow rate to the left chamber, while q_2 is the return flow rate of the right chamber. The relationship between the spool valve displacement and the load flow dictates that

$$\begin{aligned} q_1 &= k_{q1}x_v\left[s(x_v)\sqrt{p_s - p_1} + s(-x_v)\sqrt{p_1 - p_r}\right] \\ q_2 &= k_{q2}x_v\left[s(x_v)\sqrt{p_2 - p_r} + s(-x_v)\sqrt{p_s - p_2}\right] \end{aligned} \tag{10.27}$$

where the $s(x_v)$ is a function defined as follows:

$$s(x_v) = \begin{cases} 1, & x_v \geq 0 \\ 0, & x_v < 0 \end{cases} \tag{10.28}$$

and

$$\begin{aligned} k_{q1} &= C_d w_1 \sqrt{\rho/2} \\ k_{q2} &= C_d w_2 \sqrt{\rho/2} \end{aligned} \tag{10.29}$$

where p_s is the supplied pressure, p_r is the return line pressure, x_v is the spool displacement of the solenoid proportion valve, C_d is the discharge coefficient, w_1 and w_2 are the spool valve area gradients, and ρ is the fluid density. It should be noted that the parameter q_i found in Eqs. (10.26) represents the internal leakage, while q_{e1} and q_{e2} seen in Eqs. (10.26) denote the external leakages of the left chamber and right chamber, respectively. They are determined through Eqs. (10.30).

$$\begin{aligned} q_i &= c_i\left(p_1 - p_2\right) \\ q_{e1} &= c_{e1}p_1 \\ q_{e2} &= c_{e2}p_2 \end{aligned} \tag{10.30}$$

where c_i is the internal leakage coefficient and c_{e1} and c_{e2} are the external leakages of the left chamber and right chamber, respectively. Note also that the effective bulk modulus β_e is primarily affected by the chamber pressures involved as Eq. (10.31).

$$\begin{aligned} \beta_{e1} &= (c_1 + p_1)\left[\frac{1}{c_2} - \ln\left(1 + \frac{p_1}{c_1}\right)\right] \\ \beta_{e2} &= (c_1 + p_2)\left[\frac{1}{c_2} - \ln\left(1 + \frac{p_2}{c_1}\right)\right] \end{aligned} \tag{10.31}$$

where β_{e1} and β_{e2} are the input chamber and output chamber effective bulk moduli of the cylinder, respectively. c_1 and c_2 are constants.

The displacement of the shaft of the solenoid valve x_v is related to the valve's voltage input u, via a second-order system:

$$\ddot{x}_v + 2\zeta\omega_v\dot{x}_v + \omega_v^2 x = K_v\omega_v^2 u \tag{10.32}$$

where ζ, ω_v, and K_v are the damping ratio, natural frequency, and gain of the valve dynamics, respectively.

The dynamic equations of the electro-hydraulic system depicted in Fig. 10.12 can now be formulated by using a set of differential functions through recasting Eqs. (10.24)–(10.32) and the following state variables: $x_1 = x$, $x_2 = \dot{x}$, $x_3 = p_1$, $x_4 = p_2$, $x_5 = x_v$, and $x_6 = \dot{x}_v$,

$$\begin{cases} \dot{x}_1 = x_2 \\ \dot{x}_2 = \dfrac{1}{m}\left[(x_3 - x_4)A - B_e x_2 - K_e x_1 - F_L - F_F\right] \\ \dot{x}_3 = \dfrac{(c_1 + x_3)\left[1/c_2 - \ln(1 + x_3/c_1)\right]}{V_1 + Ax_1}\{k_{q1}x_5 \\ \quad \left[s(x_5)\sqrt{p_s - x_3} + s(-x_5)\sqrt{x_3 - p_r}\right] \\ \quad - Ax_2 - c_i(x_3 - x_4) - c_{e1}x_3\} \\ \dot{x}_4 = \dfrac{(c_1 + x_4)\left[1/c_2 - \ln(1 + x_4/c_1)\right]}{V_1 - Ax_1}\{-k_{q2}x_5 \\ \quad \left[s(-x_5)\sqrt{p_s - x_4} + s(x_5)\sqrt{x_4 - p_r}\right] \\ \quad + Ax_2 + c_i(x_3 - x_4) - c_{e1}x_4\} \\ \dot{x}_5 = x_6 \\ \dot{x}_6 = K_v\omega_v^2 u - 2\zeta\omega_v x_6 + \omega_v^2 x_5 \end{cases} \tag{10.33}$$

In real electro-hydraulic system, there are limits on the displacement of the piston and a dead zone in the valve spool. Such physical constraints are modeled using the state variables as follows:

$$x_1 = \begin{cases} x_{\min}, & x_1 \le x_{\min} \\ x_1, & x_{\min} < x_1 < x_{\max} \\ x_{\max}, & x_1 \ge x_{\max} \end{cases} \tag{10.34}$$

$$x_5 = \begin{cases} x_v^{\min}, & x_5 \le x_v^{\min} \\ x_5, & x_v^{\min} < x_5 < x_v^{\text{dmin}} \\ 0, & x_v^{\text{dmin}} \le x_5 \le x_v^{\text{dmax}} \\ x_5, & x_v^{\text{dmax}} < x_5 < x_v^{\max} \\ x_v^{\max}, & x_5 \ge x_v^{\max} \end{cases} \tag{10.35}$$

The initial position of the piston is set at the middle of the cylinder and according to its effective length, the parameter can be determined as: $x_{\max} = -x_{\min} = 0.2$ m. Similarly, assuming that the valve is symmetrical and the initial position of the spool is also at the middle of the valve sleeve, so the value is obtained as: $x_v^{\max} = -x_v^{\min} = 3$ mm, $x_v^{\text{dmax}} = -x_v^{\text{dmin}} = 0.5$ mm. In addition, other parameters involved that are determined by structure of system or working environment are listed in Table 10.3.

When a specific piston motion trajectory $x_0(t)$ or a force $F_0(t)$ is desired, the objective of designing the controller is therefore to generate a series of output $u_0(t)$ to the solenoid valve through adjusting the original input signal $u_i(t)$ to achieve the tracking of x_1 and the control of $(x_3 - x_4)$ subjected to the exertions of the unknown external disturbance F_L and the nonlinear friction force F_F and to maintain the displacement of the piston with limited fluctuation.

Table 10.3 The parameters used in the model

Parameters	Value	Parameters	Value
M	9.0 kg	c_i	1097 mm³/(s•MPa)
B_e	2000 N/(m/s)	c_{e1}	120 mm³/(s•MPa)
K_e	10 N/m	c_{e2}	120 mm³/(s•MPa)
A	645 mm²	K_v	0.5
c_1	99.993 MPa	ω_v	534 rad/s
c_2	0.0733	ζ	0.48
V_1	1.29×10^5 mm³	σ_0	5.77×10^6 N/m
V_2	1.29×10^5 mm³	σ_1	2.28×10^4 N/m/s
k_{q1}	2.38×10^{-5} m$^{5/2}$/kg$^{1/2}$	α_0	370 N
k_{q2}	2.38×10^{-5} m$^{5/2}$/kg$^{1/2}$	α_1	217 N
P_s	21 MPa	α_2	2318 N/m/s
P_r	0.1 MPa	v_{sk}	10 N/m

Design of Time-Frequency PID Controller

A TFC controller concept is proposed with incorporating PID control to realize the TFPID approach taking full advantages of both the PID and TF method. Figure 10.13 shows the schematic of the TFC approach. In the figure, r represents the desire or reference output, and y is the actual output of the system, which represents the system performance. The output y also implies the internal information and reflects the essence of the system. Traditional controllers, be them feedback or feedforward, always operate r and y directly to export a series signal to the system. In contrast, the results of the synthesis operation of r and y must be analyzed and processed by the time-frequency method, thus producing several sub-signals to be computed by their corresponding controller. In other words, the TFC scheme includes several controllers at the same time, all working concurrently and constituting a controller set.

The TFC is expected to design an exclusive controller for each signal component $x_i(k)$, $i = 1, 2, \ldots, n$, so that the parameter of each controller can meet the performance requirement of the system in regard to the particular frequency response. Hence, the performance of system would be optimal in all frequency domain using the controller subsets. Most important, definitive physical interpretation can be made in the TFC design process.

The TFPID controller concept incorporates the characteristics of P, I, and D control with the TF framework. The TFPID design concept is found in Fig. 10.14. Traditional PID controller often operates on the error e, while TFPID applies the DWT to sequences $\boldsymbol{e}(k)$ first to resolve the high frequency error e_H and the low frequency error e_L. The high frequency information indicative of nonlinearity, measurement noise, and external disturbances are carried by e_H, whereas the main working frequency information like phase lag is contained in e_L. The electro-hydraulic system is typically working in the low frequency 0-to-2 Hz range, so the error signal needs be multi-level wavelet decomposed, and the component belonging to the lowest frequency range can be taken as the low frequency error e_L, while the rest be synthesized to be the high frequency error e_H.

The parameters of each single controller need to be designed based on the principle of PID method with each PID controller considering a specific frequency band. Considering the digital control signal $u(k)$ at time k

$$\begin{aligned} u(k) = {} & K_{PL} e_L(k) + K_{IL} \sum_0^k e_L(k) T + K_{DL} \frac{e_L(k) - e_L(k-1)}{T} \\ & + K_{PH} e_H(k) + K_{IH} \sum_0^k e_H(k) T + K_{DH} \frac{e_H(k) - e_H(k-1)}{T} \end{aligned} \tag{10.36}$$

where K_{PL}, K_{IL}, and K_{DL} represent the coefficients of P, I, and D for the low frequency error e_L, K_{PH}, K_{IH}, and K_{DH} represent the controller coefficients for e_H, and T is the sampling period.

Fig. 10.13 Schematic of the time-frequency control approach

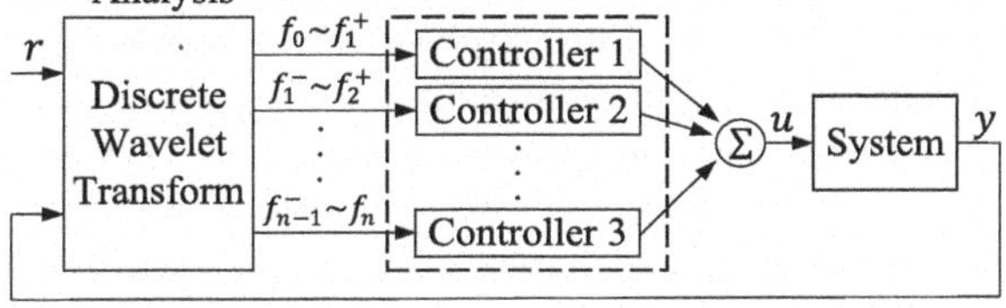

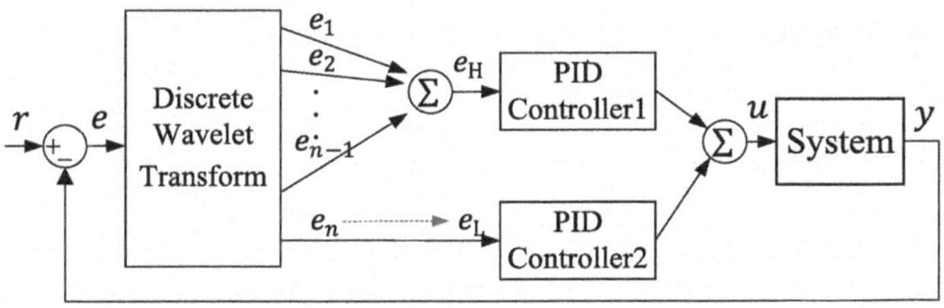

Fig. 10.14 Schematic of the TFPID control approach

To improve the precision in controlling step response and suppressing overshoot, a simple algorithm of separated integral method is introduced to the two PID controllers, rendering the following controller output:

$$u(k) = K_P e(k) + K_I \sum_{0}^{k} e_I(k)T + K_D \frac{e(k) - e(k-1)}{T} \tag{10.37}$$

The corresponding TFPID controller output is therefore

$$\begin{aligned} u(k) = {} & K_{PL} e_L(k) + K_{IL} \sum_{0}^{k} e_{LI}(k)T + K_{DL} \frac{e_L(k) - e_L(k-1)}{T} \\ & + K_{PH} e_H(k) + K_{IH} \sum_{0}^{k} e_H(k)T + K_{DH} \frac{e_H(k) - e_H(k-1)}{T} \end{aligned} \tag{10.38}$$

and the constraint conditions of the errors are as follows:

$$e_I = \begin{cases} e_{\min}, & e \le e_{\min} \\ e, & e_{\min} < e < e_{\max} \\ e_{\max}, & e \ge e_{\max} \end{cases} \tag{10.39}$$

$$e_{LI} = \begin{cases} e_{\min}, & e_L \le e_{\min} \\ e_L, & e_{\min} < e_L < e_{\max} \\ e_{\max}, & e_L \ge e_{\max} \end{cases} \tag{10.40}$$

As the TFPID controller works with two concurrent PID controllers, the number of parameters involved is therefore twice than the traditional PID controller, thus making TFPID controller more powerful and flexible. In theory, more parameters the controller has, more difficult the design task is. But with each parameter of the PID corresponding to specific response of the system, it is not demanding to design the specific TFPID to meet the control goal. Although parameters are fixed once the controller is designed, but because the constituent of different frequency in the signal is always varying in time, so the output signal $u(k)$ is dynamical with different and changing frequency components. Hence, the TFPID is adaptive to the variation in the frequency domain.

Results and Discussions

In order to validate the control method, MATLAB Simulink is employed to design the optimal parameters and a hardware-in-the-loop test bench is used to evaluate the TFPID design against the PID controller.

Extensive research has been conducted to identify the mother wavelet to be used in the study. The Daubechies wavelet is chosen for its properties of orthogonality and compact support. In addition, based on the rule that a wavelet with a large number of filter coefficients can match the characteristic features in a time series with greater efficiency, db3 wavelet of six filter coefficients is applied to decompose the error. Considering the working frequency of the system and the computational capacity of the microcomputer, the size of the error signal buffer (the number of observations in the time series) N_f is set to 512 and to be six-level decomposed.

The simulation model is built based on the model in Eq. (10.33) and the control algorithm is compiled using the S-Function. To get the minimum step response time with the condition of restraining overshoot to be less than 1%, a set of coefficients of traditional PID controller are obtained in Table 10.4. They are afterward commissioned as the references for designing the TFPID controller. The top and bottom limits of the error are chosen to be $e_{\max} = -e_{\min} = 0.0005$ m.

Per the account on PID given before, the control system will be more stable and responding faster by increasing the proportional and derivative coefficients for the low frequency error signal and decreasing the integral coefficient for the high frequency at the same time. Moreover, the wind-up phenomenon can also be eliminated. Enhancing the integral action by removing the derivative action to the high frequency component of the error signal can restrain the oscillation of high frequency. Table 10.5 summaries the parameters of the TFPID controller design.

The system model is compiled by VeriStand software and running in real-time on mainframe PXIe-8135. The control algorithm is edited in LabVIEW and downloaded to an embedded hardware device named NI myRIO-1900 as the controller prototype. Signal transmission between the virtual electro-hydraulic system and the controller is via a signal cable and the information can be uploaded to the computer at the same time. Figure 10.15 shows the test bench.

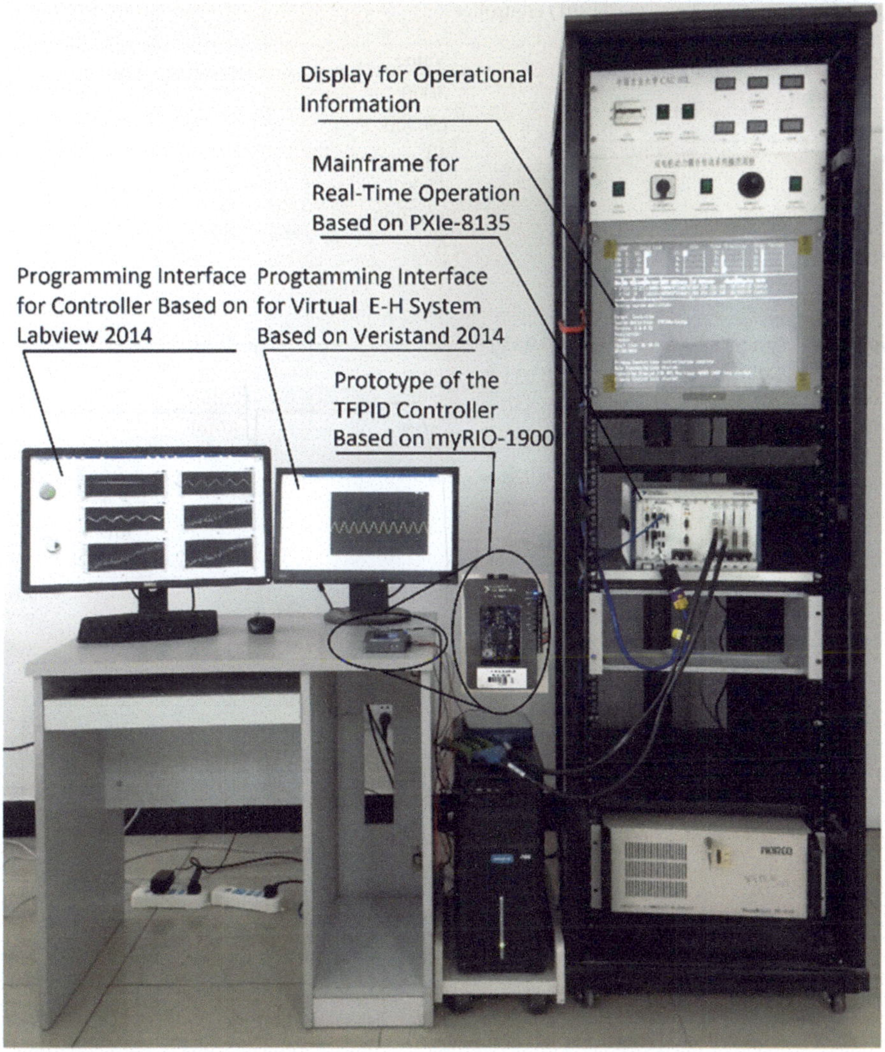

Fig. 10.15 Hardware-in-the-loop test bench for simulation of the electro-hydraulic system control

Table 10.4 Coefficients of the PID controller

Coefficient	Value
K_P	3000
K_I	60,000
K_D	14

Table 10.5 Coefficients of the TFPID controller

Coefficient	Value
K_{PL}	3000
K_{IL}	60,000
K_{DL}	14
K_{PH}	1500
K_{IH}	100,000
K_{DH}	0

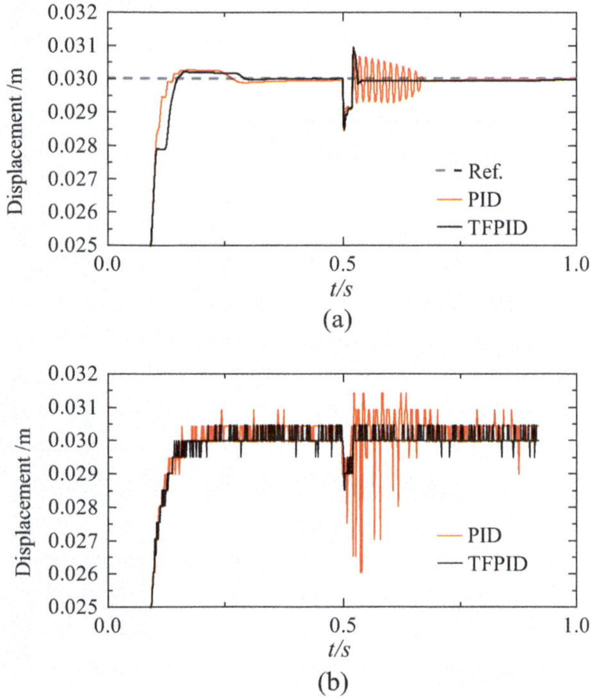

Fig. 10.16 System step response. (**a**) Simulation result. (**b**) Experimental result

Three different groups of comparison tests are performed to demonstrate the validity of the new controller design. The step responses of the two controllers are seen in Fig. 10.16, where Fig. 10.16a is the simulation result using Simulink and Fig. 10.16b is the experimental result using the test bench. A reference step signal from 0 to 0.03 m is given to the controller at the beginning and an impulse load of 10,000 N is applied at $t = 0.5$ s for 0.02 s. The step response times of the two control methods are similar, thus similar dynamic performances. However, there is a steady-state error seen in the simulation result after stepping due to the presence of the dead zone in the valve. In addition, when the high frequency oscillation occurs triggered by the impact, the system under the traditional PID control starts to oscillate and regains stability after a while. By contrast, the system restores dynamic stability immediately under the control of TFPID, thus demonstrating the robustness of the design.

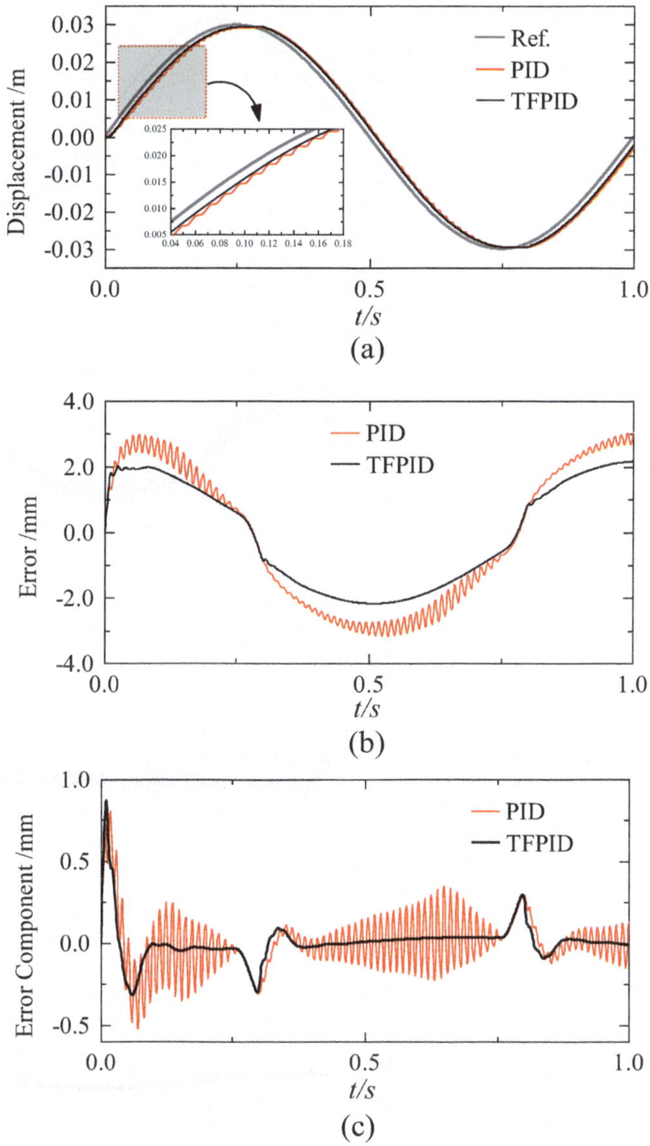

Fig. 10.17 Simulation result of sinusoidal tracking. (**a**) Results of sinusoidal tracking. (**b**) Error of sinusoidal tracking. (**c**) High frequency error component of sinusoidal tracking

The results of sinusoidal tracking are also presented as simulation and experiment in Figs. 10.17 and 10.18, respectively. Figure 10.17a gives the actual displacements of the piston in the hydraulic cylinder under the two control methods, Figure 10.17b shows the error between the actual displacement and the reference and Fig. 10.17c presents the high frequency components decomposed from the error signal.

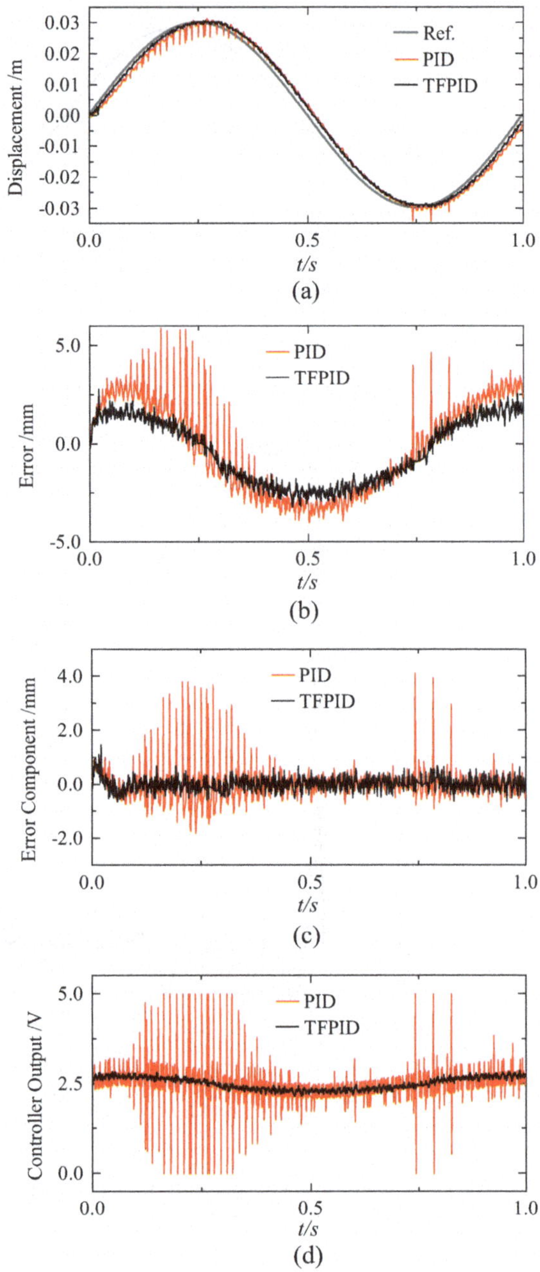

Fig. 10.18 Experimental result of sinusoidal tracking. (**a**) Results of sinusoidal tracking. (**b**) Error of sinusoidal tracking. (**c**) High frequency error component of sinusoidal tracking. (**d**) Controller output of sinusoidal tracking

As shown in Fig. 10.17a, the system can track the sinusoid faster under the TFPID control without the high frequency oscillations displayed in the one controlled by the PID. Such oscillations are manifested as system chatter as so

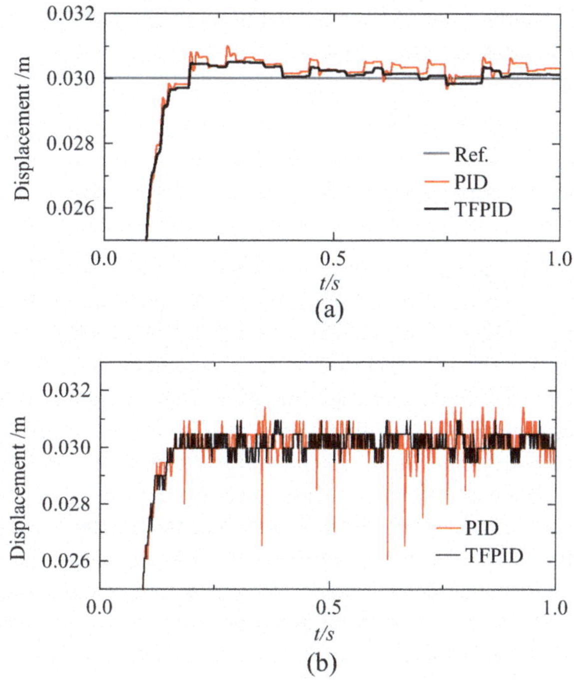

Fig. 10.19 Results of control with disturbance. (**a**) Simulation result. (**b**) Experimental result

prominently observed in Fig. 10.17b. The feasibility of the TFPID controller design is further demonstrated in Fig. 10.17c where chatter in the high frequency component is resolved. Similar observations can also be made with the corresponding experimental results found in Fig. 10.18 The controller out in Fig. 10.18d attests in unambiguous terms to the same conclusion that the TFPID controller is significantly better than the PID in giving less violent output.

A random interference signal is added to simulate disturbance. The signal is 50 Hz in frequency with an amplitude of 2% of the step applied to y. The corresponding results are given in Fig. 10.19. It is evident that the TFPID control method is able to maintain the piston close to the reference position with unremarkable transitions. This further implies that TFPID works well at anti-interference. It is more robust than the PID control to the interference of high frequency disturbance.

In overall, the P, I, and D control methods were examined to explore the feasibility of considering the frequency response of an error signal as a low and a high subband. The principle of the TFC theory was also discussed. A TFPID controller was then developed and applied to control a specific nonlinear system that is a hydraulic cylinder controlled by a solenoid proportional valve. Optimal control parameters were found using the system dynamic model derived by considering various nonlinearities including friction, dead zone, leakage, and the variable effective bulk modulus. A hardware-in-the-loop test was subsequently implemented to evaluate the performance and quality of the novel TFPID controller design. It was shown that the TFPID control demonstrates excellent performance in anti-interference, stability, and dynamic response better than that of the traditional PID controller design.

10.4 Summary

In this chapter, we briefly introduced the principle of the time-frequency control approach put forward in recent and its applications in agricultural machines. Firstly, the feature of automated agricultural machines was discussed to make clear the problem to be addressed in these actuators. It concludes that real-time analysis for feedback signals both in time and frequency domain is necessary to decide the system running status, then the controller can be specially designed to meet high control requirements of agricultural machines.

The current state of affairs of time-frequency control was reviewed and the principle and detail algorithm of the control method was described. It was argued that the discrete wavelet transform (DWT) developed in recent decades for analyzing feedback signal is feasible for conducting time-frequency control.

At last, we developed two time-frequency controller designs, namely TF-FxLMS controller and TFPID controller. The TF-FxLMS controller was used to stabilize the rotating speed of the cutting disc for sugarcane harvester, while the TFPID controller was employed to improve the dynamic performance of the electric-hydraulic system. Their control results were discussed, which demonstrated excellent performance in the respective agricultural machines considered.

References

Dassanayake, A. V., & Suh, C. S. (2007a). Machining dynamics involving whirling part I: Model development and validation. *Journal of Vibration & Control, 13*(5), 475–506.

Dassanayake, A. V., & Suh, C. S. (2007b). Machining dynamics involving whirling part II: Machining motions described by nonlinear and linearized models. *Journal of Vibration & Control, 13*(5), 507–526.

Hung, Y., Lin, F., Hwang, J., Chang, J., & Ruan, K. (2015). Wavelet fuzzy neural network with asymmetric membership function controller for electric power steering system via improved differential evolution. *IEEE Transactions on Power Electronics, 30*(4), 2350–2362.

Khan, M., & Rahman, M. A. (2008). Implementation of a new wavelet controller for interior permanent-magnet motor drives. *IEEE Transactions on Industry Applications, 44*(6), 1957–1965.

Lin, C. M., & Li, H. Y. (2012). A novel adaptive wavelet fuzzy cerebellar model articulation control system design for voice coil motors. *IEEE Transactions on Industrial Electronics, 59*(4), 2024–2033.

Lin, F. J., Shen, P. H., & Kung, Y. S. (2005). Adaptive wavelet neural network control for linear synchronous motor servo drive. *IEEE Transactions on Magnetics, 41*(12), 4401–4412.

Lin, F. J., Shieh, H. J., & Huang, P. K. (2006). Adaptive wavelet neural network control with hysteresis estimation for piezo-positioning mechanism. *IEEE Transactions on Neural Networks, 17*(2), 432–444.

Liu, M. K., & Suh, C. S. (2012). On controlling milling instability and chatter at high speed. *Journal of Applied Nonlinear Dynamics, 1*(1), 59–72.

Liu, M. K., & Suh, C. S. (2013). Synchronization of chaos in simultaneous time-frequency domain. *Applied Mathematical Modelling, 37*(23), 9524–9537.

Parvez, S., & Gao, Z. (2002). A novel controller based on multi-resolution decomposition using wavelet transforms. *Technical Papers*.

Parvez, S., & Gao, Z. (2005). A wavelet-based multiresolution pid controller. *IEEE Transactions on Industry Applications, 41*(2), 537–543.

Suh, C. S., Khurjekar, P. P., & Yang, B. (2002). Characterisation and identification of dynamic instability in milling operation. *Mechanical Systems & Signal Processing, 16*(5), 853–872.

Sun, W., Chen, Z., & Yuan, Z. (2000). Adaptive PID controller based on band-wise design using wavelet. *Acta Scien-tiarum Naturallum Universitatis Nankalensls, 33*(2), 48–52.

Tan, K. H. (2016). Squirrel-cage induction generator system using wavelet petri fuzzy neural network control for wind power applications. *IEEE Transactions on Power Electrinics*.

Tolentino, J. A. C., Silva, A. J., Velasco, L. E. R.; et al. (2012). *Wavelet PID and wavenet PID: Theory and applications*. INTECH Open Access Publisher.

Tsotoulidis, S., Safacas, A., & Mitronikas, E. (2013). Multiresolution PID control of brushless DC motor in fuel cell electric vehicles. In *International conference on renewable energy research and applications*. IEEE.

Wang, X., & Suh, C. S. (2017). Precision concurrent speed and position tracking of brushed dc motors using nonlinear time-frequency control. *Journal of Vibration & Control, 23*(19), 3266–3291.

Wu, X., Qin, J., Ma, S., Zhang, W., & Song, Z. (2019). Design and experiments of coaxial contra-rotating sugarcane base cutter with time-frequency control. *Applied Engineering in Agriculture, 35*(1), 1–8.

Wu, X., Song, Z., Du, Y., Mao, E., & Suh, C. S. (2017). A time-frequency pid controller design for improved anti-interference performance of a solenoid valve applicable to hydraulic cylinder actuation. *Vibration Testing and System Dynamics, 1*(4), 281–294.

Yang, C.-L., & Suh, C. S. (2021). A general framework for dynamic complex networks. *Vibration Testing and System Dynamics, 5*(1), 87–111.

Zekri, M., Sadri, S., & Sheikholeslam, F. (2008). Adaptive fuzzy wavelet network control design for nonlinear systems. *Fuzzy Sets & Systems, 159*(20), 2668–2695.

Zhang, Z., & Suh, C. S. (2022). Underactuated mechanical systems – A review of control design. *Vibration Testing and System Dynamics, 6*(1), 21–51.

Chapter 11
Applied Unmanned Aerial Vehicle Technologies: Opportunities and Constraints

Yongjun Zheng, Shenghui Yang, and Shijie Jiang

11.1 Introduction

Unmanned Aerial Vehicles (UAVs) are a kind of unmanned aircraft operated by radio remote control equipment and self-contained program control devices or by on-board computer completely or intermittently. With the increased demand and the advancement of technology, agricultural UAV, which is specifically configured for agricultural operation, has been developed and used extensively. Now agricultural UAVs have been gradually employed in modern farms to increase production efficiency.

This chapter will firstly demonstrate the classification of agricultural UAVs. Then, the main application field of these UAVs will be illustrated with current study examples in detail. Next, the benefits and limitations of agricultural UAVs will be analyzed and finally the development trend of the technology applied for such UAVs will be discussed with some study instances.

11.2 Classification of the UAV Configured for Agricultural Operation

As shown in Fig. 11.1, agricultural UAVs can be classified based on different considerations.

On the basis of fuselage features, agricultural UAV can be divided into fixed-wing UAV, rotor UAV, parafoil UAV, flapping-wing UAV, etc. Fixed-wing UAV and

Y. Zheng (✉) · S. Yang · S. Jiang
College of Engineering, China Agricultural University, Beijing, China
e-mail: zyj@cau.edu.cn

S. Ma et al. (eds.), *Sensing, Data Managing, and Control Technologies for Agricultural Systems*, Agriculture Automation and Control,
https://doi.org/10.1007/978-3-031-03834-1_11

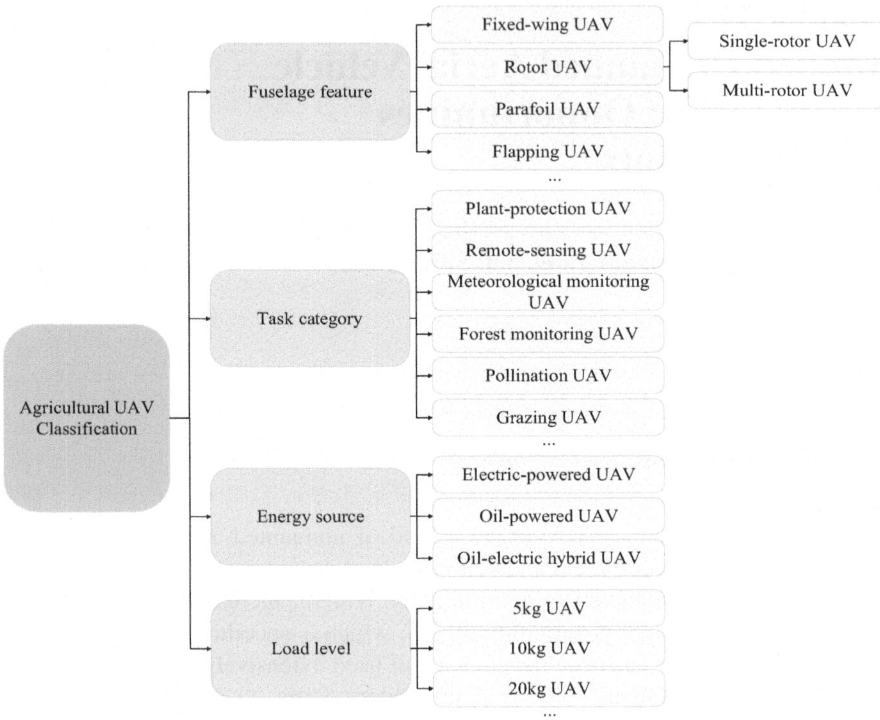

Fig. 11.1 Classification of agricultural UAV

rotor UAV are the most widely used types in agriculture. Moreover, according to the number of rotors, rotor UAV includes single-rotor UAV (or unmanned helicopter) and multi-rotor UAV. For instance, if a UAV has six rotors, it belongs to multi-rotor UAV and is called a six-rotor UAV.

In terms of task categories, agricultural UAV contains plant-protection UAV, remote-sensing UAV, meteorological monitoring UAV, forest monitoring UAV, pollination UAV, grazing UAV, etc. Plant-protection UAV, specifically for chemical spray and fertilization, and remote-sensing UAV, specifically for photographing and spectra acquisition, are commonly utilized in agriculture.

In the light of energy source, agricultural UAV consists of electric-powered UAV, oil-powered UAV, and oil-electric hybrid UAV. For electric-powered UAV, batteries, such as lithium batteries, are used as the unique energy source. This kind of UAV is eco-friendly and has low costs. Meanwhile, it has a relatively low-level request for an operation. However, small load and short endurance are its significant limitations. For oil-powered UAV, load and wind resistance are increased, but control robustness is lower with higher operation level request needed. Oil-electric hybrid UAV is a new type of UAV, integrating the advantages of the other two types, having good expectations and corresponding techniques are developing.

In consideration of load level, agricultural UAV comprises 5, 10, and 20 kg UAV, etc.

11.3 Main Applications of Agricultural UAV

11.3.1 Low Altitude Remote Sensing

Low-altitude remote sensing is a comprehensive technology of remote detection and recognition of objects and land based on the principle of electromagnetic radiation. By carrying different devices such as cameras and spectrometers, agricultural UAV can collect the images and spectra of a certain area, which are then further analyzed to obtain different parameters, such as Red-Green-Blue (RGB) bands, Normalized Difference Vegetation Index (NDVI), Leaf Area Index (LAI), and characteristic wavebands. Lastly, proper modeling approaches are selected and exploited to establish the relationship between the parameters and target output.

At present, low-altitude remote sensing in agriculture mainly focuses on crops, correspondingly containing two aspects, one, agricultural investigation and estimation, the other, the monitoring of plant diseases, pests, and natural disasters.

Agricultural Investigation and Estimation

Yield Estimation

Yield is directly related to the agricultural economy and farmer income. Distinct measures can be taken based on the result of yield estimation during crop growth. Therefore, it is greatly necessary to accurately estimate yields. Traditional estimation approaches are labor intensive, which rely on manual operation, and require people to randomly collect samples in fields.

Agricultural UAV is now employed to quantify phenotypic variances in crops by one or combinations of image or spectra features (Zhou et al., 2021). Many crops such as sugarcane, soybean, apple, and wheat have been studied and analyzed to obtain yields with common analysis methods as followed:

1. Conventional image processing: Conventional image processing involves in point-based operation and group-based operation. As listed in Table 11.1, besides Binarization and Histogram Analysis, point-based operation includes Brightness Mapping, Addictive Operation, Flip Operation, Scale Operation, Logarithm Operation, Index Operation, etc. For group-based operation, Template Convolution Operation, Filtering, Morphological Operation (expansion and corrosion), Force Field Transformation, etc. are the related spots.
2. Conventional spectral processing: Spectral technology contains Near Infrared Spectroscopy (NIR) technology and Hyperspectral Imaging Technology (HIT).

Table 11.1 Commonly used algorithms for conventional image processing

Step name	Processing algorithm	
Pre-processing	Grayscale	Float algorithm, integer algorithm, shift algorithm, average value algorithm, green only
	Binarization	Bimodal algorithm, P-parameter algorithm, iterative method, Otsu method
	Histogram analysis	
Point-based operation	Brightness mapping	
	Addictive operation	
	Flip operation	
	Scale operation	
	Logarithm operation	
	Index operation	
	Hough transform	
Group-based operation	Template convolution operation	Mx operator, my operator
	Filtering	Average value filter, maximum value filter, minimum value filter
	Morphological operation	Expansion, corrosion, filling
	Force field transformation	

NIR is an electromagnetic wave between visible light and mid-infrared light. The wavelength range is from 780 nm to 2526 nm. The main source of NIR is the absorption of frequency doubling and frequency combining of the vibration of hydrogen-containing groups (C-H, N-H, H-O). The reflection information has the composition and molecular structure of the majority of the organic compounds.

HIT is a combination of imaging technology and spectral detection. It forms dozens or even hundreds of narrow bands for each spatial pixel through dispersion to achieve continuous spectral coverage. At the same time, the spatial characteristics of targets are imaged. Thus, spectral information can reflect the internal structure of samples (e.g. molecular composition and quality components), while image information can reflect the external quality characteristics of samples (e.g. size, shape, and defects).

The spectral processing method is the same for both spectroscopy and hyperspectral imaging technologies. . The process flows in general are: pre-processing for noise reduction, dimension reduction, and sample set partition followed by model development and evaluation indicator establishment. For each step, the widely used algorithms are shown in Table 11.2.

3. Artificial intelligence (AI) algorithms: With the improvement of AI algorithms, neural networks and machine learning are becoming more popular due to their better accuracies, such as deep learning (DP), PLSR, SVM, Decision Tree (DT), Random Forest (RF), and Gradient Boosting Regression Tree (GBRT).

Table 11.2 The widely used algorithms for spectral processing

Step	Processing algorithm	
Pre-processing	S-G smoothing, standard normal variate (SNV), multiplicative scatter correction (MSC), first derivative, second derivative, wavelet transform (WT), min-max normalization, z-score normalization	
Dimension reduction	Principal component analysis (PCA), uninformative variable elimination (UVE), minimum noise fraction (MNF), successive projections algorithm (SPA), competitive adaptive reweighted sampling (CARS)	
Sample set partition	Random sampling (RS), Kennard-stone (KS), sample set partitioning based on joint x-y distance (SPXY)	
Model development	Classification	Partial least squares discrimination analysis (PLSDA), support vector machine (SVM), linear discriminant analysis (LDA), artificial neural network (ANN), K-nearest neighbor (KNN)
	Regression	Partial least squares regression (PLSR), multiple linear regression (MLR), principal component regression (PCR), support vector regression (SVR)
Evaluation indicator establishment	Classification	Correct classification rate (CCR), kappa coefficient
	Regression	Determination coefficient (R^2), root mean square error (RMSE), reactive plasma deposition (RPD)

Generally, regardless of which methods, the core of yield estimation is to find an appropriate relationship between image/spectrum features and target yields. The following is an example of apple yield estimation based on image processing by neural networks (Li et al., 2021).

Example 11.1: Yield Estimation Method of Apple Tree Based on Improved Lightweight YOLOv5

1. Hardware and experiments: As shown in Fig. 11.2, the experiment was conducted in Yantai, Shandong Province, China (N 37°16′, E 120°64′). A quad-rotor UAV was used to carry (1) a camera and a Raspberry Pi Camera V2 for image acquisition and (2) a Raspberry Pi for image storage. The flight altitude of the UAV was 1.5 m and the distance from the UAV to apple trees was 1.2 m. Images of the apples in the first day, eighth day, and fifteenth day after coloration were collected in different light conditions. White background was used to reduce noise. Meanwhile, the apples from the sampled trees were weighed when harvested.

2. Software and data processing: Fig. 11.3 shows the procedure of image processing. First, data cleaning was conducted to reduce the influence from repeated and fruitless images. Then, the dataset containing all kinds of apple tree images was divided into six subsets: front-lighting, side-lighting, and back-lighting condi-

Fig. 11.2 The UAV and the experiment setup for apple yield estimation. *1*. Raspberry Pi Camera V2. *2*. Raspberry Pi. *3*. Quad-rotor UAV. *4*. White background *5*. Apple trees. (**a**) The UAV configured for apple yield estimation. (**b**) The test field. (**c**) The test scheme

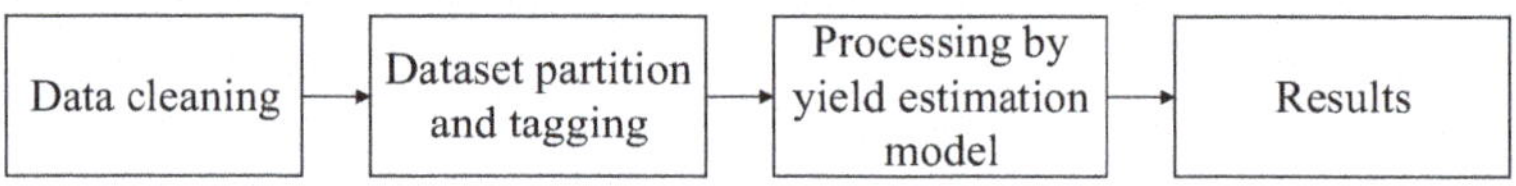

Fig. 11.3 The procedure of image processing

tion with and without white background, respectively. Each subset had 100 images and the target apples were labeled by rectangles in LabelImg software manually to obtain coordinates and species. Next, the labeled datasets were used as the input of the yield estimation model to estimate the final yield.

As shown in Fig. 11.4a, the yield estimation model included two modules, Fruit Recognition Module and Yield Fitting Module. For the former, YOLOv5 network was utilized and optimized. Fig. 11.4b shows the optimization, which was marked by the red squares. Pooling Block Attention Module (PBAM) was specifically developed and added into the network with the Depthwise-Pointwise-Batch_normalisation-Leaky_relu (DPBL) unit to deal with both large network parameters and lacking attention preference. Then, the number of fruits and the area of each rectangle tab were obtained as the input of Yield Fitting Module. Deep neural network (DNN), shown in Fig. 11.4c, was exploited as Yield Fitting Module due to the nonlinearity between the input layer parameters and actual yield. Three fully connected layers, a ReLU layer and a Sigmoid layer were added to train the model between the input and the actual yield.

3. Results: Fig. 11.5 shows the recognition of apples, indicating that the Fruit Recognition Module had a performance of more than 0.85 accuracy. Different light conditions with and without background could be adapted. Furthermore, Table 11.3 demonstrates that without the white background, the errors were less than 13% and with the white background, errors could be further reduced to less than 7%. This proposed approach showed a UAV application for in-site apple yield estimation, which is able to reduce manual labor and increase work efficiency.

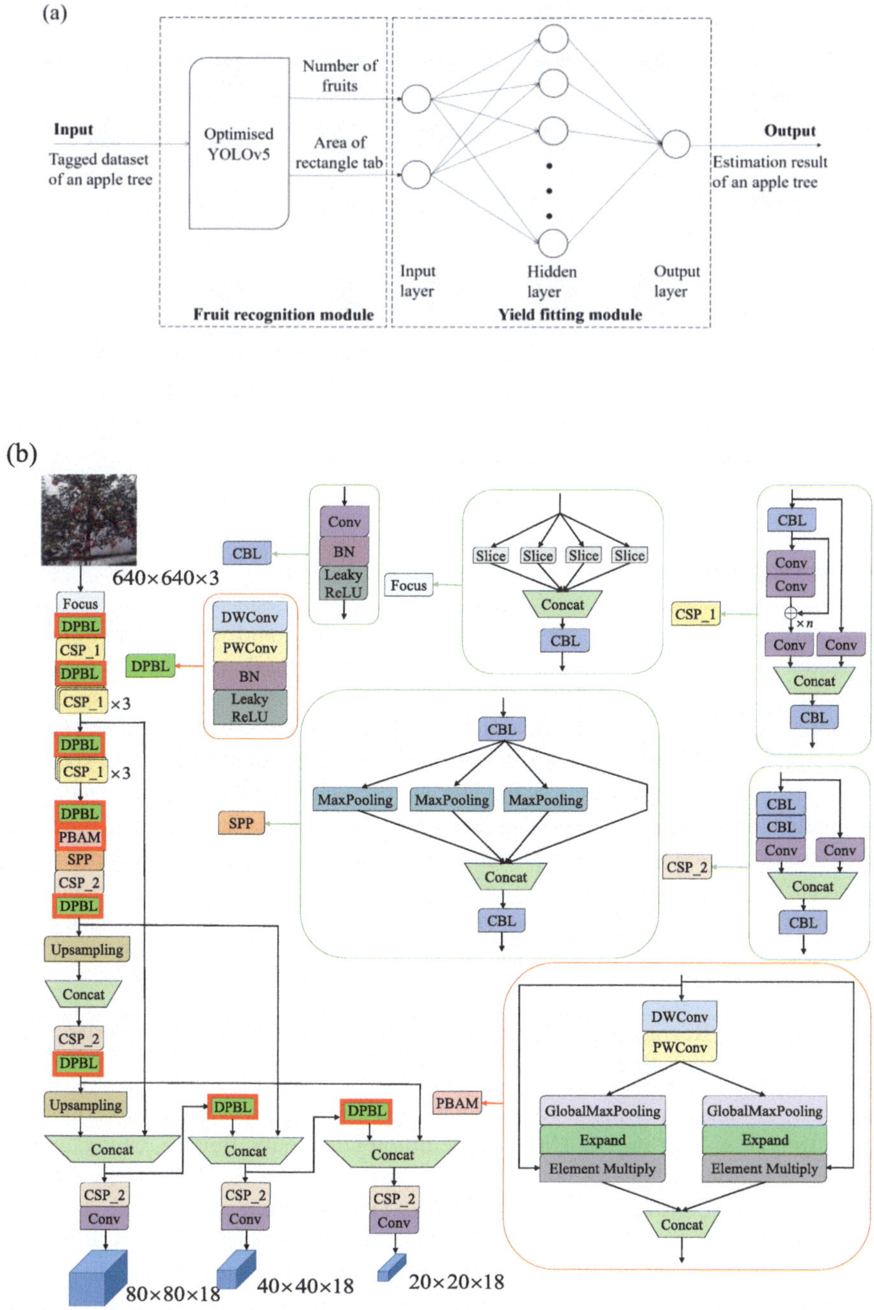

Fig. 11.4 The structure of the apple yield estimation model. (**a**) The entire structure of the yield estimation model. (**b**) The optimized YOLOv5 network. (**c**) The structure of Yield Fitting Module

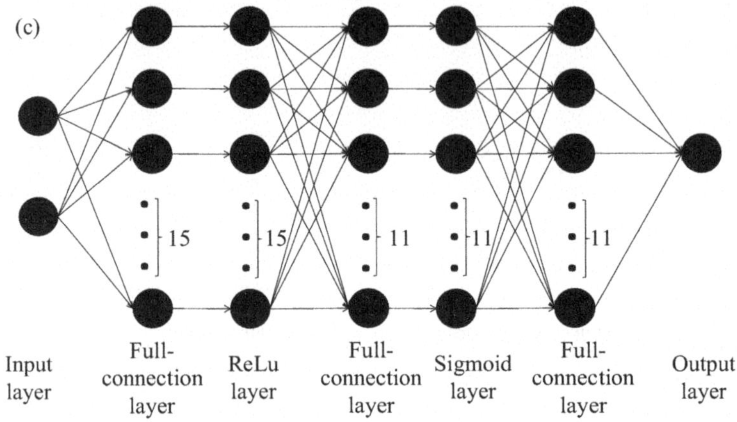

Fig. 11.4 (continued)

Nutrient Estimation

Nutrient content, which can be represented by nitrogen and water, is an important indicator for crop growth. Different quantities of nutrient determine different measures needed. Agricultural UAVs can be utilized to collect images or spectra of crops to estimate the nutrient content. Global Navigation Satellite System (GNSS) is generally employed to record the position of the crops in their corresponding areas. For ground data collection, the nutrient content from the crops was measured manually such as by using plant nutrient sensors. Then, images or spectra data from UAV are processed to find out certain significant features. At last, a proper model between the features and nutrient content can be established for future analysis. This method is non-destructive and will not require chemical analysis to save labors and time. Example below shows a research of nutrient estimation in corn (Wu et al., 2020).

Example 11.2: Recognition Method for Corn Nutrient Based on Multispectral Image and Convolutional Neural Network

1. Hardware and experiment: As shown in Fig. 11.6, the experiment was conducted in Tongzhou Test Station, China Agricultural University, Beijing, China, and the area was divided into 18 small sections (from N1-1 to N5-3 and from CK-1 to CK-3). In addition, a quad-rotor UAV, DJI Phantom, was used. A Global Positioning System (GPS) module and a Downwelling Light Sensor (DLS) were mounted to record position, flight altitude and flight speed during image data collection. Moreover, RedEdge-M was used as the multispectral camera (8 cm/pixel if the height is 120 m), while a YLS-D plant nutrition tester was utilized for ground data collection of nitrogen and leaf moisture. During the experiment, UAV was flown over all the small sections to collect image data with a flight altitude of about 20 m.

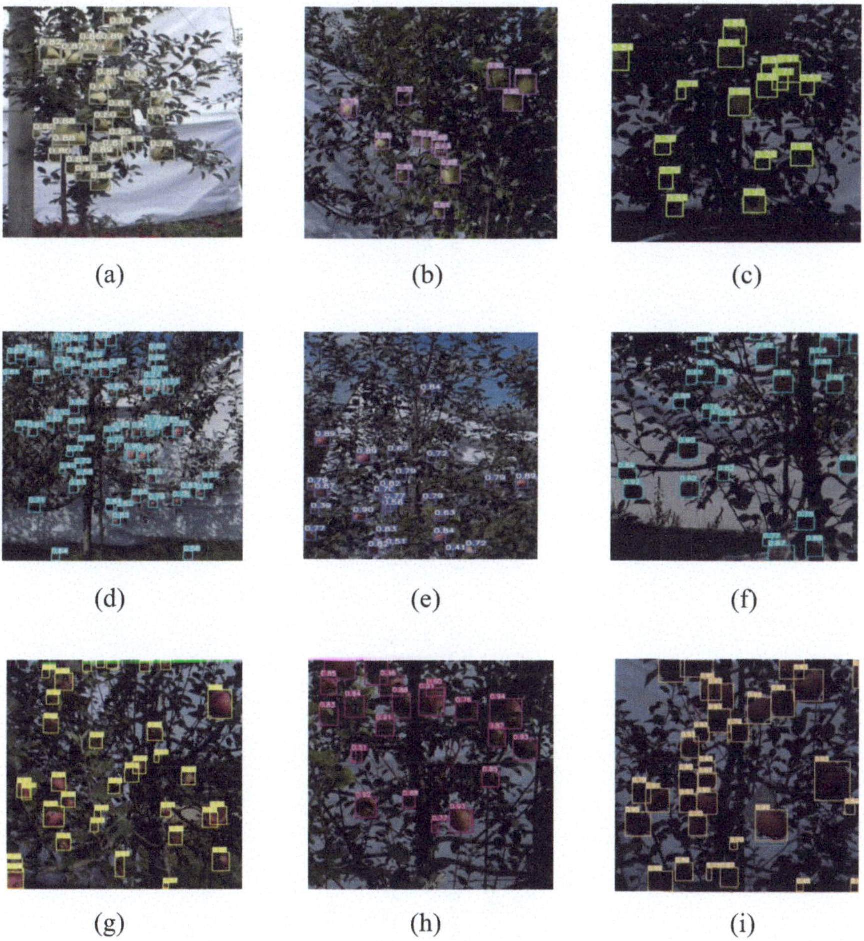

Fig. 11.5 The recognition of apples by the Fruit Recognition Module. (**a**) First day after coloration in the front-lighting condition. (**b**) First day after coloration in the side-lighting condition. (**c**) First day after coloration in the back-lighting condition. (**d**) Eighth day after coloration in the front-lighting condition. (**e**) Eighth day after coloration in the side-lighting condition. (**f**) Eighth day after coloration in the back-lighting condition. (**g**) Fifteenth day after coloration in the front-lighting condition. (**h**) Fifteenth day after coloration in the side-lighting condition. (**i**) Fifteenth day after coloration in the back-lighting condition

The configuration of the used computational station was: an Intel (R) Xeon CPU E5-2620 v4 @2.10GHz, a 256GB memory, a NVIDIA Tesla K40C graphic card with the operation system Ubuntu16.04.

2. Software and data processing: Based on the measurement from the nutrition tester and actual growth, the experiment area was classified into three grades (Table 11.4). Each grade had 800 five-channel multispectral images (4000

Table 11.3 Relative error with and without the white background

	Group 1	Group 2	Group 3	Group 4	Group 5	Group 6	Group 7	Group 8	Group 9	Group 10
Relative error δ/% with the white background	3.05	4.25	−4.17	3.75	−5.81	4.10	−6.13	−5.39	5.41	4.28
Relative error δ/% without the white background	−10.34	9.72	12.15	9.83	11.13	−10.54	−12.71	8.28	11.67	−9.37

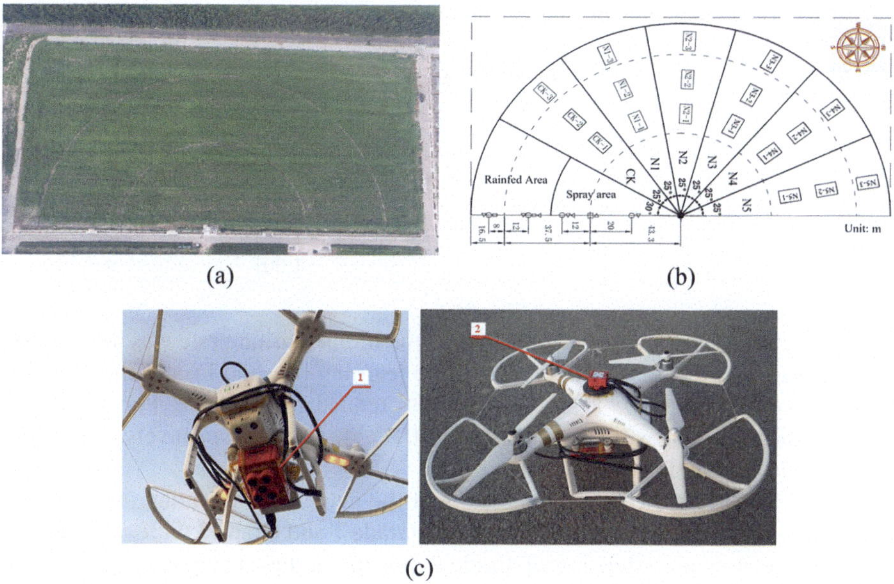

Fig. 11.6 The experiment site, UAV, and sensors used in the corn nutrient estimation experiment. (**a**) The test area. (**b**) The partition of the test area. (**c**) The UAV, DJI Phantom, and the sensors used in the tests. *1*. Multispectral camera *2*. GPS module and DLS

single-channel images in total). A total of 530 images were randomly selected from these 800 images, including 480 images as the training set and 50 images as the validation set. Furthermore, the images of Red, Green, and Blue (RGB) channels were composited to be color pictures, so that a color picture library was established.

ResNet18 network was used to estimate the corn nutrients. As shown in Fig. 11.7, it had four kinds of different ResNet modules and their output channels were 64, 128, 256, and 512, respectively. The output size of the images from the input layer was 224 × 224 × 5 (for five-channel multispectral images) or 224 × 224 × 3 (for RGB color pictures). A pooling layer, "pool" in Fig. 11.7, was for maximum pooling with the step length of 2, while the other pooling layer, "spatial_avg" in the figure, was for global average pooling. Finally, a full-connection layer, "fc 3" in Fig. 11.7, was used to match the network output and labels. Before training the network, each pixel of the images was normalized from [0,255] to [−1,1] by Eq. (11.1). TensorFlow1.3.0 was exploited as the training environment.

$$x_{\text{new}} = \frac{x}{255 \div 2} - 1 \tag{11.1}$$

where x = pixel of image

Table 11.4 The classification of the test area based on the information from the YLS-D plant nutrition tester

Grade	Area	Nitrogen (mg/g)	Leaf moisture (g/cm^2)
0	CK	14.76	1.062
0	N3	14.67	1.091
1	N1	15.09	1.095
1	N2	15.01	1.141
1	N4	15.19	1.109
2	N5	16.19	1.001

3. Results: In terms of color pictures, when the initial learning rate was 0.03, batch size was 6, and epoch number was 4, the learning rate and loss of the network performed better (Fig. 11.8a, b). The loss tended to be stable. For 5-channel multispectral images, these values were 0.05, 4, and 4, respectively, for stable loss (Fig. 11.8c, d).

Table 11.5 shows the validation results of the network, illustrating that the network performed well on the identification of Grade 1 and Grade 2 with ≥94% accuracy. The error of identification of Grade 0 was the highest because the growth of the corn in the area of CK and N3 was diverse with various nitrogen distributions. In general, the average accuracy of the proposed approach was more than 80%, indicating its potential for further application.

Figure 11.9 shows the performance of the proposed approach in testing data comprised of 208 5-channel images. All the information (i.e. image number, classified results with its corresponded nutrient content, and position) could be stored in a text file (.txt file). According to the classification result, the accuracy was about 80.3% (167/208 correctly classified). This study indicated that agricultural UAVs can be used to analyze the nutrient content of crops and the results are promising. This application would reduce both costs and labor as well as improve operation efficiency.

Surveying and Mapping

Surveying and mapping are important applications in agriculture. Agricultural UAVs can monitor and survey terrain, crop height, and affiliated facilities in fields. Sensors such as millimeter wave radar (MWR) and Lidar (light detection and ranging) are commonly used for long distance measurement. These sensor data usually is in the form of point clouds with huge amount and hence extracting the exact effective information using different processing is required. Several methods employed to process point cloud data are as followed:

1. Filtering: This step is required as the first step to remove the noises (outliers) in point clouds. Typical filtering methods include Bilateral Filtering, Gaussian Filtering, Conditional Filtering, Direct Filtering, Random Sampling Consensus Filtering, Voxelgrid Filtering, Kalman Filtering, etc.

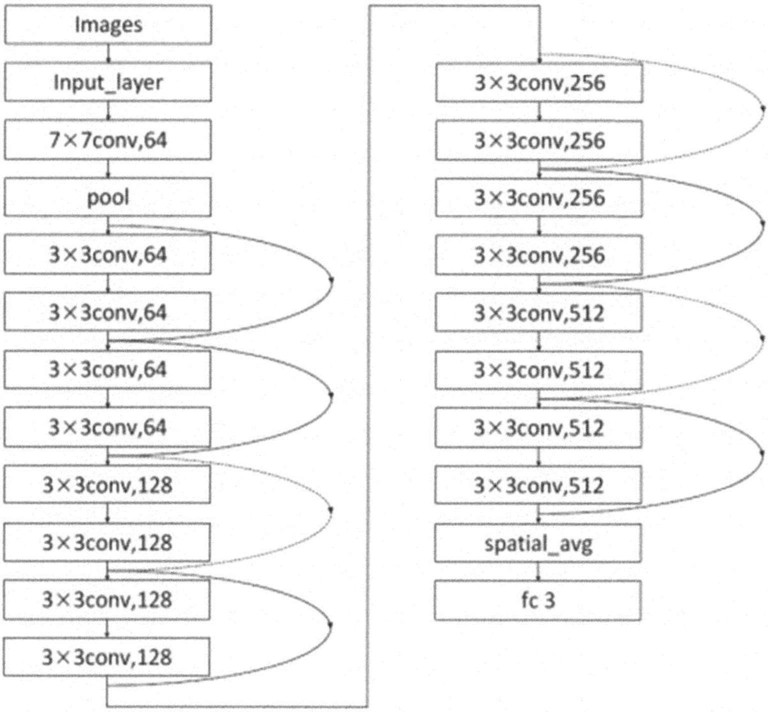

Fig. 11.7 The structure of the ResNet18 network established in the study, where "7 × 7conv,64" means 64 7 × 7 × 5 (for multispectral images) or 7 × 7 × 3 (for color pictures) convolutional kernels with the step length of 2 and "3 × 3conv, 64," "3 × 3conv, 128," "3 × 3conv, 256," and "3 × 3conv, 512" have the similar meaning

2. Key point extraction: After filtering, the extraction of key points can greatly decrease the amount of data, subsequently reducing the computational costs. Common algorithms comprise Intrinsic Shape Signature (ISS), Normal Aligned Radial Feature (NARF), Harries, Scale-Invariant Feature Transform (SIFT), Kaze, etc.
3. Feature description: In addition to the position of key points, other features to describe the three-dimensional characteristics of point clouds are generally demanded too, which include curvature, normal and texture features. The common approaches consist of Eigenvalue Analysis, Curvature and Normal Calculation, Point Feature Histogram (PFH), 3D Shape Context, etc.
4. Segmentation and classification: Point clouds can be segmented and classified based on the key points and features. In terms of segmentation algorithms, it is popular to use K-means, Hough Transform, Normalize Cut (Context based), Region Growing, Random Sampling Consensus Extraction (RANSAC), Global Plane Optimization (GPO), Connectivity Analysis, etc. For classification methods, point-based, segmentation-based and deep-learning-based algorithms are utilized in most cases.

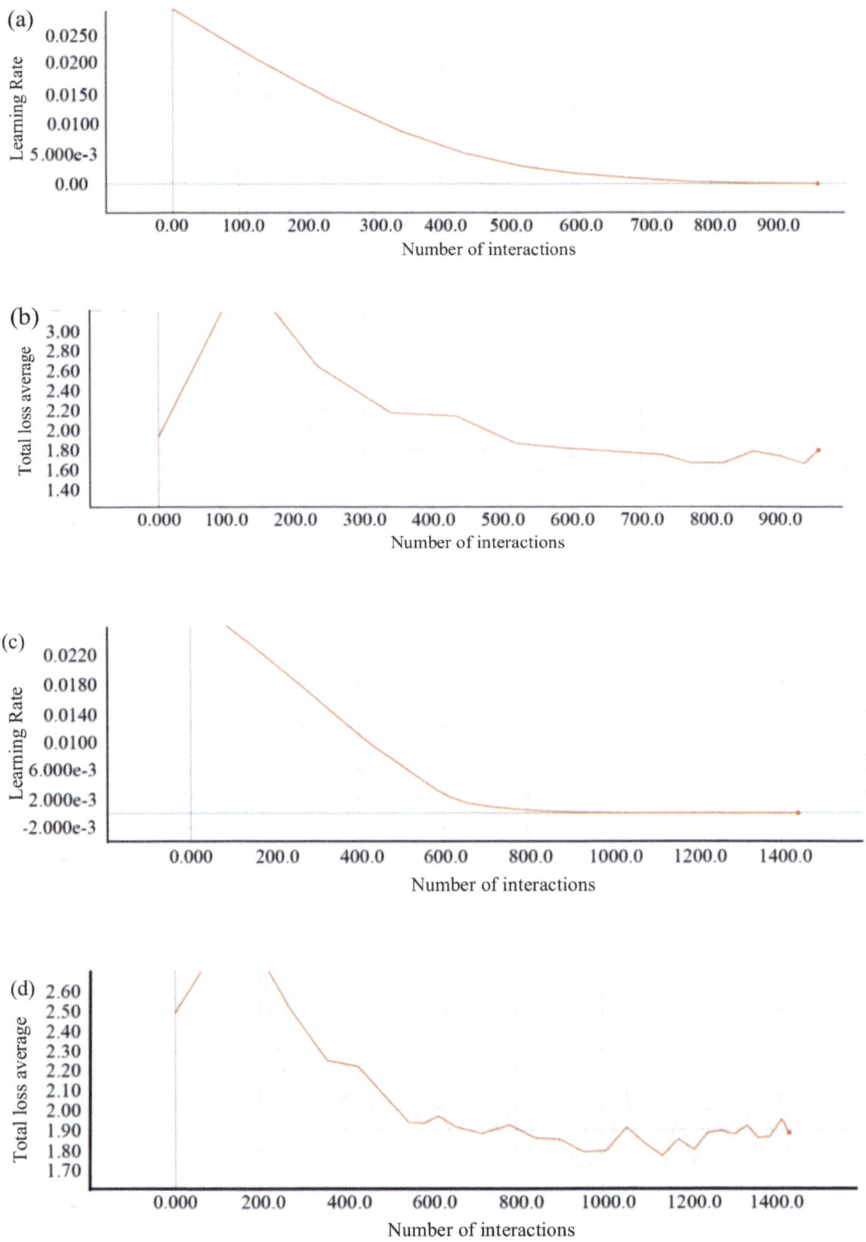

Fig. 11.8 The learning rate and loss of ResNet18 during training. (**a**) The learning rate of color pictures during training. (**b**) The loss of color pictures during training. (**c**) The learning rate of multispectral images during training. (**d**) The loss of multispectral images during training

Table 11.5 The validation results of the network

Grade	Quantity	Number of times identified as Grade 0	Number of times identified as Grade 1	Number of times identified as Grade 2	Accuracy (%)
0	50	40	4	6	80
1	50	3	47	0	94
2	50	1	0	49	98

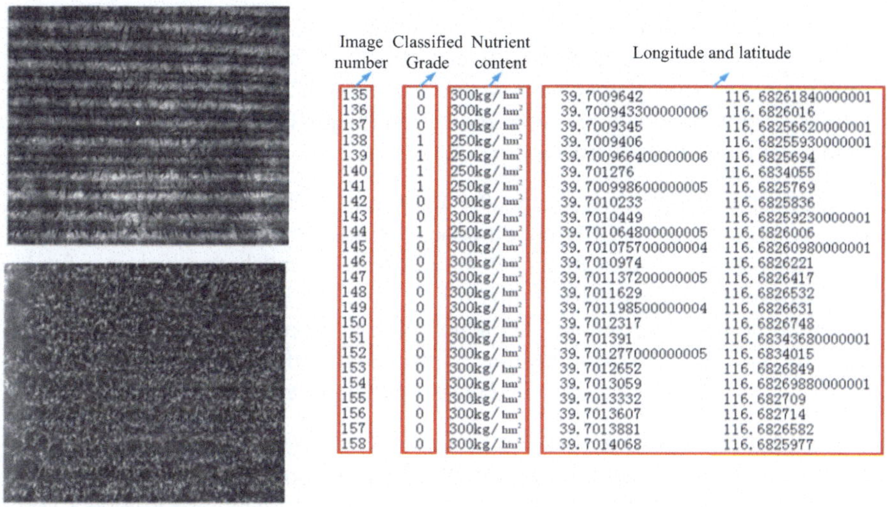

(a) (b)

Fig. 11.9 Proposed approach tested on testing set and their classification results. (**a**) The collected images. (**b**) The classification results

5. Point cloud registration: Point cloud registration contains coarse registration and fine registration. Coarse registration is the registration of point clouds in the condition that the relative position and attitude of point clouds are completely unknown. It can provide a good initial value for fine registration. The current common algorithm of automatic coarse registration includes exhaustive-search based (e.g. 4-Point Congruent Set) and feature-matching based (e.g. PFH) algorithms. Conversely, fine registration is to minimize the spatial difference between point clouds on the basis of coarse registration. Iterative Closest Point (ICP) and its varieties (e.g. robust ICP, point to plane ICP, and point to line ICP) are the most widely used algorithms for fine registration.
6. SLAM optimization: Simultaneous Localization and Mapping (SLAM) is now a trending discussion topic. There are many developed SLAM methods, i.e. ICP, Likelihood Field, Normal Distribution Transform (NDT), Gaussian Fields, and Power Iteration Clustering (PIC). Meanwhile, many libraries such as Ceres

(least square optimization library by Google) and General Graph Optimization (g2o) have been developed for subsequent optimization.

7. Three-dimensional reconstruction: This is an approach to reproduce a scene based on point clouds. Several approaches have been developed to perform reconstruction such as Poisson Reconstruction and Delaunay triangulation. The structure of a scene can be divided into multiple layers to be recognized and reconstructed according to the geometric characteristics (point, line, and plane). In agriculture, 4D real-time reconstruction of crops or fields (including stereo coordinates and time) is frequently implemented.
8. Point cloud data management: After acquiring point cloud data and for storage, they need to be managed efficiently and effectively so that the users can search and inquire existing point clouds whenever required. The usual methods adopted include point cloud compression, point cloud index (e.g. K-D tree and Octree), and point cloud Levels of Details (LOD).

After point cloud processing, specific algorithms will be developed to use these extracted point clouds for different purposes. In order to simplify the processing of point clouds, Point Cloud Library (PCL) has been established, which is a fundamental tool for point cloud processing.

In addition, high-throughput 3D phenotyping of crops is another aspect of surveying by agricultural UAVs. Similar to the yield and nutrient estimation, crop images are collected and then different measures below can be taken to process them:

1. Image mosaic: Image mosaic is an image processing method to stitch multiple images into a single large image. Examples of image mosaic comprise Digital Surface Model (DSM) for trees and buildings and Digital OrthophotoMap (DOM) for farmland. Image features can be further recognized and extracted after image mosaic.
2. Conventional image/spectrum processing: This method is similar as stated in Yield Estimation section, which the crop image features are extracted and relate with the response variables (e.g. plant responses).
3. Structure from Motion: Structure from Motion (SFM) is an offline algorithm based on unordered images. Static targets, especially crops and trees in agriculture, can be reconstructed in three dimensions if images cover the targets with sufficient overlaps.
4. Artificial intelligence algorithms, as mentioned in Yield Estimation section.

The following example demonstrates the wheat growth reconstruction (Wang, 2019).

Example 11.3: Wheat Growth Surveying Based on the Lidar Point Clouds Collected by a UAV

1. Hardware and experiment: Fig. 11.10 shows the devices and the experiment sites for the surveying of wheat growth. A quad-rotor UAV carrying an Inertial Measurement Unit (IMU) for position data, a RPLidarA2 laser ranger, and a GPS module was utilized to collect data at two wheat fields on the west campus,

China Agricultural University, Beijing, China. The UAV flew along the central line of the fields by manual control with flight altitude 2 m. The data collection lasted less than 10 min.

2. Software and data processing: Firstly, a position-velocity equation based on GPS and IMU was developed (Eq. 11.2). Secondly, closed-loop correction of centralized Kalman Filter was used to solve the decline of navigation accuracy and enhance the robustness of the system for long-time working.

$$Z(t)=\begin{bmatrix} \lambda_{INS}-\lambda_{GPS} \\ L_{INS}-L_{INS} \\ h_{INS}-h_{INS} \\ v_{\mathrm{eINS}}-v_{\mathrm{eGPS}} \\ v_{\mathrm{nINS}}-v_{\mathrm{nGPS}} \\ v_{\mathrm{uINS}}-v_{\mathrm{uINS}} \end{bmatrix}=H(t)X(t)+V(t) \tag{11.2}$$

Furthermore, the Noise Variance Matrix and the Measurement Noise Variance Matrix were optimized by Particle Swarm Optimization (PSO) and the Fitness Function was developed as shown in Eq. (11.3):

$$S=Var\left(\mathrm{X}_{Lo}\right)+Var\left(\mathrm{X}_{La}\right)+Var\left(\mathrm{X}_{Ev}\right)+Var\left(\mathrm{X}_{Nv}\right) \tag{11.3}$$

As demonstrated in Fig. 11.11, in order to calculate in the same coordinate system (real world coordinates), the three-dimensional coordinates of point clouds were transformed from the Lidar to the IMU followed by the GPS. Next, Morphological Opening Operations (MOO) were performed (corrosion followed by expansion) to remove non-surface points and only keep the surface points. Finally, Nearest Neighbor Interpolation (NNI) was conducted for reconstruction using MATLAB after MOO and due to the requirements of accuracy, re-description, and calculation efficiency.

3. Results: Fig. 11.12 and Table 11.6 present the reconstruction result of the wheat fields, indicating that the growth estimated by the UAV had a small error (maximum 8.2%), which may be caused by the effect from wind speed and manual remote control. This result indicated that reconstruction by UAVs in agriculture is able to get accurate results in shorter time and less labors are needed.

Breed Estimation

Accurate breed estimation is a hard point in agriculture, while remote sensing by agricultural UAVs can deal with it. Different breeds may have different graphic and spectral characteristics. Hence, hyperspectral analysis is one of the effective approaches, having two steps: spectral processing and image texture extraction.

Fig. 11.10 The equipment and the experiment sites for the wheat growth reconstruction experiment. (**a**) The devices used in the test. (**b**) RPLidarA2 used for the test. (**c**) Test area 1. (**d**) Test area 2. 1. The quad-rotor UAV 2. The IMU 3.RPLidarA2 4.GPS module

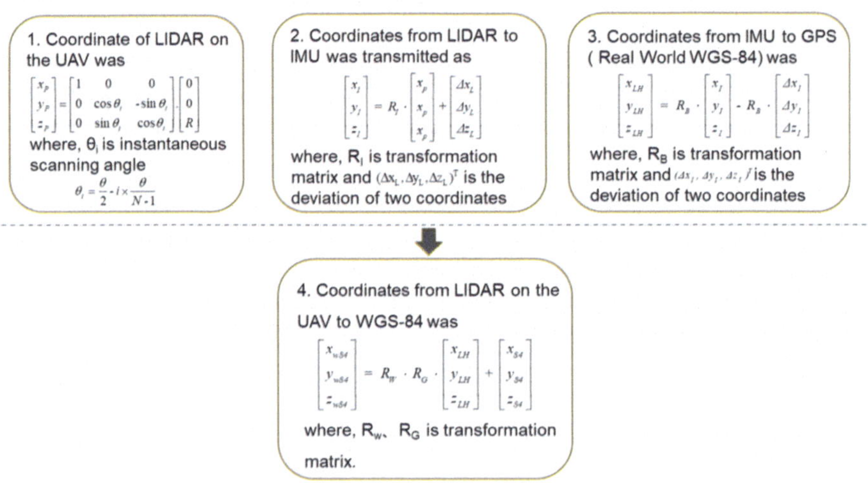

Fig. 11.11 Coordinate Transform from the Lidar to the IMU then to the GPS

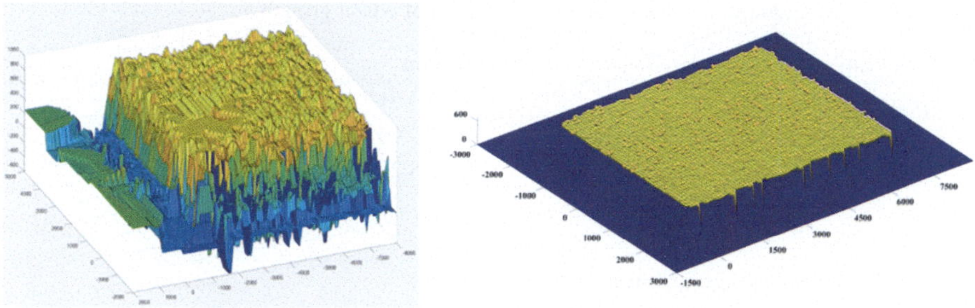

(a) (b)

Fig. 11.12 The reconstruction results of the both wheat fields 1 and 2. (**a**) The reconstruction result of test area 1. (**b**) The reconstruction result of test area 2

In terms of spectral processing, as mentioned in Table 11.2, filters can be applied to divide original spectra into different subranges. Then, sensitive wavebands can be found out by the combination of several algorithms, such as Canonical Correlation Analysis (CCA), Principal Component Analysis (PCA), Clustering, Envelope Division (ED) and Regression Analysis (RA).

Meanwhile, image texture information can be extracted by using the methods in Table 11.7.

Finally, the breed recognition model between the sensitive wavebands with the image textures and breeds is able to be established.

Monitoring of Plant Diseases, Pests, and Natural Disasters

Plant diseases, pests, and the impact of natural disasters must be minimized to ensure high yield of produce and for personnel safety. Frequent monitoring is necessary for early warning and prevention. Agricultural UAVs have been employed for this monitoring in fields, which is capable of decreasing labor costs and increasing work efficiency and accuracy.

Similar to monitoring crop changes, images and spectra data are the main data used to recognize the variations within a certain area, due to distinct reflections and features of different conditions. Thus, the methods mentioned in Table 11.1 and Table 11.2 and artificial intelligence algorithms (neural networks and machine learning) are also suitable for this type of application. An example of monitoring and classification of Huanglongbing (HLB, a kind of citrus disease) by the hyperspectral data from agricultural UAVs is detailed (Lan et al., 2019b).

Example 11.4: Monitoring and Classification of Citrus Huanglongbing Based on UAV Hyperspectral Remote Sensing

1. Hardware and experiment: As shown in Fig. 11.13, the experiment was conducted in Luobo County, Huizhou, Guangzhou Province, China (N23°29′57.81″—

Table 11.6 The measurement results of reconstruction of wheat growth

Target wheat	Value/m
Real max height/min height	0.78/0.61
Reconstructive max height/min height	0.81/0.66
Error	3.8%/8.2%

Table 11.7 Typical texture extraction methods

Method category	Typical method name
Statistics	Gray level co-occurrence matrix (GLCM)
Geometrical feature	Chessboard feature
	Structure feature
Model	Random field model
	Fractal model
Signal processing	Wavelet transform (WT)
	Autoregressive texture feature
	Tamura texture feature
Structural analysis	Syntactic texture description
	Mathematical morphologic

N23°29′59.31″, E114°28′8.39″—E114°28′12.26″). The citrus variety was Citrus Shatangju. The row spacing was 4 m and the column interval was 2.5 m. A Nano-Hyperspec spectrometer was mounted on a six-rotor UAV, DJI Matrice 600 Pro to collect UAV hyperspectral data, while an ASD FieldSpec HandHeld spectrometer was used for the ground data acquisition. A total of 30 trees were selected as healthy tree and the spectral data of three leaves of each tree were collected. For HLB-infected trees, all the infected trees were selected with the spectral data of three leaves for both apparent and inapparent symptoms were acquired, respectively. In addition, the level of HLB infection was determined by indoor lab tests. The region of interest (ROI) of both the healthy and infected leaves was established.

2. Software and data processing: First, the relative spectral reflectance of the citrus canopies from the UAV collected data was calculated by Eq. (11.4)

$$\rho_1 = \frac{\mathrm{DN}_1}{\mathrm{DN}_2}\rho_2 \tag{11.4}$$

where DN_1 and DN_2 were the radiation brightness of canopies and the calibration board, respectively; ρ_1 was the relative spectral reflectance of canopies and ρ_2 was the spectral reflectance of the calibration board.

Fig. 11.13 Experiment site and flight route of DJI Matrice 600 Pro

There were 479 samples in total and the ratio of the modeling set to the validation set was 3:1. Then, the raw data was processed following the procedures illustrated in Fig. 11.14, including pre-processing, feature extraction, and modeling. For modeling, conventional machine learning methods comprising K-Nearest Neighbor (KNN) and SVM for classification. For KNN, Euclidean Distance and Cosine Similarity were performed as the discriminative basis, while in SVM, linear kernel function, radial basis function (RBF) kernel and polynomial kernel function were tested for comparison.

3. Results: KNN had a low accuracy to determine original spectra and inverse-logarithm spectra (about 8.40%). For SVM, the accuracy of the function with secondary kernel was 94.7%. In terms of overall effect, the accuracy was more than 90%, suitable for precise determination of HLB.

11.3.2 Plant Protection

Representative Enterprises and Types

The number of enterprises in developing agricultural UAVs is increasing worldwide and some typical ones are introduced as below:

DJI

DJI is a famous UAV company in China. It starts from consumer UAVs and then expands to multi-application fields. It has occupied more than 80% of the global UAV market and more than 70% of the Chinese market so far. At present, according to DJI official website, it has several products for plant protection such as MG-1p, T10, T16, T20, and T30 (latest).

As shown in Fig. 11.15 and Table 11.8, T30 is a kind of electric-powered six-rotor UAV. Its intelligent route mode can plan routes independently prior to each operation. Meanwhile, T30 uses a novel targeting technology, which can adjust six

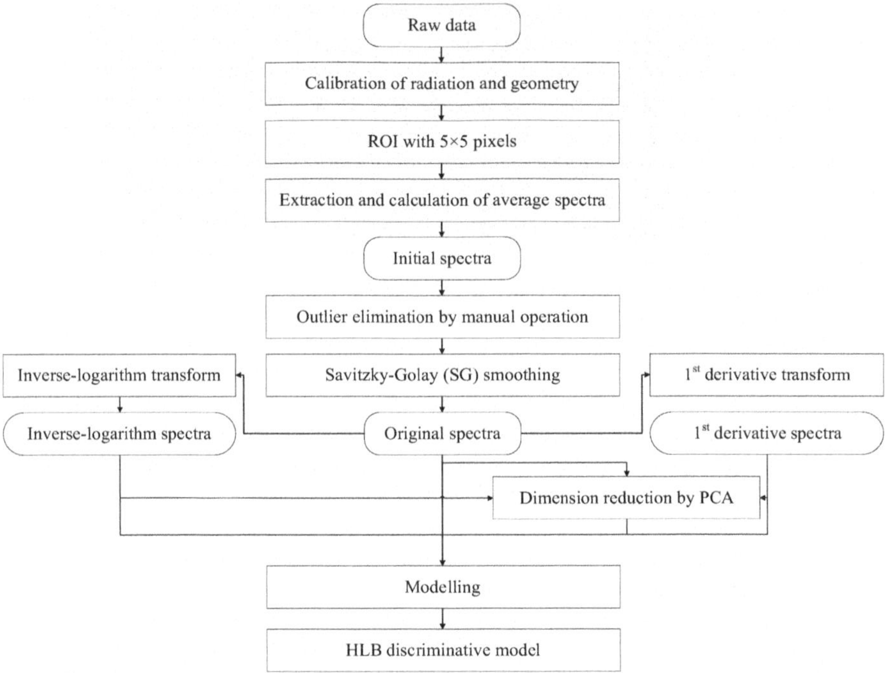

Fig. 11.14 The procedure flow of hyperspectral data processing

arm angles and make spray droplets penetrate through thick canopies at an oblique angle to ensure the uniform adhesion of liquid spray. Moreover, 16 nozzles are installed under the main body for spray application including pesticides, herbicides, and fertilization.

XAG

XAG is another well-known UAV company in China founded in the year 2007. Since 2013, it began to have agricultural UAV and products have been marketed to more than 40 countries and regions around the world with research and development centers and experimental bases in many countries. They have two series types of products: P-type as the main product (P80 as the latest product) and V-type as the new product in the year 2021. P-type is the main product, while V-type is a kind of new product in 2021.

As shown in Fig. 11.16 and Table 11.9, P80 is an electric-powered four-rotor UAV with the load capability of 40 kg. It is equipped with a new intelligent control system (Superx4) and can be used for spraying and surveying. Meanwhile, it can follow AI prescription maps to spray.

Fig. 11.15 T30 plant-protection UAV

Table 11.8 Major specification of T30

Name	Value/unit
Size	2858 mm × 2685 mm × 790 mm
Weight	26.3 kg
Maximum power consumption	13,000 W
Maximum speed	7 m/s
Maximum starting altitude	2000 m
Tank volume	30 L

YAMAHA

YAMAHA is a company renowned for agricultural UAV products. It has developed unmanned helicopters as the main products for a long time since the downwash of unmanned helicopters is steadier than that of multi-rotor UAVs. One example of their unmanned helicopters is from the Fazer-series, which has been marketed for about 30 years. In recent years, YAMAHA has also gradually launched multi-rotor UAVs for agricultural purposes with the representative product as YMR-08.

According to YAMAHA official website, FAZER-R as indicated in Fig. 11.17 is the latest product with its major specification listed in Table 11.10. It is an oil-powered UAV with 32 kg workload and can spray 4 hectares of farmland for each flight. Due to its flexible flight control system, FAZER-R can be adjusted to different workloads for different missions.

Meta Robotics

Meta Robotics is a robot enterprise located in Yishan, Quanbei, South Korea and has introduced Vandi-series, a type of plant-protection UAV for agriculture, including Vandi-A1, Vandi-C1, Vandi-A10, Vandi-B10, etc.

According to their official website, Vandi-A1 (Fig. 11.18 and Table 11.11) is the representative product, which is an electric-powered sixteen-rotor UAV. It is suitable for the spray of liquid and granular chemicals for different field conditions such as large-scale rice field.

Fig. 11.16 P80 plant-protection UAV

Table 11.9 Major specification of P80

Name	Value/unit
Size	2460 mm × 2487 mm × 564 mm
Weight	32 kg
Maximum speed	10 m/s
Maximum starting altitude	4000 m
Tank volume	40 L

Main Research Topic

In general, the main research focuses on examining the spray effect from plant-protection UAVs in pest-disease control as affected by different operation parameters which include the flight altitude, speed, and spray pressure on droplet deposition. Relationship between these factors and deposition values is studied. Moreover, some research also considered environmental variables such as wind on drift as related to spray deposition.

Furthermore, increasing studies have focused on the downwash airflow (or wind-field flow) generated by propellers, since it can strongly affect the distribution of droplets. These studies involve the development of downwash simulation (e.g. Computation Fluid Dynamics, CFD) and verification (e.g. windspeed measurement system; Particle Image Velocimetry, PIV) approaches.

Meanwhile, the chemicals efficacy in controlling pest and disease as well as optimizing the fertilizer use in fields is another research topic. Different kinds of substances have been developed for the chemical use with desired outcome of long-lasting substance but less toxic.

The example below indicates a study of droplet detection and the adjustment of spray swath to enhance the spray effect of a six-rotor plant-protection UAV (Zheng et al., 2017a, b).

Example 11.5: Droplet Detection and the Optimization of Spray Effect of a Six-Rotor UAV

1. Hardware and experiments: As shown in Fig. 11.19a, b, a Lidar, LMS512-20100, was utilized for droplet detection and a six-rotor UAV, JF01-10, was utilized for spraying. Two nozzles were mounted on the UAV with constant spray pressure. The Lidar was connected to a personal computer via Ethernet as a detection

Fig. 11.17 FAZER-R UAV

Table 11.10 Major specification of FAZER-R

Name	Value/unit
Size	2782 mm × 770 mm × 1078 mm
Real payload	32 kg
Main propeller size	3115 mm
Engine	2 cylinders with 4 stroke
Fuel	Conventional unleaded gasoline
Tank volume	16 L × 2

system and specific algorithms were developed to extract and calculate effective spray swath. Then, the detection system was employed to adjust the nozzle interval for uniform droplet distribution (Fig. 11.19c).

Then, the UAV with adjusted nozzle interval was operated to spray water onto corns at different heights (i.e. 1, 1.5, 2 m) and speeds (2, 4, and 6 m/s). Three different growth stages of corns (represented by corn height, 1.20 m, 1.53 m, and 2.08 m) were investigated. For each kind of corn height, water-sensitive papers were put on the top, middle, and bottom canopy layers (Fig.11.19d) for the analysis of spray effect. Ten collection points in the direction of flight were selected and the interval between each collection point in the same side was 5 m (Fig. 11.19e), then the spray effect was analyzed.

2. Software and data processing: Fig. 11.20 shows the algorithm developed for the Lidar detection system, containing 4 main steps. The details of the algorithm can be found in the study by Zheng et al. (2017a). The system could finally calculate the value of a two-dimensional spray swath. Based on the actual request from the company (Viga UAV Co. Ltd.), the spray swath at a hovering height of 1.5 m was tested.

3. Results: Fig. 11.21 shows the results of the spray swath, indicating that after the adjustment, the droplet distribution was more uniform rather than concentrated in two pieces with even the smaller spray swath.

Penetration rate can be calculated by using Eq. (11.5), in which d_{max} is the maximum deposition value of the three collection layers and d_{min} is the minimum one, while $\overline{d}$ is the mean of deposition values of the three collection layers. Figure 11.22 shows the spray effect of the UAV with adjusted spray swath based on penetration

Fig. 11.18 Vandi-A1 UAV

Table 11.11 Major specification of Vandi-A1

Name	Value/unit
Size	1350 mm × 1350 mm × 777 mm
Maximum landing load	30 kg
Maximum speed	18 km/h
Spray height	3-4 m
Maximum working time	8 min
Tank volume	10 L

rates and their regression models (between the operation parameters and penetration) at three kinds of growth stages were indicated in Eq. (11.6).

$$p = \frac{d_{\max} - d_{\min}}{\bar{d}} \tag{11.5}$$

$$\begin{aligned} b_1 &= 0.7566 + 3.5829h - 1.1324v - 0.0316vh - 1.1165h^2 + 0.137v^2 \\ b_2 &= 0.9155 - 1.3471h + 0.5524v + 0.2324vh + 0.2238h^2 - 0.0999v^2 \\ b_3 &= 5.9194 - 3.5275h - 0.983v + 0.3652vh + 0.5251h^2 + 0.0669v^2 \end{aligned} \tag{11.6}$$

where h is UAV flight height, v is UAV flight speed, and b_1, b_2, and b_3 are the penetration rate of 1.2 m height stage, 1.53 m height stage, and 2.08 m height stage, respectively.

Penetration rate, p, should be as small as possible to indicate the deposition difference in three layers is little. According to Fig. 11.22, the optimal operation parameters were 1 m and 4 m/s for 1.2 m height stage corns, 2 m and 2 m/s for both 1.53 m height stage and 2.08 m height stage corns. Moreover, the R^2 of all the regression models (Eq. 11.6) was greater than 0.9, which indicated the high reliability and predictability of the models.

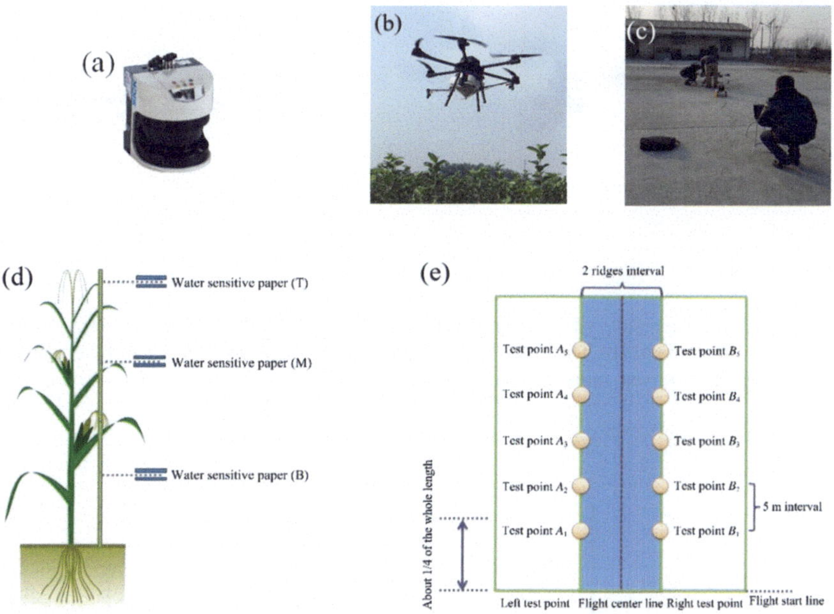

Fig. 11.19 Devices used and the experiment setup of spray effect tests. (**a**) The Lidar, LMS512-20100. (**b**) The six-rotor UAV, JF01-10. (**c**) Adjusting the nozzle distance of JF0-10. (**d**) The layout of the water-sensitive paper in canopies for spray effect tests. (**e**) The sampling points of the water-sensitive paper in fields for spray effect tests

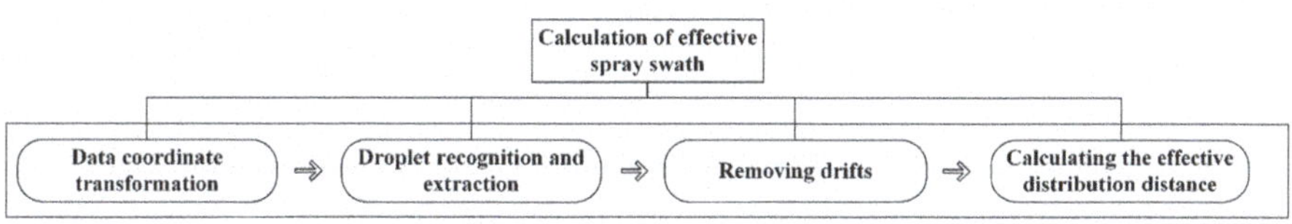

Fig. 11.20 Algorithm developed for the droplet detection system

11.4 Agricultural UAV Advantages and Limitations

11.4.1 Advantages

1. High Efficiency

Currently, the flight speed of plant-protection UAVs can reach up to 8 m/s and accommodate liquid tanks of larger than 10 L or even 30 L. This means that the UAVs are able to spray wide areas with less time cost, which is about 0.067 to 0.134 hectare/min (Lan et al., 2019a). Furthermore, many agricultural UAVs have incorporated several smart functions such as automatic route planning, one-click take-off, autonomous flight, RTK differential positioning and breakpoint continuous spraying or photographing. Some of the electric-powered UAVs also have the functions of terrain following, autonomous obstacle avoidance and

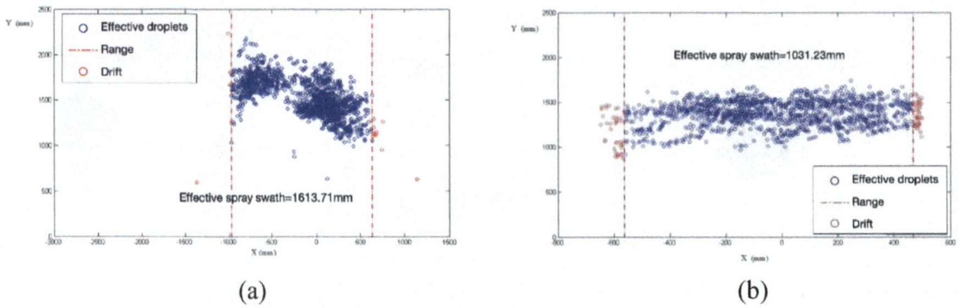

Fig. 11.21 Detection results of the spray swath before and after adjustment. (**a**) The UAV spray swath before adjustment. (**b**) The UAV spray swath after adjustment. Note: The blue points are effective droplets, the red ones are outliers, and the red lines are the boundary of effective distribution

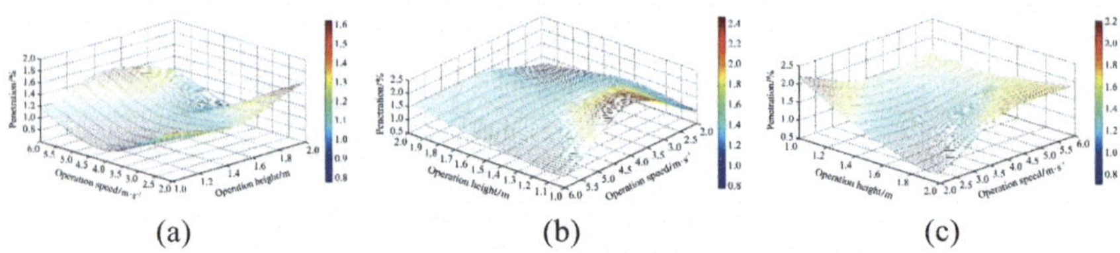

Fig. 11.22 3D surfaces of operation height, operation speed and penetration for the three growth stages. (**a**) 3D surface for 1.2 m height stage. (**b**) 3D surfaces for 1.53 m height stage. (**c**) 3D surfaces for 2.08 m height stage

night flight. Therefore, compared with manual remote control and manual spray, agricultural UAVs are highly efficient with less labor intensive.

2. Superior Terrain Adaptability

Since agricultural UAVs can be operated at fast speed, they are particularly suitable for small-sized fields, scattered fields and densely populated agricultural areas (Lan et al., 2019a). In addition, UAVs operation is not affected by terrain, which is sometimes a limited factor by ground vehicles. Though tall obstacles such as buildings, trees and telegraph poles can bring hazards when operating the agricultural UAVs, the developed terrain following and obstacle avoidance technology will help to adapt these surroundings variation. Therefore, agricultural UAVs are considered to be high terrain adaptability.

3. Acceptable Target Adaptability

Plant-protection UAVs can be operated and performed well in almost all crops irrespective of high-stalk crops, trees, or small plants. In orchards, plant-protection UAVs have been utilized for upper canopies, improving the spray effect from top to middle parts of trees based on strong downwash (Zheng et al., 2020a, b).

Meanwhile, for crops with narrow row spacings and canopy closure in later growth stage, it is reliable for plant-protection UAVs to spray pesticides using strong airflow, which is in accessible by ground vehicles.

4. Separation of Human from Chemicals

During operation, plant-protection UAVs can be controlled by ground remote controllers or on-board controllers, ensuring the users to keep a safe distance from the sprayed area. This can prevent a series of risks caused by human exposure to chemical drift and help to maintain spraying safety.

11.4.2 Limitations

1. Balance Between Operation Parameters

For the same crops, the change of operation parameters may lead to diverse results of pest-disease control or resolutions of images. For instance, a high flight altitude and fast speed may give a large spray swath but low droplet penetration, while low flight altitude and speed probably result in good penetration but a small swath and excessive concentration of droplets. Thus, before spraying, the relationship between operation parameters and spray effect should be studied to find an optimized setting.

2. Different spray effects on different crops

The spray effect of plant-protection UAVs correlates exceedingly with target features such as shape, area, height and leaf density, whereas the chemicals used for crops are distinct. For instance, in terms of the same height, there is a dosage variation between apple tree seedlings and corns. Hence, before operation, key spray indices (such as spray pressure, nozzle distance) for UAV spray should be first investigated.

3. Comparison with manned aircrafts

Manned aircrafts for agriculture are generally fixed-wing planes and have three main advantages over agricultural UAVs. Firstly, they have a greater load, usually in tons or 100 kg, which is much higher than that of UAVs with about 5 to 40 kg. Secondly, the endurance of manned aircraft is noticeably longer, regularly in hours and kilometers, while agricultural UAVsjust run in minutes and meters. Thirdly, due to the smooth surface and unchanging section of wings from manned aircrafts, the downward pressure generated is more concentrated and uniform than that of plant-protection UAVs, especially rotor UAVs, resulting in a more stable spray.

4. Specific pesticide required

Conventionally, water is added to ordinary pesticides to attain right concentration of the liquid mixture, matching with the flight parameters of plant-protection UAVs in order to maintain the total amount of pesticides applied per unit area. Nonetheless, if the concentrations of some pesticides are not properly attained with the flight parameters, severe chemical damage to crops may occur during spraying. Meanwhile, it is possible to block the spraying system of UAVs if the pellet of ordinary pesticides is too big and rough. Therefore, specialized pesticide for aviation spray is required.

11.5 Expectation of Agricultural UAV

11.5.1 Existing State-of-the-Art Technology

The current advanced technology associated with agricultural UAVs is mainly concentrated on autonomous operation, including the following three aspects:

Navigation and Obstacle Avoidance

During actual operation, agricultural UAVs will probably encounter some complex obstacles. Thus, autonomous planning and navigation for obstacle avoidance based on the real-time obstacle positions is an effective need for them to complete subsequent tasks without human intervention. To do this, algorithms of surrounding perception are developed such as A* algorithm, Particle Swarm Optimization (PSO), Artificial Potential Field (APF), geometric reinforcement learning, evolutionary algorithm and Rapidly exploring Random Tree (RRT). The algorithms should be first optimized before application based on the exact demand and flight characteristics of an agricultural UAV. The following example demonstrates a brief introduction of an algorithm in UAV obstacle avoidance during navigation (Zheng et al. 2020c).

Example 11.6: Obstacle Avoidance Path Planning Algorithm for Multi-Rotor UAVs

1. System structure and environment: A quad-rotor UAV named Carto F4 with 5 kg maximum load mounted with a Lidar scanner, Rplidar S1 were used in the experiment. Meanwhile, a PIXHAWK flight controller was installed for the detection and flight control of the UAV, and NVIDIA TX2, a high-performance computing module, was added for the processing of point clouds and obstacle avoidance logic algorithm.

Figure 11.23 illustrates the structure of the UAV system. The software for NVIDIA TX2 was based on the Robot Operating System (ROS) and distributed software architecture to decouple and simplify the development. Moreover, the software of PIXHAWK was provided by Px4, which could exchange data with the API interface provided by Px4 through Mavlink protocol. As shown in Fig. 11.24, tall trees and street lamps were considered as the obstacles and the UAV was flown through this area. The flight altitude was strictly restricted to half of the tree height.

2. Software and methods: Fig. 11.25 delineates the procedures of the developed methods with four steps. Multiple scenarios were established for simulation vali-

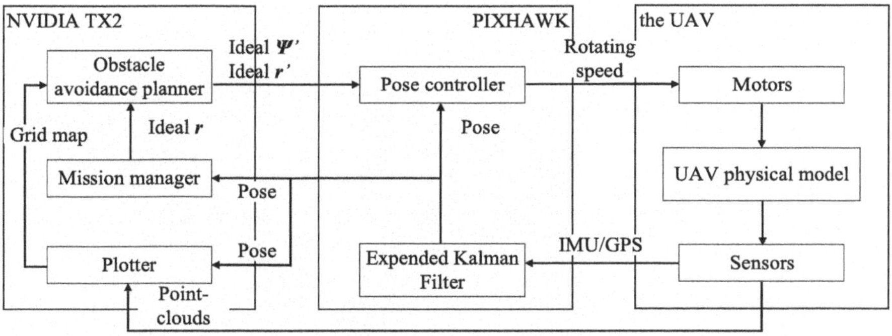

Note: ψ is the yaw angle of the UAV, rad, and $\boldsymbol{r}$ is the position vector of the UAV, m.

Fig. 11.23 Structure of the UAV system for navigation and obstacle avoidance systems. Note: ψ is the yaw angle of the UAV, rad, and $\boldsymbol{r}$ is the position vector of the UAV, m

dation. The dimension of the probability map was firstly reduced to 2 dimensions (Fig. 11.26a) and the optimized RRT-Connect algorithm was performed. Next, the full path was optimized based on Minimum Snap method, especially including the expansion of the 2D probability grid map (Fig. 11.26b) to avoid path interference.

3. Results: Fig. 11.26c shows the result of navigation and obstacle avoidance systems. Based on the simulation verifications, the optimized RRT-Connect algorithm could reduce the re-planning time of obstacle avoidance paths by 23.69%. The effective planning time of obstacle avoidance paths was less than 0.33 s and the average tracking speed of obstacle avoidance paths was more than 1.12 m/s. The proposed method could achieve the real-time planning of obstacle avoidance paths in different complex scenes, effectively improving the efficiency and stability of path planning. It provides a feasible technical solution for UAV obstacle avoidance in the actual operation environment such as farms, orchards and forests.

Terrain Following During Flying

Terrain following functions from agricultural UAVs is established based on sensor measurement that allows them to follow the outline of the terrain automatically and keeps a constant height difference from the ground to them during operation. This function enables the UAV to adapt to different terrains to obtain effective and accurate spray or data acquisition. The determining factor of this technology is the precise measurement of the height between UAVs and ground. Table 11.12 listed the popular approaches for this technology followed by an example study (Example 11.7 by Xu et al., 2021).

Fig. 11.24 Flight conditions and obstacles

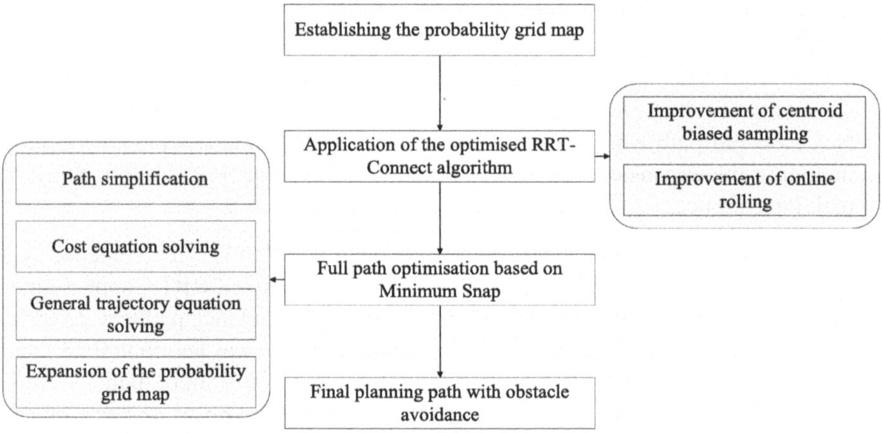

Fig. 11.25 The procedures of the developed methods for navigation and obstacle avoidance systems

Example 11.7: Detection of Crop Heights by UAVs Based on the Adaptive Kalman Filter

1. System structure and environment: A quad-rotor UAV, DJI Phantom 3 was used in the study. A data acquisition module (Fig. 11.27) known as NRA24 MWR (Nanoradar Technology Co, Ltd. Hunan, China) was utilized to measure the relative altitude. An IMU, JY901 (Junyue Intelligent Control Technology Co, Ltd., Guangdong, China) was utilized to measure the flight angles and acceleration. Meanwhile, a global positioning unit, ATK-S1216F8-BD (SkyTraq Technology, Inc. Taiwan, China) was used to obtain the UAV real-time positions. A digital transmission radio, XROCK V3 (Xili innovation Electronic Technology Co, Ltd., Zhejiang, China) was selected as the wireless transmission module due to

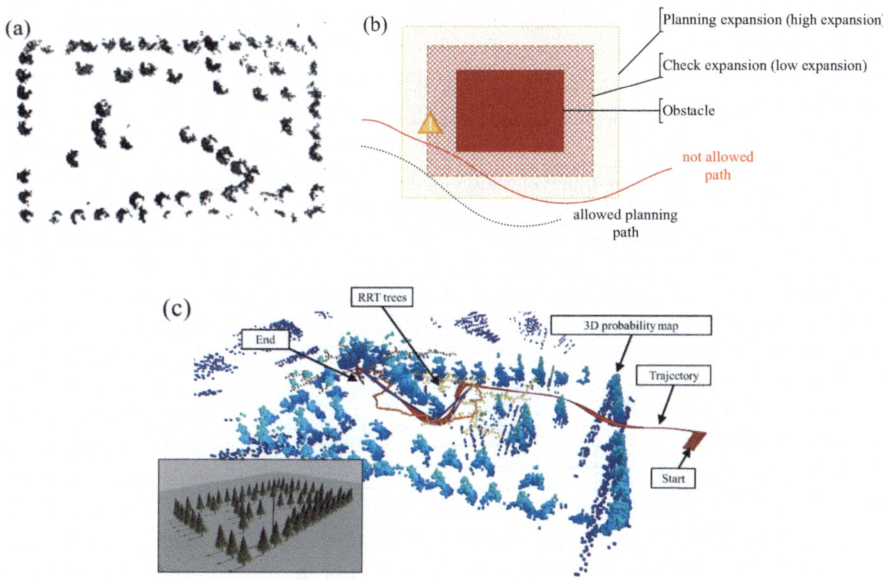

Fig. 11.26 Processing steps and planning results. (**a**) 2D probability grid map. (**b**) Interference avoidance of the path after grid-map expansion. (**c**) The result of navigation and obstacle avoidance

its long transmission distance. In addition, a microcontroller, STM32F103 (Xingyi Electronic Technology Co, Ltd., Guangdong, China) was exploited as the slave computer for the fusion and processing of sensor data, while the host computer was a laptop for data storage and picture display. The height measurement tests were conducted in a cornfield (with 5 m and 10 m flight altitude), while the terrain following test was based on stairs.

2. Software and methods: Fig. 11.28 illustrates the flow of the algorithm. The proposed data fusion algorithm was based on the Adaptive Kalman Filter (AKF). The data from the MWR, IMU, and GPS was used to obtain the parameters of both the state and measurement equations. Then, the data from the MWR was modified by the angle information from the IMU. Meanwhile, the modified data and the data from the GPS and IMU were asynchronously fused by the AKF to get the final appropriate height results. Based on the real-time height results, the UAV could adjust its flight altitude.

3. Results: The tests demonstrated that: (1) when compared with the direct detection by the MWR, the error of detection was reduced by 0.035 m, and (2) when compared with the real corn heights, the error of detection was 0.02 m. Furthermore, as indicated in Fig. 11.29, the terrain was profiled properly, hence, the proposed algorithm showed a good ability and accuracy of terrain following.

Table 11.12 Popular approaches for agricultural UAV terrain following technology

Method category	Name
Hardware	Real-time kinematic GPS (RTK GPS)
	Millimeter wave radar (MWR)
	Lidar
Off-line height acquisition	Known digital elevation models (DEMs) by existing software such as ASTER GDEM30m and ALOS12.5 m
	Manual measurement
Real-time height acquisition	Sensor scanning during flying

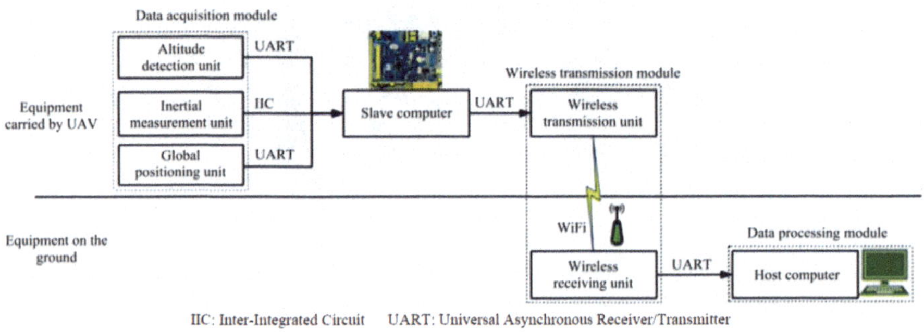

Fig. 11.27 Structure of the terrain following system. *IIC* inter-integrated circuit, *UART* universal asynchronous receiver/transmitter

Collaborative Flying and Operation

Agricultural UAVs can significantly enhance work effectiveness and efficiency, especially for plant protection when operating in a large scale. The greater the number of UAVs is, the higher operational efficiency is. However, current commercial products are mainly concentrated on single automatic operation. Some of them are even manually telecontrolled, which restricts the advantages of agricultural UAVs. Hence, multi-agent collaboration is a development trend and it is primary to develop appropriate policy of collaboration. Task allocation is the key for collaborative flying and several typical algorithms are listed in Table 11.13. Most of the studies focus on the theoretical solutions of this type of problem. An example of collision avoidance of multiple UAVs is as followed (Liu et al. 2021).

Example 11.8: Developed Algorithm, DPIO-SA, and GA, for Task and Location Assignment

In this research, tasks were first assigned based on DPIO-SA algorithm. Each task would be executed by multiple UAVs. Then, the locations of the UAVs in the formation were arranged. Since there was no constraint in the location assignment, GA was utilized to solve the location assignment that was only simple crossover and mutation operations. Figure 11.30 demonstrates the flow of task and location assignment.

11.5.2 Development Trend of Agricultural UAV

The conception of smart agriculture has been proposed and is developing as an eco-friendly, effective and high-yield agricultural pattern. This includes not only the precision perception, control, decision-making and communication, but also agricultural internet of things, agricultural e-commerce, food traceability, agricultural tourism and agricultural information services. Current and future agricultural UAVs can undoubtedly play an expert role in smart agriculture due to their increasing intelligent technologies. Based on the existing limited techniques, the following aspects serve as the development trend that must be advanced.

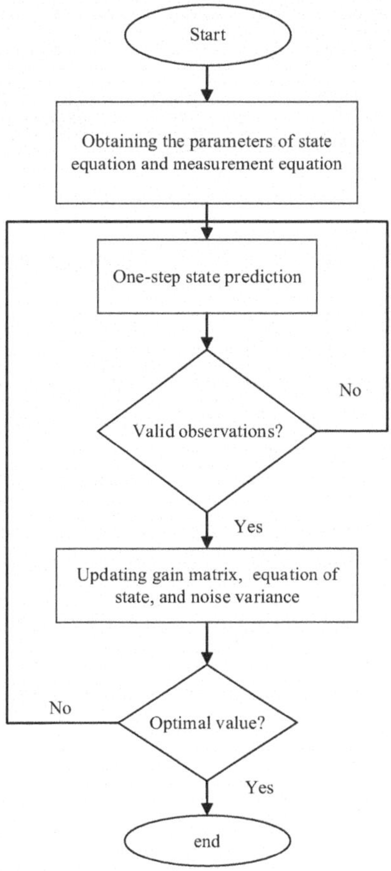

Fig. 11.28 Flow of the proposed algorithm

Fig. 11.29 Terrain following results

Strategy of Aviation-Ground Collaboration

Stereo operation, which is defined as the simultaneous operation of both agricultural UAVs and unmanned ground systems, is developing (Lan et al., 2019a). It is especially suitable for large-scale farms and orchards, which is able to solve the problem of low droplets penetration through densed canopies. Current research has mainly worked for theoretical development and its validation but less practical application. Hence, more studies should be conducted on investigating:

- how a UAV can cooperate with other UAVs, ground vehicles and satellites;
- how a UAV group can cooperate with other groups of UAVs and ground vehicles;
- what specific strategies (e.g. the variation of flight height and speed for different places) can be used by an operation agent in real-time based on previous test results;
- how to administer these operation agents to achieve safe flight.

Table 11.13 Typical algorithms for collaborative flying

Algorithm		Task/function
Simulated Annealing (SA) algorithm and Particle Swarm Optimization (PSO)	SA-PSO	Path planning
	Discrete DPSO-SA	Task allocation
	NPSO	Optimal locations
Genetic Algorithm (GA)		Location assignment
Pigeon Inspired Optimization (PIO)		Parameter optimization and collision avoidance
Contract Network Algorithm (CNA)		Local task scheduling
Artificial Potential Field (APF)		Collision avoidance
Optimal Reciprocal Collision Avoidance (ORCA)		Collision avoidance

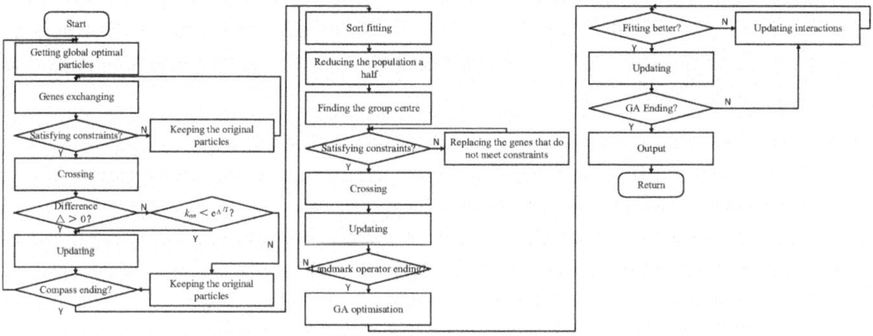

Fig. 11.30 Flow of task and location assignment. Note: T is the initial temperature, k_{nn} is the temperature attenuation factor, and △ is the difference between the new fitting and the old fitting

Data Link and Multi-agent Communication

At present, intelligent communication among multiple agents has been studied more in military and industry but less in agriculture. The data links for agents and users should be capable and robust. Therefore, it is required to study the data link and communication of agricultural UAVs about:

- the approaches to ensure the capacity and robustness;
- the way to achieve multi-agent communication and agent-user communication;
- the design and optimization of data link structures;
- the methods of data store and tele-transmission.

Single UAV Control Based on Targeting and Ground Effect

Control strategy is very important for single UAV. According to Sect. 11.3, though studies related to spray effect have been conducted, two necessary issues have not been considered: autonomous operation by toward-target and the ground effect on the stability of UAVs. Hence, the following three sections may need further explorations:

- the mechanism of the ground effect on agricultural UAVs;
- the approaches of stable control to eliminate the ground effect on agricultural UAVs;
- the control methods and strategies of target localization and toward-target operation.

Capability of Endurance and Load

According to Sect. 11.2, oil-powered agricultural UAVs can load more but are not eco-friendly and stably controlled, while the opposite is for battery-powered UAVs. Oil-electric hybrid agricultural UAVs make the balance between the two, which are still in a developing stage. Hence, it is needed to expand the endurance and load of agricultural UAVs by studying:

- the materials, structures and capacity of batteries to have superior performance;
- the optimization of fuselage and propeller structures and the motor performance to make better aerodynamics;
- energy switching mode, stable control and the comprehensive performance of the power system of oil-electric hybrid agricultural UAVs;
- stable control approaches for oil-powered agricultural UAVs.

Law of Downwash-Droplet-Crop Interaction

In fact, the two key variables crucial for droplet distribution are downwash and spray parameters (pressure, nozzle intervals, nozzle types, etc.). In terms of a certain agricultural UAV, spray parameters have been set by manufacturers and are usually unchanged, so mainly the downwash determines the droplet distribution. High-efficient use of downwash can greatly reduce drift and increase distribution uniformity, subsequently decreasing chemical use and environmental pollution. Although some research has been involved in downwash, they do not focus on the interaction between downwash and droplet penetration in crop canopies. When droplets penetrating through crop canopies, downwash results in both crop shaking and droplet motion. Therefore, future studies related to spray effect may be involved in:

- the principle of downwash transmission in crop canopies;
- the interaction principles of (1) crops with droplets, (2) droplets with downwash and (3) downwash with crops;
- the model between these interaction and spray effect for practical use.

11.6 Summary

This chapter introduced the development of unmanned aerial vehicles configured for agricultural-related operation. The classification of such type of UAVs was firstly demonstrated. Then, the main fields of application of these UAVs, including low altitude remote sensing and plant protection, were illustrated and relevant applied techniques were contained. Each sub-field of application was explained with a study instance and the prevalent types of plant-protection UAVs were shown. Furthermore, both advantages and disadvantages of the UAVs were analyzed. Finally, expected development directions were discussed with current study examples.

References

Lan, Y. B., Chen, S. D., Deng Ji, Z., Zhou, Z. Y., & Ou, Y. F. (2019a). Development situation and problem analysis of plant protection unmanned aerial vehicle in China. *Journal of South China Agricultural University, 40*(5), 217–225. (in Chinese with English abstract). https://doi.org/10.7671/j.issn.1001-411X.201905082

Lan, Y. B., Zhu, Z. H., Deng, X. L., Lian, B. Z., Huang, J. Y., Huang, Z. X., & Hu, J. (2019b). Monitoring and classification of citrus Huanglongbing based on UAV hyperspectral remote sensing. *Transactions of the Chinese Society of Agricultural Engineering (Transactions of the CSAE), 35*(3), 92–100. (in Chinese with English abstract). https://doi.org/10.11975/j.issn.1002-6819.2019.03.012

Li, Z. J., Yang, S. H., Shi, D. S., Liu, X. X., & Zheng, Y. J. (2021). Yield estimation method of apple tree based on improved lightweight YOLOv5. *Smart Agriculture, 3*(2), 1–15. (in Chinese with English abstract). (Online available) https://kns.cnki.net/kcms/detail/10.1681.S.20210713.1118.002.html

Liu, X., Han, Y., & Chen, J. (2021). Discrete pigeon-inspired optimization-simulated annealing algorithm and optimal reciprocal collision avoidance scheme for fixed-wing uav formation assembly. *Unmanned Systems, 9*(2), 1–15.

Wang, G. (2019). *Three-dimensional ground reconstruction technology based on lidar data of unmanned system*. China Agricultural University.

Wu, G., Peng, Y. Q., Zhou, G. Q., Li, X. L., Zheng, Y. J., & Yan, H. J. (2020). Recognition method for corn nutrient based on multispectral image and convolutional neural network. *Smart Agriculture, 2*(1), 111–120. (in Chinese with English abstract).

Xu, P. F., Wang, H. C., Yang, S. H., & Zheng, Y. J. (2021). Detection of crop heights by UAVs based on the Adaptive Kalman Filter. *International Journal of Precision Agricultural Aviation, 4*(1), 52–58.

Zheng, Y. J., Chen, B. T., Lyu, H. T., Kang, F., & Jiang, S. J. (2020a). Research progress of orchard plant protection mechanization technology and equipment in China. *Transactions of the Chinese Society of Agricultural Engineering (Transactions of the CSAE), 36*(20), 110–124. (in Chinese with English abstract). https://doi.org/10.11975/j.issn.1002-6819.2020.20.014

Zheng, Y. J., Jiang, S. J., Chen, B. T., Lyu, H. T., Wan, C., & Kang, F. (2020b). Review on technology and equipment of mechanization in Hilly Orchard. *Transactions of The Chinese Society of Agricultural Machinery*. (in Chinese with English abstract), *v51*(11), 8–27. https://doi.org/10.6041/j.issn.1000-1298.2020.11.001

Zheng, Y. J., Yang, S. H., Lan, Y. B., Clint, H., Zhao, C. J., Chen, L. P., Liu, X. X., & Tan, Y. (2017a). A novel detection method of spray droplet distribution based on LIDARs. *International Journal of Agricultural and Biological Engineering, 10*(4), 54–65.

Zheng, Y. J., Yang, S. H., Zhao, C. J., Chen, L. P., Lan, Y. B., & Tan, Y. (2017b). Modelling operation parameters of UAV on spray effects at different growth stages of corns. *International Journal of Agricultural and Biological Engineering, 10*(3), 57–66.

Zheng, Z., Yang, S. H., Zheng, Y. J., Liu, X. X., Chen, J., & Su, D. (2020c). Obstacle avoidance path planning algorithm for multi-rotor UAVs. *Transactions of the Chinese Society of Agricultural Engineering (Transactions of the CSAE), 36*(23), 59–69. (in Chinese with English abstract). https://doi.org/10.11975/j.issn.1002-6819.2020.23.007

Zhou, J., Zhou, J., Ye, H., Ali, M. L., & Nguyen, H. T. (2021). Yield estimation of soybean breeding lines under drought stress using unmanned aerial vehicle-based imagery and convolutional neural network. *Biosystems Engineering, 204*(5), 90–103.

Chapter 12
Robotic Tree Fruit Harvesting: Status, Challenges, and Prosperities

Long He, Azlan Zahid, and Md Sultan Mahmud

12.1 Introduction

Up to today, majority of tree fruit crop production operations highly depend on seasonal human labor. Many critical activities are not only labor-intensive, but also highly time-sensitive. With the increasing concerns on the labor shortage and associated high labor cost, harvesting as the most labor-intensive operation in tree fruit production has been attracting more and more attention. Improving harvesting efficiency and reducing the dependence on human workers have been the major motivation for developing new harvesting technologies. In recent decades, automation technologies, especially the auto-guidance for field tractors have been investigated widely. However, for specialty crops including tree fruit crops, the application of automatic technologies has lagged due to the complexity of field operations and inconsistency of crop systems. Three different harvesting technologies have been investigated in tree fruit harvesting, including harvest assist platform, massive mechanical harvesting, and robotic harvesting. Harvest assist platforms have significant improvement in harvesting efficiency (Schupp et al., 2011; Zhang et al., 2016), while large amount of human labor is still needed. Mechanical harvesting based on the shake-and-catch concept to conduct massive but non-selective harvesting led to higher harvesting efficiency but may cause more bruise to the fruits (He et al., 2017; Ma et al., 2018). Robotic harvesting as a selective harvesting method is showing potential of replacing human hand picking (Silwal et al., 2016; Hohimer et al., 2019). Two major components with robotic harvesting are fruit detection with machine vision system and fruit picking with robotic maneuvering mechanisms and

L. He (✉) · A. Zahid · M. S. Mahmud
Department of Agricultural and Biological Engineering, Pennsylvania State University, Fruit Research and Extension Center, Biglerville, PA, USA
e-mail: luh378@psu.edu

S. Ma et al. (eds.), *Sensing, Data Managing, and Control Technologies for Agricultural Systems*, Agriculture Automation and Control,
https://doi.org/10.1007/978-3-031-03834-1_12

arms. Tree architecture is another core factor relating to the canopy–robot interaction. With the adoption of narrow tree canopy system especially two-dimensional trellis trained tree systems, robotic harvesting technologies showed more promising compared to the traditional trees. The interaction between harvester and tree canopy requires optimal path planning to avoid obstacles to reach the targeted fruits.

12.2 Tree Fruit Industry and Current Challenges

12.2.1 Overview of Tree Fruit Industry in USA

The tree fruit industry is an important component of the nation's agricultural sector that contributes about 25% of the market share ($18 billion) among all specialty crops produced in USA (USDA-ERS, 2018). Production of major tree fruits in USA is shown in Fig. 12.1. Citrus fruits are the top fruit crops in world trade in terms of highest worth (FAOSTAT, 2016), are one of the most famous fruit commodities widely accepted for their flavor and nutritional facts. Fresh and processed (e.g., mainly juice) are the two major markets of the US citrus fruits. The fruits mainly used for fresh consumption are grown in California, Arizona, and Texas, where Florida covers almost the entire processed citrus fruit market for orange juice. California produced about 51% of total citrus fruits in the USA in 2018–2019 season where Florida accounted for 44% of the total production and remaining 5% shared by Texas and Arizona (USDA-NASS, 2019a). A total of 7.94 million tons of citrus fruits (valued $3.35 billion) produced in 2018–2019 was 31% higher than 2017–2018 season (USDA-NASS, 2019a). Apples are the second most produced fruits after orange and most valuable fruit crops in the USA. Apples are commercially grown in 32 states, but Washington is by far the largest producer accounting

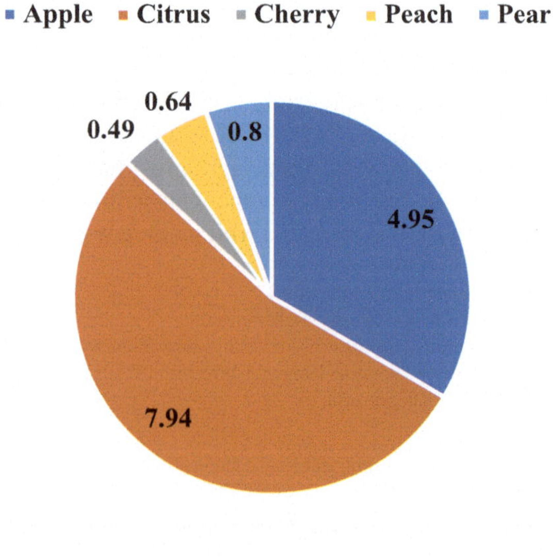

Fig. 12.1 Production of major tree fruits (per million tons) in USA (USDA-NASS, 2019a, b)

for 70% of the total apple production. New York, Michigan, Pennsylvania, and California are the next four top producers producing a significant amount of apples every year (U.S. Apple Association, 2018). Nearly 7500 growers produced around 4.95 million tons of apple (valued $3.01 billion) on an approximated 130.3 hectare of land in 2018–2019 (USDA-NASS, 2019b). Conversely, pears are mainly grown in six states of USA including California, Michigan, New York, Oregon, Pennsylvania, and Washington. Of these states, California, Oregon, and Washington are producing majority of the pear production every year. Pears contributed $429 million to the economy by producing a total production of 0.8 million tons in 2018–2019 season (USDA-NASS, 2019b). Peaches are the fourth most produced tree fruits in the USA, producing 0.64 million tons in 2018–2019 which is valued $511 million. Peaches are commercially grown in 20 states where California is the largest producer and supplied about 56% of the US fresh peach fruit and nearly 96% of processed peaches (USDA-NASS, 2018). Other top producing states are South Carolina, Georgia, and new Jersey. Contrarily, almost 90% of sweet cherry mainly produced in three states (i.e., Washington, California, and Oregon) and 74% of tart cherry produced by Michigan alone (USDA-NASS, 2018). The US cherry growers produced a total of 0.34 million tons of sweet cherry (valued $638 million) and 0.15 million tons of tart cherry ($57 million) in 2018–2019 (USDA-NASS, 2019b). Despite the significant increasing of production for tree fruits in the past decade because of the proper orchard managements, the fruit industry in USA is facing tremendous challenges due to high dependency on farm labors resulting increasing costs of production (Fennimore & Doohan, 2008; Calvin & Martin, 2010).

Among the costs associated with production of tree fruits, the harvesting operation (e.g., only picking and hauling) itself is accounting for 11–26% of the total production costs. Cost of harvesting is varying from one fruit to another and also depends on the size of the fruit orchards. Citrus fruits such as orange are costing for $926 per acre for only fruit picking and hauling which is about 11% of the total production cost (University of California Cooperative Extension, 2015). Conversely, the picking and hauling cost for apple is much higher than citrus fruits accounting for 26% of the total production cost where the harvesting cost is $1320 per acre (University of California Cooperative Extension, 2014). Similar to apple harvesting cost, peach requires $1339 for picking and hand sorting of one acre orchard (University of California Agriculture and Natural Resources Cooperative Extension, 2017a, b). Conversely, pear fruit accounts for $1780–$1969 per acre which is about 20–25% of total production cost (University of California Agriculture and Natural Resources Cooperative Extension, 2018). Aside from citrus, apple, peach, and pear, cherry fruit accounts for $720–$960 per acre for picking by using hand (University of California Agriculture and Natural Resources Cooperative Extension, 2017a, b).

12.2.2 Challenges and Opportunities for Fruit Harvesting

Harvesting of tree fruits (i.e., apples, citrus, cherries, peaches, and pears) is the process of gathering a ripe fruit from the orchards which highly depends on labor workforce and is becoming less feasible due to the decreasing trend of farm labor in agriculture with increasing cost of production. Although a rapid development in agricultural automation has been progressed in the twentieth century, tree fruit harvesting is still largely dependent on manual labor due to lack of efficient and effective harvesting methods. Most of the developments reported in the last few years are in prototype phase and not fully feasible to the large scale orchard condition due to lower efficacy and efficiency, unreliable performance, and high cost (Zhao et al., 2016). Among the tree fruits, the apple industry alone is accounting for \$1150–\$1700 per acre for manual harvesting (e.g., handpicked) by seasonal labors (Gallardo et al., 2010). Therefore, a large number of seasonal workers is required every year for only tree fruit harvesting considered as the top labor-intensive task in orchard management. Fruit growers of Washington State utilized about 36,425 seasonal labors in the peak harvesting month (i.e., September) for only apple harvesting (Washington State Employment Security Department, 2013), accounting for one-third of the annual variable costs combining tree pruning and thinning (Gallardo et al., 2010). Conversely, increasing demand for seasonal workforce in fruit industries is pretending the high uncertainty of the farm labor availability in the near future (Calvin & Martin, 2010). Tree fruit industries in the USA are hiring a major portion of seasonal labors from migrant Latino populations which is also following decreasing trend in the past few years (Gonzalez-Barrera, 2015) gaining serious concern of fruit growers for harvesting in the upcoming years. Contrarily, most of the tree fruits are picked by hand of farm labors using a ladder and bag that pose a high risk of back strain and musculoskeletal injuries because of hand lifting, repetitive hand actions, and awkward postures while picking fruits (Fathallah, 2010). The main reason for the musculoskeletal injury is ascending and descending of ladders with heavy loads. Ladder-caused injuries accounted for about \$21 million compensation in the year between 1996 and 2001, which was 50% of all compensations claimed in the fruit industry of Washington State over the time frame (Hofmann et al., 2006). Considering labor injury issues during fruit picking at high locations, labor assist systems (i.e., mechanical platforms) were commercialized that help the pickers by raising up and by raising the bins close to them; however, adoption of these technologies is not widespread among tree fruits growers in the USA (Robinson et al., 2013). A total of nearly 11% fruit growers utilized mechanical labor support systems for harvesting operation in Washington State (Gallardo & Brady, 2015). Contrariness between the mechanical labor assisted systems and the previous orchard design and tree architecture was referred to as the most noteworthy obstacle to their utilization and brought a significant compatibility problem in the tree fruits harvesting (Duraj et al., 2010). To address the challenges associated with labor shortage, risk of labor injuries, limitation of labor assisted systems, and also to reduce the harvesting cost and saving time, the development and application of

Fig. 12.2 Illustration of the evaluation of tree fruit harvesting methods, from left to right are: manual picking, harvest assist platform, mechanical shake-and-catch, and robotic harvesting

automatic or robotic harvesting is utmost importance and essential considering innovations in developing advanced sensors, horticultural advancement, and evaluation of mechanical technologies in the past decades. Figure 12.2 illustrates the evaluation of tree fruit harvesting methods from manual picking to robotic harvesting.

12.3 Overview of Robotic Harvesting Technologies

Beside robotic harvesting, using harvest assist platforms for harvesting tree fruits can be back to the 1990s. Peterson and Miller (1996) developed a harvest aid by placing two pickers strategically under a tree canopy, whose primary task was to pick and drop apples onto a padded catching surface. The machine was modified for narrow inclined trellises that allowed pickers' free movement to optimize their picking time, field tests demonstrated the potential to improve worker productivity up to 22% and effectively remove culls in the orchard (Peterson & Bennedsen, 2005). However, apple damage incidence was unacceptably high, requiring refinements on the handling components.

Vibratory or shaking is the most widely used mechanical harvesting method to transmit kinetic energy to fruiting branches, thus to generate a detaching force on the fruit–stem interface and removes fruit from the tree (Erdoğan et al., 2003). During shaking, a tree will respond differently to different excitation frequencies and amplitudes and fruit removal occurs when the induced detachment force exceeds the pedicel fruit tensile strength (Markwardt et al., 1964). Upadhyaya et al. (1981) studied a single degree of freedom model to describe the response of a tree to impact input and found that 50–60% of the mechanical energy was converted to kinetic energy when impact excitation was used. Savary et al. (2010) developed a simulation tool for predicting the interaction between a tree and the shaker using finite element analysis. Experimental results revealed that the resultant acceleration of the tree would increase with the increase of shaking frequency. Du et al. (2012) conducted a series of dynamic tests to find the energy responses of a sweet cherry tree to vibratory excitations in both laboratory and orchard environments. They found that the energy delivery efficiency and its distribution pattern were heavily related to tree structure. Recently, a localized multi-layer shake-and-catch harvesting system was developed and tested in the apple orchards, which found the posiblility of reducing mechanical-induced damage to the fruits (He et al., 2019). While mechanical harvesting is non-selective harvesting and more precise harvesting should be applied, such as robotic harvesting.

12.3.1 Concept of Robotic Harvesting

The use of robots in tree fruit production is primarily associated with decreasing labor availability and increasing associated costs. An agricultural robot can be defined as an integration of sensing, computing, and manipulation systems to execute pre-defined tasks including thinning, pruning, and harvesting (Kondo & Ting, 1998). In the production cycle of the tree fruits, harvesting is the most labor-intensive operation. As fruit harvesting is time sensitive operation, a large seasonal workforce of skilled labor is required, which is a concern for the fruit growers due to decrease in the labor availability. In addition, the harvesting labor accounts for the significant portion of the variable production cost. Thus, robotic harvesting is an alternate solution to address the issue of labor availability and associated costs and timeliness. The robotic harvester can be classified into two categories: bulk (mass) harvesting and selective (ripe/ready) harvesting. The selective harvesting in which only harvesting the ripened fruits received more attention from the researchers. As a result, several robotic tree fruit harvesting systems have been developed for harvesting various types of fruits including apples (Silwal et al., 2016), citrus (Mehta & Burks, 2014), and cherries (Tanigaki et al., 2008), but no commercial success has been achieved yet. With the recent advances in sensing, controlling, and computing capabilities, the robotic tree fruit harvesting is becoming a possible long-term technology to ensure the sustainability of the tree fruit industry. In this section, a general overview of the different components along with some recent efforts for developing an integrated robotic system for tree fruit harvesting is presented, followed by detailed discussion on the core technologies in the next section (Fig. 12.3).

12.3.2 Robotic Harvesting Review

In recent years, many researchers have worked on development of integrated robotic harvesting system for different tree fruits including apples, citrus, litchi, and cherry. However, these robotic systems are still in the development phase. A universal robotic system may not be feasible for different tree fruits as the harvesting principles vary for different fruits due to the challenging features, e.g., canopy characteristics, and fruit attributes such as size, shape, and weight. Different robotic systems

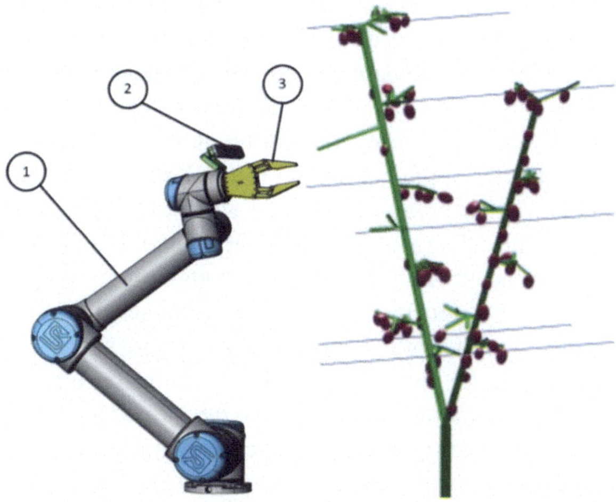

Fig. 12.3 Illustration of the Integrated Robotic Tree Fruit Harvester (Apple). Components: (1) manipulator, (2) camera vision system, and (3) end-effector tool (gripper)

Fig. 12.4 Example of three robotic apple picking systems. From left to right: FFRobotics (multi-layer linear motion with three-finger gripper), Abundant Robotics (parallel robotic arm with vacuum gripper), and Washington State university (serial robotic arm with three finger gripper)

were developed implementing various combinations of integrating different types of sensing systems with different types of manipulators and end-effectors to facilitate the robotic harvesting for tree fruits. Among the high valued tree fruits, the robotic harvesting of apple has gained more attention. Figure 12.4 shows three different types of apple robotic harvesting systems. The modern tree canopy architecture for apple orchards such as trellis fruiting wall, v-trellis, and tall spindle makes most of the fruit visible and accessible from outside, has encouraged researchers for automated apple harvesting. The features of apple fruit including shape, size, color (esp. red varieties) are easier to detect and the other attributes such as hard nature of apple fruits help robotic harvesting as the end-effector could pick it without damaging/bruising. An apple harvesting robot was developed by Silwal et al. (2016) using a seven DoF robotic system integrated with a three tandem fingers gripper end-effector and over-the-row time of flight-based color camera. For establishing the controls, the developed system used the global view system to take the images at the start of each harvesting cycle. The developed system was able to detect 90–100% of the fruits; however, the harvesting/picking success was 84% with an average speed of 6.1 s per fruit. Onishi et al. (2019) developed a robotic apple harvester using deep learning for fruit detection. The system comprised of a six DoF robotic arm integrated with a stereo camera and gripper end-effector. The system was able to detect 90% of the fruits with average harvesting cycle time of 16 s per fruit. However, the gripper made four turns to twist break the peduncle, resulting in higher harvesting time. Another apple harvesting robot developed by Baeten et al. (2008) consists of a six DoF manipulator integrated with soft gripper end-effector (vacuum operated) having the camera attached in the center (hand-in-eye configuration). The fruit detection accuracy was 80% (diameter range 6–11 cm) and average harvesting speed was 9 s per fruit. However, a better sensing of the environment is essential to avoid the contact of the soft gripper with the sharp limbs. Also, the communication between the vision and control unit could be improved to reduce the harvesting time. Bulanon and Kataoka (2010) developed a prototype for robotic apple harvesting by integrating an RGB camera with a laser sensor. The single fruit detection accuracy was 100% and the picking success was as high as 90%, with an average detachment time of 7.1 s per fruit. However, the study was conducted in laboratory environment, and further investigations are still required to confirm the performance in the field conditions. FFRobotics (2020) developed a commercial robotic apple harvester and claimed to have the fruit detection 95% in high-density orchards with a bruise free fruit picking accuracy as 90%. However, the collision with limbs and trellis wire still needs to be addressed.

Some other tree fruits gained attention for robotic harvesting including citrus, cherry, peach, and litchi. Mehta et al. (2014) developed an integrated citrus harvesting robot with a position controller. The system consists of a seven DoF manipulator equipped with a gripper and RGB cameras. The system was able to harvest 95% of the fruits on the tree with harvest cycle as 8 s per fruit. The error in the end-effector positioning was observed less than the fruit diameter, however, with average position accuracy of about ±15 mm, could only be suitable for medium to large size citrus varieties. Harrell et al. (1990) reported the harvesting success rate as 50% with harvest cycle time of 36 s per fruit for citrus harvester. Energid (2020) has developed a prototype for citrus harvesting. The system comprised of two DoF (for aiming and extension) and a camera system (for detection), while no picking end-effector was attached for grasping. The developed prototype was able to pick 50% of the citrus fruit and the average harvesting cycle time was 3 s per fruit. Robotic cherry harvesting has also gained attention of the researchers. Tanigaki et al. (2008) developed a cherry harvesting robot, comprising a four DoF manipulator integrated with a vacuum end-effector and 3D vision sensor having red and infrared laser diodes. For all detected cherries on the tree, the average harvesting cycle time was 14 s per fruit and the harvesting success with and without peduncle attached to cherry was 83% and 66%, respectively. The robot prototype was tested on a model cherry tree in the laboratory, however considering the delicacy of cherry fruit, a more sophisticated end-effector is essential to test the system performance in the field conditions on real trees. Some efforts for the integrated robotic systems for peach and litchi harvested are also reported. Yu et al. (2018) developed a prototype of an autonomous peach harvester. The system consists of a six DoF manipulator

Table 12.1 Recent developments for tree fruit harvesting robots

Crops	Robotic system	Harvesting speed	Harvesting success (%)	References
Apple	6 DoF arm, stereo camera	16 s per fruit	90	Onishi et al. (2019)
Apple	7 DoF arm, color camera, time of flight-based 3D camera	6.1 s per fruit	84	Silwal et al. (2016)
Apple	Color CCD camera, laser range sensor	7.1 s per fruit	90	Bulanon and Kataoka (2010)
Apple	5 DoF arm, stereovision imaging sensor	7.3 s per fruit	67	Hohimer et al. (2019)
Apple	5 DoF arm, color CCD camera, pressure sensor	15 s per fruit	77	Zhao et al. (2011)
Apple	6 DoF arm, high-frequency light camera	9 s per fruit	80	Baeten et al. (2008)
Citrus	7 DoF arm, color CCD (charge coupled device) camera	8 s per fruit	95	Mehta et al. (2014)
Citrus	2 DoF platform, color camera	3 s per fruit	50	Energid (2020)
Cherry	4 DoF arm, 3D vision sensor	14 s per fruit	83	Tanigaki et al. (2008)
Peach	6 DoF arm, color camera	Not reported	90	Yu et al. (2018)

integrated with a gripper end-effector and three RGB cameras and a laser sensor for peach detection, measuring distance, and obstacle avoidance. The fruit detection success was 90% with a tracking speed of 40 fps. However, the system was greatly affected by the illumination conditions, which resulted in lower detection accuracy. Similarly, Xu et al. (2011) reported a virtual prototype for litchi harvesting robot consisting of a five DoF manipulator but further research is required for the integrated system development. A summary of the recently developed robotic tree fruit harvester is presented in Table 12.1. The reviewed integrated robotic harvesters for various fruits are still in the development phase. An interdisciplinary approach is needed to address the engineering, horticultural, and economical issues, to make a substantial progress toward the adoption of robotic tree fruit harvesting in the orchard environment.

Different metrics could be used to determine the performance of the integrated robotic systems. Bac et al. (2014) present eight different performance measuring indicators including fruit localization success, false-positive fruit detection, detachment success, harvest success, harvest cycle time, damage rate, number of fruits evaluated in a test, and detachment attempt ratio. However, in general the performance of the harvesting robots as reported by researchers could be determined using two metrics including: harvesting success, which refers to percentage of the successfully picked from the available total fruits on the tree, and harvesting speed, which refers to the amount of time required to complete the harvesting cycle (sensing, reaching, and detaching) for a single fruit. The integrated harvesting robots developed for different tree fruits greatly differ from each other as the design requirements vary for different fruits, depending on the fruit and canopy characteristics and thus could not be compared directly. However, the metrics used to determine the performance are similar. The figure is presented to better understand the status of harvesting robots for different tree fruits and also provide the understanding on how the fruit and canopy characteristics could affect the harvesting success and harvesting speed.

12.4 Core Technologies in Robotic Harvesting

As shown in Fig. 12.1, the core components of the robot include a camera based sensing system to detect the environment including fruits, leaves, and branches, an efficient computing and processing algorithm to extract the useful information from the environment, a mechanical manipulation system for reaching the target fruit location, an end-effector tool to harvest/pick the target fruit, and a conveyer system to place the harvested fruit into a container/bin. The process of robotic tree fruit harvesting begins from detecting the fruit using a camera vision system and finding the location of the fruit so that the mechanical manipulation system could reach target fruit and an effector tool could detach it from the tree. With advancement in the imaging and sensing and technologies, numerous studies have reported different vision-based techniques for getting useful information for fruit feature extraction including color, size, shape, and texture, etc., localization, and tracking (Silwal et al., 2016; Tabb et al., 2006). Environmental sensing or fruit detection could be done using a single viewpoint or multiple viewpoints, however, the vision system has certain challenges due to various factors including heavy occlusion by the leaves, fruit clustering, unpredicted tasks, unstructured environment, and variable lighting conditions (Zhang et al., 2019). The second step in the robotic fruit harvesting is to approach the fruit using a mechanical manipulation system. This step primarily involves the optimal trajectory planning to position the end-effector at required location and orientation, and sequencing or prioritization of fruit harvesting to minimize the path length, time, energy or parameters that affect the performance of a robot (Silwal et al., 2017). The manipulator degrees of freedom (DoF) is critical for precise positioning and orientation of the end-effector. In general, most widely adopted manipulators for agriculture usually have five or more DoFs. The target could be approached in two different ways. The first way refers as visual surveying, which involves detecting the fruit coordinates in the 2D image and continuously changing the manipulator joint positions to keep the fruit at the same image coordinates at all time (Ringdahl et al., 2019). The second approach is using a global camera system, which involves mounting a camera at a fixed position to take images at the beginning of a harvesting cycle to estimate the position of all fruit in the camera view. The approaching path could be established using the inverse kinematics for each initial and final position of the end-effector; however, an accurate calibration between vision system and manipulator is essential for reaching the target precisely (Zhang et al., 2019).

12.4.1 Machine Vision for Fruit Harvesting

Robotic tree fruit harvesting requires two major tasks to be done; one is to accurately recognize the fruit in the tree and second is to be detaching the fruit without having any damage on it or any particular part of the tree. An illustration of machine vision based automatic tree fruit detection is presented in Fig. 12.5. Machine vision uses advanced sensors (i.e., cameras) that captures the images, processing hardware and software algorithms to automate visual inspection or detection and localization tasks and accurately/precisely guide the end-effectors to successfully harvest the fruits from the tree branches. For robotic fruit harvesting, the fruit automatic detection and localization have been conducted mainly by using machine vision techniques. Camera sensors are used to capture the images from the trees, which is considered as the first step toward fruit detection as well as fruit harvesting.

Camera Sensors for Fruit Harvesting

Camera is an optical instrument used to record visual important features in the form of image or video signals to distinguish fruits from leaves, trunks, branches, and other neighboring objects in the real-time orchard condition. A camera lens takes all the light beams skipping around and utilizes glass to divert them to a single point, making a sharp picture of the objects. Four types of cameras are used in fruit recognizing so far including black and white, color, spectral, and thermal cameras, and three types of cameras are used for fruit localizing including color, stereovision, and time-of-flight cameras. A color camera uses filtering to look at the light in its tree primary colors including red, green, and blue. After recording all three primary colors, the camera combines them to create the full spectrum. Color camera captures light across three wavelength bands in the visible spectrum (400–700 nm). Spectral camera uses multiple electromagnetic spectrum bands (e.g., near-infrared: 750–900 nm; hyperspectral: 400–1100 nm in steps of 20 nm, and so on) and usually

Fig. 12.5 An illustration of machine vision based automatic tree fruit detection

go beyond color camera to collect objects information. Conversely, thermal camera detects temperature by capturing different levels of infrared light using wavelength of 1–14 μm to distinguish between objects. Apart from single camera lens, stereovision camera is consisted of two or more lenses with separate image sensors to see the same object that can provide 3D structure of the object. Conversely, time-of-flight camera measures the distance between the camera sensor and the object for each point of the image by calculating the time difference between emission and return of an artificial light signal, after being reflected by the object. Favorable circumstances and drawbacks of various camera sensors are discussed in Table 12.2.

Earliest studies dated back in the late 1980s initiated the application of black and white cameras for fruit detections aiming to ensure first step respecting to the development of automatic fruit harvesting system (Whittaker et al., 1987), however, successes were not sufficient to move forward because of the sensor's limitations and inability to acquire useful color information/features. Color is most prominent features for tree fruit detection, especially for ripe fruit detection (e.g., red apple,

Table 12.2 Advantages and disadvantages of different camera sensors used for tree fruit detection and localization

Camera types	Advantages	Disadvantages
Black and white	• Less affected by lighting condition • Relatively cheaper in price	• Only provide black and white color; are not suitable for distinguishing multiple objects
Color	• Provide color information about fruits, leaves, trunks, branches, and background • Easily to find features • Less expensive	• Highly sensitive to the illumination variations • Only provide 2D information of the objects
Spectral	• Acquired both color and spectral information • Able to distinguish differences between similarly colored objects	• Time consuming • Large data storage capacities are required • Costly and complex in operation
Thermal	• Not affected by the color of the fruits • Does not require an illumination source and possible to use under low-light condition	• Limited operation time (narrow range) during day • Size of the fruits greatly affected the performance
Stereovision	• Ability to capture three-dimensional images • Robust enough for real-time field applications	• Susceptible to lighting condition • Computationally expensive • Depth range is highly dependent on the baseline distance
Time-of-flight	• Data can be acquired at night or even in low-light conditions • Provide 3D image of the objects that help to localize fruits • Able to extract the distance information of the object • High precision at long range measurement	• Low pixel resolution • Most of the sensors are affected by direct sunlight • High cost

yellow orange, dark yellow peach), which is not possible to extract from black and white camera specifying the need to use color cameras. Color cameras introduced in the early 1990s provides the first time opportunity to detect fruit based on color features along with geometric and texture information. Success of the color cameras is adequate when the ripe fruits color is different than leaves, branches, and background (e.g., red apples, yellow citrus, and yellow pear fruits in green background). The sensor performs poorly when the fruit color is same as the leaves or background considering only color information. Another problem noticed that the color camera is highly susceptible to the illumination variations and make the sensor unsuitable in the orchard condition. Spectral camera sensors came up in addressing the color similarity problem between fruit and background by providing spectral information along with special information about fruits, leaves, branches, or other objects (Kondo et al., 1996). Potential of spectral camera has been delineated for fruit detection using different wavelengths considering the appearance of different fruits. However, major limitation is reported for the longer data acquisition and processing time, especially using hyperspectral camera (Kim & Reid, 2004) that forged the spectral sensor difficult and challenging for real-time detection. Thermal cameras also utilized for fruit detection aiming to solve the color similarity problem between fruits, other objects and background, but performance of this types of sensors is greatly affected by fruit size and direct sunlight exposure. The accuracy of the thermal camera is lower in shaded and high canopy density area because there is not any significant temperature difference existing between fruits and other objects including leaves, branches, and background in those regions. Aside from the fruit recognition sensors, the stereovision and time-of-flight cameras are mainly used for fruit localization. Stereovision camera measures the position of the target objects from the camera sensor by performing the stereo matching of multiple images acquired using various cameras installed in various arms. However, performance of this vision system is affected by illumination variation, wind speed and direction, and efficient of the hardware component (Plebe & Grasso, 2001). Another major limitation is long computational time and complexity. Time-of-flight camera introduced for the fruit localization due to its faster data acquisition and processing speed. In the last few years, time-of-flight cameras showed promising potential for fruit localizing which is also suitable for outdoor orchard environment especially using an RGB-D (Red, Green, Blue-Depth) camera (Fu et al., 2020), which provides the RGB information along with depth and infrared information; however, direct sunlight exposure can affect the accuracy of the sensor.

Fruit Detection and Classification for Harvesting

The first step of camera vision system for fruit harvesting is the image acquisition stage where images are captured from the tree fruit orchards. After the image has been captured, different processing methods have included feature extraction and classification can be applied to the pre-processed image to detect fruits from the leaves, branches, and other objects background. Color is one of the most valuable features used in image processing based detection to differentiate fruits from other neighboring objects (i.e., fruits, foliage, or branches) presence in orchard environment. Distinguishing oranges from the natural background was the first attempt toward developing robotic harvester using color features and detected 75% of the fruit pixels successfully showed the potential of applying color features for fruit detection (Slaughter & Harrell, 1989). The accuracy of the fruit detection was improved in the later years up to 88.0% using only color features (Bulanon et al., 2002; Qiang et al., 2014), however, fruit detection accuracy based on color features is greatly affected by illumination variation, fruit variety, fruit maturity level, and uncontrollable orchard environment. Illumination variation during image capturing can provide different light intensities; therefore, it would be very difficult and challenging to detect fruits under uncontrollable lighting environment using color features. Geometric features mainly considering the size and shape of the fruits are being used to address the color feature problems especially when the green fruits need to be detected from green leafy background. These types of features are also less susceptible to illumination variations which make it suitable for real-time orchard condition unless the blurred image caused during data acquisition due to high wind velocity. Lu et al. (2018) detected green immature citrus fruit using geometric features and achieved 82.3% of precision rate. Performance of geometric features (i.e., searching circles) is boosted up to 85% of accuracy for detecting the green apples from green background when they were visible in the captured images, but the occluded apples caused the false-positive detection (i.e., considered leaves, stems, and branches as fruits) (Linker et al., 2012). Conversely, iterative Circular Hough Transform (CHT) and blob analysis based geometric features provided over 90% of accuracy for "Jazz" and "Fuji" apples detection in clearly visible and partially occluded apples on tall spindle architecture canopy trees (Silwal et al., 2014). However, the major problem using the geometric features is the occlusion of fruits, which results in the poor performance due to alter in size, shape, and other geometric characteristics of the tree fruits. Textures are another important feature which is not affected by the surface color so it can also be used to detect fruits from the similar color background (i.e., leaves and stems). Tree fruits generally have smoother surfaces compared to the leaves, branches, stems, and other objects. Detection of fruits using texture features isolates the surfaces with the homogeneous texture and afterward distinguishes the edges of the isolated surface (Zhao et al., 2005). Performance of these types of features for fruit detection is not so high when only the texture features are used. Considering a novel Eigen Fruit approach and blob analysis, a Gabor wavelet based texture analysis was utilized to detect green citrus and achieved an accuracy of 75.3% with 27.3% false detection (Kurtulmus et al., 2011). Variable illumination condition, complexity of the fruit background, and varying fruit size have tremendous effect on the texture properties of fruits reducing the accuracy of the detection (Zhao et al., 2005; Kurtulmus et al., 2011). Combining texture features and other features (i.e., color and geometric) can enhance the accuracy up to 89% while detecting "Golden Delicious" and "Jonagold" apples (Stajnko et al., 2009). Besides color, geometric, and texture features, a 3D shape of the fruit was reconstructed (Fig. 12.6) for improving the detection accuracy, but the

Fig. 12.6 Original acquired image (upper left image), color segmented version (right top image), cleaned version by applying morphological operators (bottom left image), and finally, 3D shape analysis (bottom right image). (Adopted from Rakun et al., 2011)

methodology was only justified hypothetically, and therefore ample tests were required to show its reliability in real-time orchard applications (Rakun et al., 2011).

To perform successful fruit detection from the other neighboring objects, the image classification is required after extracting valuable features from the images. Supervised classifiers have included Bayesian and K-nearest neighbor; unsupervised classifiers included K-means clustering; and soft computing methods included artificial neural network (ANN) and support vector machine (SVM) were used so far for fruit detection. Bayesian is one of the multivariate statistical classification techniques used widely for object detection/classification based on prior knowledge and probability distributions also called posterior probability theory. Bayesian discriminant was used to classify oranges considering the color information and classified 75% of fruits successfully (Slaughter & Harrell, 1989). Considering the similar method, Juste and Sevila (1992) applied a pattern classification method of Bayes's rules for citrus fruit detection and reported up to 90% of accuracy. Although the higher detection accuracy showing the potential of Bayesian classifier for fruit detection, the major drawback is that the prior probabilities information require in detection that can be affected due to the changes of color value of fruits caused by the illumination variations (Chinchuluun et al., 2007). Contrarily, K-nearest neighbor (KNN) based supervised classifier, also susceptible to illumination variable is used to classify unknown feature vector to the class by measuring the closeness measure between the obscure and each training samples. To detect the green apples in captured RGB image, a KNN classifier was used in two dataset recorded in direct illumination and diffusive light conditions and reported 85% and 95% of accuracies for correct detection (Linker et al., 2012). Another significant impediment of KNN based algorithm is huge processing time to group an obscure feature vector which makes it inadmissible for real-time field applications (Mitchell, 1997). Besides supervised classifiers, K-means clustering based unsupervised machine learning classifier is also used for fruit detection, which allocates every data point into the nearest cluster dependent on their intrinsic distance between one another. However, the performance of K-means clustering in fruit detection is not so high using different images including color and thermal especially for green apples (Wachs et al., 2010). The soft computing methods including ANN and SVM are also supervised machine learning algorithms become so popular and widely accepted for fruit detection in the orchards (Wachs et al., 2010; Qiang et al., 2014). An SVM based classifier isolates the two classes with a greatest edge between them by a hyper-straight plane to classify objects. Tao and Zhou (2017) detected apples, branches, and leaves using a multi-class SVM classification method and achieved an accuracy of 94.64%, 47.05%, and 75%, respectively, while acquired images by a Kinect V2 camera sensor. Using the same camera, Lin et al. (2019) detected citrus fruits based on SVM algorithm and reported a F1-score of 91.97% using color, gradient, and geometry features. Qiang et al. (2014) used RBF kernel function for applying a multi-class SVM classifier to detect citrus fruits from the leaves and branches by using color features and reported a detection accuracy of 92.4%. The authors identified that illumination variations, fruit occlusion, and immature fruit were the major factors reducing the classifier as well as system performance. Apart from SVM based soft computing method, an ANN based machine learning algorithm detects the fruits by learning specific patterns/model defined by the training data through the iterative training process. To develop orange picking robot, a neural network (i.e., back propagation) based machine learning algorithm along with color features was used to detect oranges from the images captured at different lighting conditions and achieved an accuracy of 87% with 15% false positive and 5% false negative (Plebe & Grasso, 2001). Additionally, Kurtulmus et al. (2014) compared three classifiers including a statistical classifier (i.e., discriminant analysis), an ANN, and an SVM performance for immature peaches detection under various illumination conditions and reported the ANN classifier performed better (82%) than discriminant analysis (80%) and SVM (62%). Despite both supervised and unsupervised machine learning classifiers showed good performances, but most recently, significant advancement and effort have been accomplished through the application of deep learning algorithms on fruit detection due to its larger learning capabilities resulting in higher performance and precision, which is based on multiple layer ANNs (Koirala et al., 2019).

Deep learning is one of the machine learning techniques that can learn the features themself from raw data and provides a hierarchical representation of the data through deeper neural networks and various convolutions. Object detection using deep learning algorithms becomes more popular due to their higher detection rate and fast detection speed in the past years which is applied in various fields of research (Gao et al., 2020). Deep learning networks including convolution neural network (CNN), region-based CNN (R-CNN), Fast R-CNN, Faster R-CNN, You Only Look Once (YOLO) network are increasingly applied in recent years for orchard management and provide an excellent framework for fruit detection (Bargoti

& Underwood, 2017; Fu et al., 2020; Gao et al., 2020). Considering rapid progress and improvement in deep learning algorithm, a Faster R-CNN model was tested and achieved an accuracy of 95% for Fuji apple detection (Gené-Mola et al., 2019). To reduce detection time and improve detection accuracy, the convolution and pooling layers of Faster R-CNN were modified by Wan and Goudos (2020), the developed model was tested for green apple and orange detection and achieved 92.51% and 90.73% of accuracies, respectively. Numerous deep learning algorithms (i.e., Yolov3, R-CNN, and VGG-16) application for apple fruit detection was reviewed and reported the detection accuracies ranged between 84% and 95% (Koirala et al., 2019). By combining Gaussian Mixture Models based semi-supervised method and deep learning method, Häni et al. (2020) developed a novel semantic segmentation-based approach for fruit detection and counting and reported the performance can be better than a single deep learning model with detection accuracies ranged from 95.56% to 97.83%. Compared to the conventional machine learning models, emerging deep learning algorithms are showing promising potential and benefit with the higher detection accuracy and the faster detection speed that are necessary for robotic fruit harvesting in real-time orchard condition.

Fruit Localization for Harvesting

Next step of detection is fruit localization, another very essential part of computer vision system for guiding robotic end-effectors to grab and detach fruit from the tree. Inaccurate fruit localization information causes failure of the end-effectors in successful fruit harvesting. Despite there are different types of challenges exist due to uncontrollable orchard condition (i.e., wind velocity, fruit occlusion, etc.), studies have conducted toward the accurate fruit localization (Bac et al., 2014). Fruit localization began with using a single black and white camera to identify fruit centroids aimed to extract 3D coordinate for grasping fruit from the branches by developing a mathematical transformation model (Parrish & Goksel, 1977). After about a decade, the color camera had been applied to identify fruit centroids by stick out the telescopic end-effector. This was made conceivable when the camera mounted at the center of the end-effector, at that point the fruit centroid in the captured image lined up with the pivot of the prismatic joint (Slaughter & Harrell, 1989). Apart from the fruit centroids, studies also conducted to localize fruit peduncle by using the color camera for the ease of fruit harvesting especially for detachment (Bulanon et al., 2001). For obtaining more precise fruit location, the laser systems were also utilized in some extend along with camera sensors where 2D location of fruit accessed via camera vision and a laser sensor used to measure the distance from the end-effector and fruit (Bulanon et al., 2004). Besides single camera applications for fruit localization, several attempts were reported using more than one camera by applying stereovision where fruits were located by triangulation. However, the main problem using a stereovision system was the correspondence problem where obtaining reference points in the practical view is difficult (Wang et al., 2013). Researchers attempted to solve the correspondence problem while using stereovision system, but they ended up with the error less than 20 mm due to densely distributed tree canopies (Si et al., 2015). Aside from stereovision, the red-green-blue-depth (RGB-D) cameras by Kinect V2 offer a new approach to extract 3D space for detecting and localizing fruits simultaneously (Fu et al., 2020). Studies reported that RGB-D camera along with advanced machine learning algorithm including Bayes classifier and Faster R-CNN can be appropriate for real-time orchard conditions with detection/recognition accuracies went from 92% to 95% and localization errors of 7.0 ± 2.5 mm, -4.0 ± 3.0 mm, and 13.0 ± 3.0 mm in x, y, and z axis direction, respectively (Zheng et al., 2018; Lin et al., 2019). On the other hand, several studies reported RealSense RGB-D camera performed better than Kinect V2 with an image resolution of 1280×720 pixel and sample frequency of 90 frames per second compared to 512×424 and 30 (Mejia-Trujillo et al., 2019). Considering the promises of RealSense RGB-D cameras shown in fruit detection and localization, we can assume that it could be an effective tool for real-time orchard applications in the future with high accurate manner.

12.4.2 Fruit Removal Dynamics and End-Effector Development

Fruit detachment is one of the major tasks in the robotic fruit harvesting. Prior to designing a fruit picking end-effector, it is necessary to investigate the dynamics for fruit detachment. The information provided by the dynamics includes picking or cutting force/torque, picking angle, and fruit detachment motion. Typically, robotic picking requires fruit detachment motions planned and performed with sufficient grasping forces applied to the target fruit (Tillett, 1993). For a human picker, an apple is detached by gently griping it with fingers and twisting it around the connection point of its stem and limb. At the same time, pickers put one finger on the connection point to minimize the movement of the connection point or the pivot. Reduced movement of the pivot point will increase the torque around this point and thus increases effectiveness of fruit detachment. Preliminary tests showed that twisting of apples by attaching the pivot could achieve more effective and efficient detachment than pulling them (Bulanon & Kataoka, 2010).

To provide baseline information for developing a conceptual robotic end-effector for apple picking, a series of fundamental physics studies for apple picking were conducted by a Washington State University (WSU) research group with mimic human picking operations (Fig. 12.7, He et al., unpublished document). These physics included the picking orientation, the applied force/torque, and the relations to the apple weight and stem length. Three flexible force sensors were mounted on three fingers of picker to measure the force applied to the apple surface during picking operation. A hand-held picking device, consisting of a gripper and a torque sensor, was built to measure the twisting torque for removing apples from the tree. Tests showed that picking apple along the peduncle direction obtained much higher

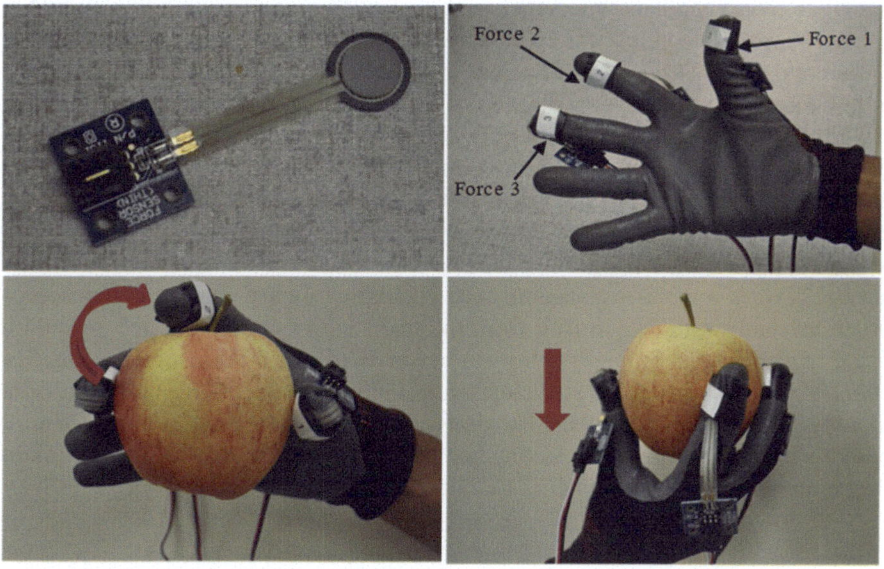

Fig. 12.7 Hand picking apple force measurement setup and method. (**a**) force sensor; (**b**) sensor equiped picking glove; (**c**) picking apple by twisting; (**d**) picking apple by pulling

picking efficiency. Force applied to the surface of apple varied from different pickers and different fingers, the force applied was from 0.43 ± 0.27 to 1.16 ± 0.33 kgf in this study, also the applied force showed positive relation to the fruit weight. The results also indicated that the detachment torque increased as the increasing of apple weight, and the picking angle increased as the increasing of apple stem length. Furthermore, Davidson et al. (2016) investigated the hand picking dynamics for robotic apple harvester design. The results indicated that each variety has different detachment force. And the study also suggested to use a tactile sensor in a robotic end-effector to potentially determine the point of fruit separation and minimize the path traveled by the end-effector during harvesting. Li et al. (2016) indicated that bending motion could improve the fruit detachment performance for apple picking. To remove a fruit from the branch, bend-and-pull picking will require less energy than straight pulling along stem growth direction. Flood (2006) designed a robotic citrus harvesting end-effector and a force control model using physical properties and harvesting motion tests.

End-effector is a critical component for a harvesting robot, which is used to detach fruits from the tree with appropriate force and motion. Designing an end-effector tool for fruit harvest can be a challenging task due to the complex canopy environment and unique fruit characteristics. The design should consider the mechanical and spatial requirements including size, shape, weight, and maneuverability, and the task object requirements including physical, horticultural, and biological properties (Kondo & Ting, 1998). Researchers in the past have put a lot of efforts on developing end-effectors to harvest different kind of crops including orange, tomato, eggplant, cucumber, and apples (Muscato et al., 2005; Whittaker et al., 1987; Van Henten et al., 2003; Hayashi et al., 2002; Davidson & Mo, 2015). Different detachment motions also have been tested in these studies, such as pulling, twisting, cutting, and combination of two. Zhang et al. (2020) did an extensive review for different robotic grippers used for agricultural applications along with their grasping and control strategies. Many picking end-effectors use either two or more fingers to grasp the fruit to detach it (Burks et al., 2005). Some of these end-effectors used air to suck the object and grip it, then use scissors to cut the peduncle to detach the object, which may cause damage to the fruit peduncle. Conversely, suction devices comprised a vacuum cup to hold the fruit and combined with appropriate mechanism to detach fruit form the tree such as cutting the peduncle with blade mounted on the fingers (Hayashi et al., 2014), or a twist motion (Yaguchi et al., 2016). Bac et al. (2017) developed a four-fingered hand with a pair of scissors mounted on top to cut the stem. The designed hand may be more suitable for fruit with longer stems, however detecting and locating the stem is a challenging task in the complex canopy environment.

12.4.3 Harvesting Robot Manipulation

Robotic Manipulators

The tree canopy–machine interaction could be interpreted as manipulation of a machine (robotic arm/manipulator) within tree canopy to reach the identified fruit locations to perform the harvesting using an end-effector. The robotic arm or manipulator is the mechanical system like a human arm, usually comprised of links connected in a series joints that perform the intended tasks in the one-two-three-dimensional space. Each joint in the manipulator has one DoF and the kinematic dexterity is directly related to the number and type of joints in the manipulator (Burks et al., 2018). The currently available industrial manipulators are designed to perform repetitive tasks with uniform objects in unconstrained workspace. Conversely, the adoption of robotic manipulators for fruit harvesting has many challenges as agriculture is a constrained dynamic environment where the target objects vary in shape, size, position, and orientation (Simonton, 1991). The successful adoption of robotic manipulators requires consideration of its working environment (Kondo & Ting, 1998; Simonton, 1991). Thus, the robotic manipulators for tree fruit harvesting should be designed considering different factors such as canopy structures, and branch density, etc., for safe operation in the unstructured agriculture environment.

In an agricultural robot, the first joint of the manipulator is connected to the base of a mobile platform, and the last joint of the manipulator is an integrated end-effector unit, which consists of a tool/gripper to perform the required task is attached. The manipulator mainly works for the positioning of the end-effector close to the target fruit and then move the harvested fruit to the collection bin/container. For tree fruit harvesting, the manipulator could be designed with various configurations, based on total DoFs, and different combination of joint types. The selection of joints configuration is critical as it affects the kinematic dexterity and spatial requirements during manipulation of the robot to attain different positions and orientations of the end-effector. Based on total DoF selection, a manipulator can be designed with different number of joints starting from three or higher. However, increasing the number of joints (DoF) exponentially increase the computation and control complexity (Choset et al., 2005). A three DoF (3 DoF) manipulator is the most common choice due to its simple design and control architecture. For a known Cartesian position of the fruit, the 3 DoF manipulator (Harrell et al., 1990) could easily reach the desired position using the inverse kinematics. However, the end-effector (gripper) could not alter the orientation due to lack of DoFs. As the fruits on a tree grow at random orientations, the manipulator should have the ability to grasp the fruit from different orientations. The manipulator performance will be decreased if the fruits are occluded behind leaves or branches and the gripper may not be able to harvest the fruit. Adding additional DoFs to the manipulator such as a four DoF (Tanigaki et al., 2008) or five DoF (Zahid et al., 2020a) could be a solution to the problem to some extent by giving the capabilities to adjust the orientation of the end-effector, but harvesting the fruits present behind the obstacles deep inside the canopy could still be problematic. To completely describe the six components of the Cartesian space including three positional (x, y, and z) components, and three angular (yaw, pitch, and roll) components, the manipulator should have six joints in its assembly. Thus, the agricultural manipulator should have at least six DoFs (Onishi et al., 2019) to attain all possible orientation and position in the workspace. However, with higher DoFs, the kinematics of the manipulator results in two poses (elbow up and elbow down) for any desired position and orientation, which can lead to a higher chance of manipulator collision with the branches at some poses, causing damage to manipulator, fruit, or tree. Another problem with six DoFs is its limitation of a single pose in the workspace, and it may not be able to avoid all the obstacles, which is essential for the safe operation of robot (Burks et al., 2018). Considering the unstructured canopy environment, the manipulator with at least one excess DoF such as a seven DoF (Mehta et al., 2014; Silwal et al., 2016) for the positioning and orientation is a better solution to avoid collisions, also known as redundant manipulators. These redundant manipulators can attain multiple orientations for any target position to avoid collisions by changing the pose to the optimal. Although the redundant manipulators improve the kinematic dexterity to grasp the fruit by attaining different orientation, it also increases the complexity for manipulation controls (Fig. 12.8).

The performance of the tree fruit robotic manipulator could also be affected by the type of joints such as prismatic, rotational, or combination of both joints, used for its assembly. Figure 12.8 shows few examples of different configurations of first three joints for a six DoF manipulator integrated with spherical wrist gripper end-effector. The first three joints, referred as Cartesian positioning (x, y, and z) links, move the end-effector in the proximity of target fruit. The last three joints, referred as wrist, alter the orientation of the end-effector for accurate positioning at the target. Each of the shown manipulator has a different workspace and spatial

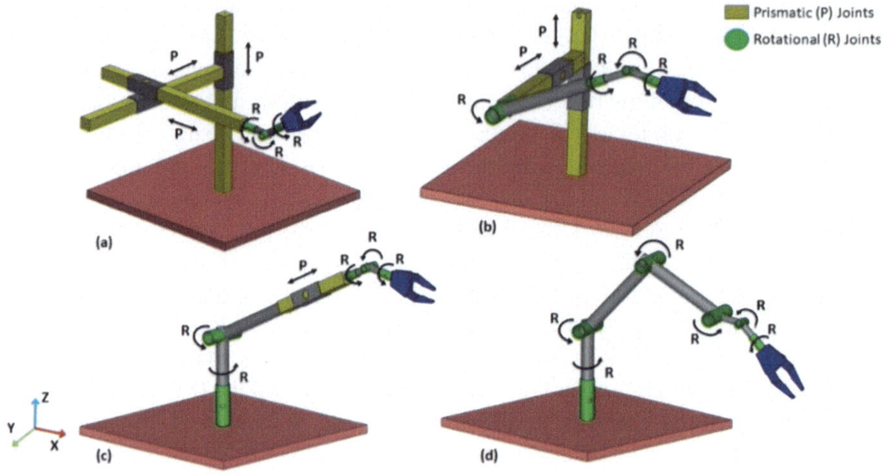

Fig. 12.8 Illustration of manipulator with different joint configurations and wrist end-effector; (**a**) Cartesian (PPP), (**b**) Cylindrical (PPR), (**c**) Spherical (RRP), and (**d**) Articulated (RRR)

requirements for manipulation. During maneuvering, each joint contributes to alter the manipulator pose and end-effector position and orientation in the canopy. When the manipulator starts maneuvering inside the canopy, the major change in position and pose of the manipulator link is due to the positioning joints and a small contribution is from the wrist joints. With greater degree of pose change, the chances for collision with branches increase within the canopy; therefore, the joints for positioning should be selected which allow the minimum change in pose of the manipulator during maneuvering. Zahid et al. (2020b) developed apple tree pruning manipulator by integrating three prismatic joints (3P DoF) with three revolute (3R DoF) joints. The integrated tree pruning manipulator showed promising results as it was able to reach all selected branches with lower pose change, which reduced the collision potential. In general, the Cartesian/prismatic joints have low pose change attributes, as the orientation of the links remains the same irrespective of the joint movement. Thus, a manipulator could be developed considering different joint types to reduce the spatial requirements. For example, the positioning joints as shown in Fig. 12.8a may perform positioning motion outside the canopy with a slight pose change and could have less spatial requirements for the maneuvering of the spherical wrist end-effector within the tree canopy for reaching target fruits. Similarly, when aiming to reach the fruits inside the canopy, the maneuvering within the tree canopy for reaching target fruits using different joint combinations as shown in the figure could affect the manipulator pose change differently. Thus, the manipulator design should consider the requirements for different tree features such as canopy sizes and structures to ensure that the end-effector reaches all positions in the canopy with minimum spatial requirements and least chances for collision with the branches.

Robotic System Control

An agriculture robot must solve multitude of problems to perform the operation such as fruit harvesting, thinning, and pruning, etc. Unlike industrial robots, where a repetitive work is performed for same objects, the target fruits are located at different position and orientation. Thus, during agricultural operation, there is no repetition of the same motion/path, and the robot needs the information about every target to perform the target-specific motion. The manipulator movement and control could be established using the information from the sensing or vision system, also referred as vision-based manipulation control. The vision-based control for the harvesting robot is essential as the manipulator could use the visual information for path planning and motion. The inefficiency of vision-based control is one of the primary factors limiting the performance of the harvesting robot. The vision-based control is categorized into two types: visual navigation or visual servo control and eye-hand coordination or global camera system (Zhao et al., 2016). The global camera system is an open-loop control system which is operated based on "3D positioning." The camera system scans the complete scene to detect all fruits and then start moving to the target fruits. The control efficiency in terms of the end-effector positioning depends on the accuracy of the vision system, calibration of manipulator and camera system (Yau & Wang, 1996). To achieve higher efficiency, the vision system may be consisting of stereovision or range sensors to precisely measure the distance to the target fruits. However, for open-loop visual control, an accurate kinematic model of the manipulator is essential for the path planning to reach target fruits. Han et al. (2012) successfully established the open-loop visual control for path planning using a color stereoscope camera and a laser sensor. The execution time for successful harvest was less than 7 s per fruit. However, one downside of the open-loop visual control is low efficiency in the situations where the fruit is under the influence of wind or movement from other reasons.

The second category of visual based control is the visual based feedback control loop, also referred as visual servo (Corke & Hager, 1998). The visual servo is a closed-loop control system which is operated based on "concurrent looking and moving" as a dynamic system. The visual servo used the image features extracted from the camera-in-hand system to control the position and orientation of the end-effector on the fly (Hashimoto, 2003). A major advantage of visual feedback control is that the performance does not rely on the accuracy of the kinematic model and the calibration of vision and manipulator system. However, one important consideration to achieve high efficiency of visual servo control is that the bandwidth of the vision controllers should match the frame rate of the visual information coming from the camera system. Zhao et al. (2011) successfully implemented the visual servo controls in an apple harvesting robot. Font et al. (2014) combined open-loop and visual servo controls. Using the open-loop control, the end-effector moves quickly in the proximity of the target fruit, followed by adjusting position and orientation through guidance from visual servo to harvest the fruit. The general comparison of these two types of control is given in Table 12.3.

Table 12.3 The comparison of two types of vision based controls

Visual control	Principle	Advantages	Limitations
Global camera control/ open-loop control	Control based on precise 3D positioning	Simple and smooth controls; higher stability region	Required high accuracy of vision system; required accurate manipulator and camera calibration
Visual servo control/ closed-loop control	Control based on manipulator-vision dynamic interaction	Real-time applications; vision-manipulator calibration not required	Problems related to local minima of unpredicted camera paths; required high bandwidth

Collision-Free Path Planning

The path planning of a harvesting robot is one of the most important components for successful operation. The path planning strategies including picking order and obstacle avoidance, etc. are essential to achieve higher harvesting efficiency as well as safety of the robot during interaction with the canopy. With the advancement in the computing theory, the path planning and controls are becoming more reasonable and efficient (Jia et al., 2020). Different path planning and harvesting order strategies are discussed by various researchers. The path of the robot can be established using the kinematic model of the manipulator, which calculates the displacement toward the target fruit position. The manipulator uses the inverse kinematics equations to establish the path using open-loop control (Yau & Wang, 1996) or visual servo control (Hashimoto, 2003). The kinematic model considers the body dimensions of the robotic manipulator and the target position, so the collisions could be possible during the operation, which could result in the damage of robot or the tree. A separate set of algorithms are required to avoid the collisions during operation. The task or harvesting order planning strategies are also studied by many researchers. Most common method is to detect and localize the target fruit and the path for each harvesting cycle starts from the home position of manipulator (Roldan et al., 2018). Researchers have also developed harvest sequencing schemes to optimize the harvest cycle time. The Traveling Salesman Problem (TSP) is widely reported scheme used for harvest sequencing. Yuan et al. (2009) implemented an algorithm to covert the apple harvesting task into a three-dimensional traveling salesman problem (TSP) to get the finite field information and then used ant colony algorithm to optimize the path planning. Some other task planning schemes were also developed by researchers such as harvesting all detected fruits in the scene (Baeten et al., 2008) and optimal harvesting sequence by moving fruit-to-fruit for reducing the cycle time (Reed et al., 2001). Plebe and Anile (2002) obtained an efficient harvesting sequence plan by converting the harvesting task into twin traveling salesman problem (TTSP). All these path planning and task planning strategies could be feasible for reaching target following the optimized path. However, the manipulator collision with branches could still be a problem and needs to be addressed as it is essential for the safe operation of the fruit harvesting robot.

The tree fruit canopies usually have complex structure with branches growing in the random direction and orientations, which limits the manipulation capabilities of the robotic manipulators. To ensure the safe and successful robotic operation, it is essential to establish the collision-free paths for the robot movement. The collision-free path refers to the movement of manipulator and end-effector toward the target fruit without hitting the branches. In the recent years, the challenges of obstacle detection and collision avoidance for tree fruit harvesting robot have gained interest from the researchers. The obstacle detection is the task performed by the machine vision system such as camera and proximity sensors, etc. The collision detection sensors can be integrated with the end-effector such as a position sensor in an apple harvesting robot (Zhao et al., 2011), Light Imaging Detection and Ranging (LIDAR) sensor in a cherry harvesting robot (Tanigaki et al., 2008), and a camera for litchi harvesting robot (Cao et al., 2019). However, the obstacle avoidance task presents more challenges. For collision-free path planning, many researchers have proposed algorithms including grid-based, neural networks, and random sampling. Grid-based algorithms such as A*, or Phi* or ant colony, etc. are suitable for multi-objective problems but computationally expensive for complex environment and could give satisfactory results with up to two or three DoF manipulators (Nash et al., 2009). With the increase in the DoFs of the manipulator, the computational complexity and planning time increase exponentially (Choset et al., 2005). As mentioned earlier, the tree fruit harvesting robot should have at least six or seven DoFs, giving manipulator the flexibility in the poses to avoid the obstacles. Most of these grid-based path planning algorithms may not be suitable for agricultural application. Conversely, the sampling-based planning approaches such as rapidly exploring random tree (RRT), RRT*, or bi-directional RRT are probabilistic-complete algorithms, i.e., if solution exists, they find path, and perform better for high dimensional complex problems and are less influenced by the DoFs of the manipulator.

Nowadays, the RRT based search algorithms are widely adopted for collision-free path planning in the agricultural environment. Nguyen et al. (2013) proposed a framework for motion and hierarchical task planning of a nine DoF manipulator for harvesting apples. The strategy was first implemented in simulation and then real-time communication between sensing and execution was successfully established in the orchard environment. The authors used seven different sampling-based planning algorithms including RRT, and RRT connect, and concluded that the RRT connect as most efficient for path planning in terms of processing time. However, the nine DoF manipulator has enough flexibility in the pose to avoid collision with branches. Cao et al. (2019) successfully also used RRT for collision-free path planning for six DoF litchi harvesting robot. The path calculated using the sampling-based algorithms is not the optimal solutions, as it has less convergence speed and more processing time. These deficiencies could be minimized by combining optimization algorithms such as genetic algorithm (GA) to reduce the path cost (LaValle, 2006). Also, the random sampling-based algorithm have longer path length due to intrinsic search properties (generating and connecting random nodes in the search space). The path smoothing method which aims to omit the unnecessary nodes (Zahid et al., 2020c) can be used to reduce the length of collision-free path. For random sampling-based search algorithms, the path planning time depends on the sampling resolution, which should be optimized, considering the required path success rate.

12.5 Conclusions and Future Directions

Robotic harvesting systems have been investigated in the past decades, the enhancements in both technologies and horticulture have bring much more promising for the adoption of these systems for agricultural applications. For tree fruit crops, tree structures in modern orchard are getting much simpler with high-density canopy systems. These tree systems are much more robot-friendly for implementing robotic harvesting system by comparing to the conventional tree systems. While even with these trees, the harvesting task is still relative complex due to the natural of biological system. A successful robotic harvesting system would be considered as accurate, robust, fast, or even inexpensive system. Therefore, the critical points for success of robotic harvesting for fruit trees are the accuracy of fruit detection, the spatial requirement of picking end-effector, and the efficiency of picking operation (time for fruit identification and the time for maneuvering the end-effector).

The current research on tree fruit harvesting robots mainly focused on developing vision systems for accurate detection and localization of the target fruits. However, the improvements in many other components including the manipulation controls, optimized harvest sequencing, and obstacle avoidance are also required. The robot path planning is critical for accurately reaching the target points. The path planning mainly involves three operations: manipulation controls, task sequencing, and collision avoidance. With the recent advancement in computing technologies and control algorithms, there are many opportunities of developing efficient controls for tree fruit harvesting robots. As discussed earlier, the vision-based manipulation control (open- and close-loop) is critical for the robotic harvest operation as the target location is unknown, the type of control scheme should be selected carefully to achieve the desired outcome. The open-loop manipulation controls could be a good control scheme, but some natural factors in the field such as wind could alter the position of the targeted fruit, making it a dynamic environment to reduce the robot performance. Conversely, the visual servo control is computationally expensive and requires highly accurate vision system for successful operation. As both types of vision-based control have advantages and limitations, a robotic harvester integrated with a combination of both open- and close-loop (global and local) manipulation controls could improve the harvesting efficiency. By doing so, the global path planning can provide the initial guideline to start the robot motion, and once the end-effector reaches the proximity of the target fruit, the manipulation can be changed to local control for accurate positioning. Using the global control scheme, the harvesting robot can have the information about the complete scene (all fruits) before the start of operation and the path can be calculated for multiple fruits simultaneously, to reduce the cycle time. Conversely, the local control can guide the end-effector to attain desired orientation to grasp the fruit, employing parallel computation.

In addition, the harvest sequencing is also essential to optimize the path length and cycle time for each fruit. The TSP-based sequencing algorithms can be a potential solution to optimize the path lengths and cycle time. Studies have been reported using TSP and other TSP variants to optimize the path length and cycle time (Yuan et al., 2009). A redundant manipulator can perform well as it has infinite pose configurations for reaching a target point. However, for redundant manipulators, optimizing the task sequencing with TSP may not be enough, require optimization of pose configuration as well, which could be solved using TSP-N-based optimal path planning (Vicencio et al., 2014). Additionally, the collision avoidance is one of the biggest challenges for tree harvesting robots. Researchers have implemented collision-free path algorithms for different robotic operations in tree fruits. The random sampling-based search algorithms such as RRT, RRT*, and bi-RRT are widely adopted to collision-free path planning because of their higher success rate. The path solutions from the random sampling algorithms are not always optimal. The recent advancement of intelligence-based optimized search algorithms such as ant colony optimization (ACO), particle swarm optimization (PSO), and genetic algorithm (GA) can provide the optimal collision-free path solutions in the constraint tree canopy environment. Adding a numerous approach poses of the manipulator to reach the target could also improve collision avoidance (Zahid et al., 2020c). Also, a redundant manipulator could perform well for collision avoidance due to its higher pose flexibility; however, additional DoFs will increase the path finding time and cost of the manipulator (Bac et al., 2017). Overall, the computation complexity of the robotic harvest operation will be increased with the addition of obstacle avoidance in the path planning scheme. Thus, fast and efficient collision-free path algorithms are required to ensure successful and safe operation. An efficient fusion of path planning algorithms, including task sequencing and obstacle avoidance is essential for successful robotic tree fruit operations.

Being an emerging technology, machine vision combined with machine learning algorithms has become a crucial factor in the development of automatic harvesting robots. The complexity of harvesting robots has been minimized to a great extent due to extensive progress in machine vision technologies, including advanced camera sensors and artificial intelligence (AI) algorithms. Time-of-flight cameras (e.g., RGB-D) have been used in recent years showing promise for fruit recognition. The potential of using all types of time-of-flight cameras is not identical. Studies reported good accuracy with RealSense RGB-D cameras, but these types of cameras have high sensitivity to outdoor illumination and could provide low-resolution images. The high-resolution cameras including but not limited to Microsoft Kinect and Zed stereo cameras might be better options instead. Overcoming image acquisition problems caused by different environmental conditions should be the key. Although scholars have carried out many studies using traditional machine learning (ML) for fruit detections, the current innovations of deep learning algorithms, including Faster-RCNN, Mask-RCNN, ResNet, and DenseNet outperformed traditional ML algorithms have been proven in different agricultural researches. The deep learning algorithms assembled with graphics processing units (GPUs) have been widely applied to increase the computing power while processing high-density data. The machine vision technology has been rigorously used in complicated and unknown plant environmental conditions for its robustness and high complexity. But at the same time, most existing machine vision-based systems are only implemented in laboratory, semi-customized, and customized environments for experimentation, resulting in a huge inconsistency between the experimental and original field conditions for fruit recognition. Due to this limitation of machine vision technologies, the adaptability of the harvesting robots to complex and unstructured environments still remains a major bottleneck problem affecting the harvesting robot's maturity and limiting the application in orchard conditions. Therefore, universal machine vision technologies need to be developed that could recognize fruits in any environmental condition.

References

Bac, C. W., Henten, E. J., Hemming, J., & Edan, Y. (2014). Harvesting robots for high-value crops: State-of-the-art review and challenges ahead. *Journal of Field Robotics, 31*(6), 888–911. https://doi.org/10.1002/rob.21525

Bac, C. W., Hemming, J., van Tuijl, B. A. J., Barth, R., Wais, E., & van Henten, E. J. (2017). Performance evaluation of a harvesting robot for sweet pepper. *Journal of Field Robotics, 34*(6), 1123–1139. https://doi.org/10.1002/rob.21709

Baeten, J., Donné, K., Boedrij, S., Beckers, W., & Claesen, E. (2008). Autonomous fruit picking machine: A robotic apple harvester. In C. Laugier & R. Siegwart (Eds.), *Field and service robotics* (pp. 531–539). Springer. https://doi.org/10.1007/978-3-540-75404-6_51

Bargoti, S., & Underwood, J. P. (2017). Image segmentation for fruit detection and yield estimation in apple orchards. *Journal of Field Robotics, 34*(6), 1039–1060.

Bulanon, D. M., & Kataoka, T. (2010). Fruit detection system and an end effector for robotic harvesting of Fuji apples. *Agricultural Engineering International: CIGR Journal, 12*(1), 203–210.

Bulanon, D. M., Kataoka, T., Ota, Y., & Hiroma, T. (2001). A machine vision system for the apple harvesting robot. CIGR E-Journal, Volume 3, December 2001.

Bulanon, D. M., Kataoka, T., Ota, Y., & Hiroma, T. (2002). AE-automation and emerging technologies: A segmentation algorithm for the automatic recognition of Fuji apples at harvest. *Biosystems Engineering, 83*(4), 405–412.

Bulanon, D. M., Kataoka, T., Okamoto, H., & Hata, S. (2004). Determining the 3-D location of the apple fruit during harvest. In *Automation technology for off-road equipment proceedings of the 2004 conference* (p. 91). American Society of Agricultural and Biological Engineers.

Burks, T., Bulanon, D., & Mehta, S. (2018). Opportunity of robotics in precision horticulture. In *Automation in tree fruit production: Principles and practice*. CAB International.

Burks, T., Villegas, F., Hannan, M., Flood, S., Sivaraman, B., Subramanian, V., & Sikes, J. (2005). Engineering and horticultural aspects of robotic fruit harvesting: opprtunities and constraints. *HortTechnology, 15*(1), 79–87.

Calvin, L., & Martin, P. (2010). *The US produce industry and labor: Facing the future in a global economy* (No. 1477-2017-4011).

Cao, X., Zou, X., Jia, C., Chen, M., & Zeng, Z. (2019). RRT-based path planning for an intelligent litchi-picking manipulator. *Computers and Electronics in Agriculture, 156*, 105–118. https://doi.org/10.1016/j.compag.2018.10.031

Chinchuluun, R., Lee, W. S., & Burks, T. F. (2007). Machine vision-based citrus yield mapping system. *Proc Florida State Hortic Soc, 119*, 142–147.

Choset, H., Lynch, K. M., Hutchinson, S., George, K., Burgard, W., Kavraki, L. E., & Thrun, S. (2005). Principles of robot motion: Theory, algorithms, and implementations. *Cambridge University Press*. https://doi.org/10.1017/S0263574706212803

Corke, P. I., & Hager, G. D. (1998). In B. Siciliano & K. P. Valavanis (Eds.), *Vision-based robot control - Control problems in robotics and automation*. Springer.

Davidson, J. R., & Mo, C. (2015). Mechanical design and initial performance testing of an apple-picking end-effector. In *ASME International Mechanical Engineering Congress and Exposition* (Vol. 57397, p. V04AT04A011). American Society of Mechanical Engineers.

Davidson, J., Silwal, A., Karkee, M., Mo, C., & Zhang, Q. (2016). Hand-picking dynamic analysis for undersensed robotic apple harvesting. *Trans ASABE, 59*(4), 745–758.

Du, X., Chen, D., Zhang, Q., Scharf, P., & Whiting, M. (2012). Dynamic responses of sweet cherry trees under vibratory excitations. *Biosystems Engineering, 111*, 305–314.

Duraj, V., Miles, J. A., Tejeda, D. J., Mitcham, E. J., Biasi, W. V., Asín, L., et al. (2010). Comparison of platform versus ladders for harvest in northern California pear orchard. In *XI International Pear Symposium* (Vol. 909, pp. 241–249).

Energid. (2020). *Citrus harvesting. Agricultural robotics*. Retrieved from https://www.energid.com/industries/agricultural-robotics

Erdoğan, D., Güner, M., Dursun, E., & Gezer, I. (2003). Mechanical harvesting of apricots. *Biosystems Engineering, 85*(1), 19–28

FAOSTAT. (2016). http://www.fao.org/faostat/en/#data

Fathallah, F. A. (2010). Musculoskeletal disorders in labor-intensive agriculture. *Applied Ergonomics, 41*(6), 738–743.

Fennimore, S. A., & Doohan, D. J. (2008). The challenges of specialty crop weed control, future directions. *Weed Technology, 22*(2), 364–372.

FFRobotics. (2020). *Robotic fruit harvestor. Apple harvesting robot*. Retrieved from https://www.ffrobotics.com/

Flood, S. J. (2006). Design of a robotic citrus harvesting end effector and force control model using physical properties and harvesting motion tests 73(5).

Font, D., Pallejà, T., Tresanchez, M., Runcan, D., Moreno, J., Martínez, D., Teixidó, M., & Palacín, J. (2014). A proposal for automatic fruit harvesting by combining a low cost stereovision camera and a robotic arm. *Sensors (Switzerland), 14*(7), 11557–11579. https://doi.org/10.3390/s140711557

Fu, L., Gao, F., Wu, J., Li, R., Karkee, M., & Zhang, Q. (2020). Application of consumer RGB-D cameras for fruit detection and localization in field: A critical review. *Computers and Electronics in Agriculture, 177*, 105687.

Gallardo, R. K., & Brady, M. P. (2015). Adoption of labor-enhancing technologies by specialty crop producers. *Agricultural Finance Review, 75*(4), 514–532.

Gallardo, R. K., Taylor, M. R., & Hinman, H. (2010). *Cost estimates of establishing and producing gala apples in Washington. Extension fact sheet FS005E*. University of Washington, School of Economic Sciences, Tree Fruit Research and Extension Center.

Gao, F., Fu, L., Zhang, X., Majeed, Y., Li, R., Karkee, M., & Zhang, Q. (2020). Multi-class fruit-on-plant detection for apple in SNAP system using Faster R-CNN. *Computers and Electronics in Agriculture, 176*, 105634.

Gené-Mola, J., Vilaplana, V., Rosell-Polo, J. R., Morros, J. R., Ruiz-Hidalgo, J., & Gregorio, E. (2019). Multi-modal deep learning for Fuji apple detection using RGB-D cameras and their radiometric capabilities. *Computers and Electronics in Agriculture, 162*, 689–698.

Gonzalez-Barrera, A. (2015). *More Mexicans leaving than coming to the US*. Retrieved from http://www.pewhispanic.org/files/2015/11/2015-11-19_mexican-immigration__FINAL.pdf

Han, K.-S., Si-Chan, K., Young-Bum, L., Sang-Chul, K., Dong-Hyuk, I., Hong-Ki, C., & Heon, H. (2012). Strawberry harvesting robot for bench-type cultivation. *Journal of Biosystems Engineering, 37*(1), 65–74. https://doi.org/10.5307/JBE.2012.37.1.065

Häni, N., Roy, P., & Isler, V. (2020). A comparative study of fruit detection and counting methods for yield mapping in apple orchards. *Journal of Field Robotics, 37*(2), 263–282.

Harrell, R. C., Adsit, P. D., Munilla, R. D., & Slaughter, D. C. (1990). Robotic picking of citrus. *Robotica, 8*(4), 269–278. https://doi.org/10.1017/S0263574700000308

Hashimoto, K. (2003). A review on vision-based control of robot manipulators. *Advanced Robotics, 17*(10), 969–991. https://doi.org/10.1163/156855303322554382

Hayashi, S., Ganno, K., Ishii, Y., & Tanaka, I. (2002). Robotic harvesting system for eggplants. *Japan Agricultural Research Quarterly: JARQ, 36*(3), 163–168.

Hayashi, S., Yamamoto, S., Tsubota, S., Ochiai, Y., Kobayashi, K., Kamata, J., Kurita, M., Inazumi, H., & Peter, R. (2014). Automation technologies for strawberry harvesting and packing operations in Japan. *Journal of Berry Research, 4*(1), 19–27. https://doi.org/10.3233/JBR-140065

He, L., Fu, H., Sun, D., Karkee, M., & Zhang, Q. (2017). Shake-and-catch harvesting for fresh market apples in trellis-trained trees. *Transactions of the ASABE, 60*(2), 353–360.

He, L., Zhang, X., Ye, Y., Karkee, M., & Zhang, Q. (2019). Effect of shaking location and duration on mechanical harvesting of fresh market apples. *Applied Engineering in Agriculture, 35*(2), 175–183.

Hofmann, J., Snyder, K., & Keifer, M. (2006). A descriptive study of workers' compensation claims in Washington State orchards. *Occupational Medicine, 56*(4), 251–257.

Hohimer, C. J., Wang, H., Bhusal, S., Miller, J., Mo, C., & Karkee, M. (2019). Design and field evaluation of a robotic apple harvesting system with a 3D-printed soft-robotic end-effector. *Transactions of the ASABE, 62*(2), 405–414. https://doi.org/10.13031/trans.12986

Jia, W., Zhang, Y., Lian, J., Zheng, Y., Zhao, D., & Li, C. (2020). Apple harvesting robot under information technology: A review. *International Journal of Advanced Robotic Systems, 17*(3), 1–16. https://doi.org/10.1177/1729881420925310

Juste, F., & Sevila, F. (1992). Citrus: An European project to study the robotic harvesting of oranges. In *Proceedings of the 3rd International Symposium on: Fruit, Nut, and Vegetable Harvesting Mechanization Jordbrugsteknik Agricultural Engineering* (No. 67, pp. 331–338).

Kim, Y., & Reid, J. (2004). Apple yield mapping using a multispectral imaging sensor. In: *International Conference on Agricultural Engineering, AgEng2004 Paper* (No. 010-PA-235).

Koirala, A., Walsh, K. B., Wang, Z., & McCarthy, C. (2019). Deep learning–Method overview and review of use for fruit detection and yield estimation. *Computers and Electronics in Agriculture, 162*, 219–234.

Kondo, N., & Ting, K. C. (1998). Robotics for plant production. *Artificial Intelligence Review, 12*(1–3), 227–243. https://doi.org/10.1007/978-94-011-5048-4_12

Kondo, N., Monta, M., & Fujiura, T. (1996). Fruit harvesting robots in Japan. *Advances in Space Research, 18*(1–2), 181–184.

Kurtulmus, F., Lee, W. S., & Vardar, A. (2011). Green citrus detection using 'eigenfruit', color and circular Gabor texture features under natural outdoor conditions. *Computers and Electronics in Agriculture, 78*(2), 140–149.

Kurtulmus, F., Lee, W. S., & Vardar, A. (2014). Immature peach detection in colour images acquired in natural illumination conditions using statistical classifiers and neural network. *Precision Agriculture, 15*(1), 57–79.

LaValle, S. M. (2006). Planning algorithms. In *University of Illinois* (1st ed.). Cambridge University Press. https://doi.org/10.1017/CBO9780511546877

Li, J., Karkee, M., Zhang, Q., Xiao, K., & Feng, T. (2016). Characterizing apple picking patterns for robotic harvesting. *Computers and Electronics in Agriculture, 127*, 633–640.

Lin, G., Tang, Y., Zou, X., Li, J., & Xiong, J. (2019). In-field citrus detection and localisation based on RGB-D image analysis. *Biosystems Engineering, 186*, 34–44.

Linker, R., Cohen, O., & Naor, A. (2012). Determination of the number of green apples in RGB images recorded in orchards. *Computers and Electronics in Agriculture, 81*, 45–57.

Lu, J., Lee, W. S., Gan, H., & Hu, X. (2018). Immature citrus fruit detection based on local binary pattern feature and hierarchical contour analysis. *Biosystems Engineering, 171*, 78–90.

Ma, S., Karkee, M., Fu, H., Sun, D., & Zhang, Q. (2018). Evaluation of shake-and-catch mechanism in mechanical harvesting of apples. *Transactions of the ASABE, 61*(4), 1257–1263.
Markwardt, E. D., Guest, R. W., Cain, J. C., & Labelle, R. L. (1964). Mechanical cherry harvesting. *Transactions of the ASAE, 7*(1), 70–74, 82.
Mehta, S. S., & Burks, T. F. (2014). Vision-based control of robotic manipulator for citrus harvesting. *Computers and Electronics in Agriculture, 102*, 146–158. https://doi.org/10.1016/j.compag.2014.01.003
Mehta, S. S., MacKunis, W., & Burks, T. F. (2014). Nonlinear robust visual servo control for robotic citrus harvesting. In *IFAC Proceedings Volumes (IFAC-PapersOnline)* (Vol. 19, Issue 3). IFAC. https://doi.org/10.3182/20140824-6-za-1003.02729
Mejia-Trujillo, J. D., Castaño-Pino, Y. J., Navarro, A., Arango-Paredes, J. D., Rincón, D., Valderrama, J., et al. (2019). Kinect™ and Intel RealSense™ D435 comparison: A preliminary study for motion analysis. In *2019 IEEE International Conference on E-health Networking, Application & Services (HealthCom)* (pp. 1–4). IEEE.
Mitchell, T. M. (1997). *Machine learning*. McGraw Hill.
Muscato, G., Prestifilippo, M., Abbate, N., & Rizzuto, I. (2005). A prototype of an orange picking robot: past history, the new robot and experimental results. *Industril Robot, 32*(2), 128–138.
Nash, A., Koenig, S., & Likhachev, M. (2009). *Incremental Phi*: Incremental any-angle path planning on grids*. Lab Papers (GRASP).
Nguyen, T. T., Kayacan, E., De Baedemaeker, J., & Saeys, W. (2013). Task and motion planning for apple harvesting robot. In *IFAC Proceedings Volumes (IFAC-PapersOnline)* (Vol. 4, Issue PART 1). IFAC. https://doi.org/10.3182/20130828-2-SF-3019.00063
Onishi, Y., Yoshida, T., Kurita, H., Fukao, T., Arihara, H., & Iwai, A. (2019). An automated fruit harvesting robot by using deep learning. *ROBOMECH Journal, 6*(1), 13. https://doi.org/10.1186/s40648-019-0141-2
Parrish, E. A., & Goksel, A. K. (1977). Pictorial pattern recognition applied to fruit harvesting. *Transactions of the ASAE, 20*(5), 822–0827.
Peterson, D. L., & Bennedsen, B. S. (2005). Isolating damage from mechanical harvesting of apples. *Applied Engineering in Agriculture, 21*(1), 31–34.
Peterson, D. L., & Miller, S. S. (1996). Apple harvesting concepts for inclined trellised canopies. *Applied Engineering in Agriculture, 12*(3), 267–271.
Plebe, A., & Anile, A. M. (2002). A neural-network-based approach to the double traveling salesman problem. *Neural Computation, 14*(2), 437–471. https://doi.org/10.1162/08997660252741194
Plebe, A., & Grasso, G. (2001). Localization of spherical fruits for robotic harvesting. *Machine Vision and Applications, 13*(2), 70–79.
Qiang, L., Jianrong, C., Bin, L., Lie, D., & Yajing, Z. (2014). Identification of fruit and branch in natural scenes for citrus harvesting robot using machine vision and support vector machine. *International Journal of Agricultural and Biological Engineering, 7*(2), 115–121.
Rakun, J., Stajnko, D., & Zazula, D. (2011). Detecting fruits in natural scenes by using spatial-frequency based texture analysis and multiview geometry. *Computers and Electronics in Agriculture, 76*(1), 80–88.
Reed, J. N., Miles, S. J., Butler, J., Baldwin, M., & Noble, R. (2001). AE—Automation and emerging technologies: Automatic mushroom harvester development. *Journal of Agricultural Engineering Research, 78*(1), 15–23. https://doi.org/10.1006/jaer.2000.0629
Ringdahl, O., Kurtser, P., & Edan, Y. (2019). Evaluation of approach strategies for harvesting robots: Case study of sweet pepper harvesting. *Journal of Intelligent & Robotic Systems, 95*(1), 149–164. https://doi.org/10.1007/s10846-018-0892-7
Robinson, T., Hoying, S., Sazo, M. M., DeMarree, A., & Dominguez, L. (2013). A vision for apple orchard systems of the future. *New York Fruit Quarterly, 21*(3), 11–16.
Roldan, J. J., del Cerro, J., Garzón-Ramos, D., Garcia-Aunon, P., Garzón, M., de León, J., & Barrientos, A. (2018). Robots in agriculture: State of art and practical experiences. In *Service robots* (pp. 67–90). https://doi.org/10.5772/intechopen.69874

Savary, S. K. J. U., Ehsani, R., Schueller, J. K., & Rajaraman, B. P. (2010). Simulation study of citrus tree canopy motion during harvesting using a canopy shaker. *Transactions of the ASABE, 53*(5), 1373–1381.

Schupp, J., Baugher, T., Winzeler, E., Schupp, M., & Messner, W. (2011). Preliminary results with a vacuum assisted harvest system for apples. *Fruit Notes, 76*(4), 1–5.

Si, Y., Liu, G., & Feng, J. (2015). Location of apples in trees using stereoscopic vision. *Computers and Electronics in Agriculture, 112*, 68–74.

Silwal, A., Gongal, A., & Karkee, M. (2014). Apple identification in field environment with over the row machine vision system. *Agricultural Engineering International: CIGR Journal, 16*(4), 66–75.

Silwal, A., Davidson, J., Karkee, M., Mo, C., Zhang, Q., & Lewis, K. (2016). *Effort towards robotic apple harvesting in Washington State*. Paper Number: 162460869. ASABE. https://doi.org/10.13031/aim.20162460869

Silwal, A., Davidson, J. R., Karkee, M., Mo, C., Zhang, Q., & Lewis, K. (2017). Design, integration, and field evaluation of a robotic apple harvester. Journal of Field Robotics, 34(6), 1140–1159. https://doi.org/https://doi.org/10.1002/rob.21715.

Simonton, W. (1991). Robotic end effector for handling. *Transaction of the American Society of Agricultural Engineers, 34*, 2615–2621.

Slaughter, D. C., & Harrell, R. C. (1989). Discriminating fruit for robotic harvest using color in natural outdoor scenes. *Transactions of the ASAE, 32*(2), 757–0763.

Stajnko, D., Rakun, J., & Blanke, M. (2009). Modelling apple fruit yield using image analysis for fruit colour, shape and texture. *European Journal of Horticultural Science, 74*(6), 260.

Tabb, A. L., Peterson, D. L., & Park, J. (2006). *Segmentation of apple fruit from video via background modeling*. ASABE Paper No. 063060. ASABE. https://doi.org/10.13031/2013.20873

Tanigaki, K., Fujiura, T., Akase, A., & Imagawa, J. (2008). Cherry-harvesting robot. *Computers and Electronics in Agriculture, 63*(1), 65–72. https://doi.org/10.1016/j.compag.2008.01.018

Tao, Y., & Zhou, J. (2017). Automatic apple recognition based on the fusion of color and 3D feature for robotic fruit picking. *Computers and Electronics in Agriculture, 142*, 388–396.

Tillett, N. D. (1993). Robotic manipulators in horticulture: A review. *Journal of Agricultural Engineering Research, 55*(2), 89–105.

U.S. Apple Association. (2018). *Apple industry statistics*.

University of California Agriculture and Natural Resources Cooperative Extension. (2017a). *Sample costs for sweet cherries*. Retrieved from https://coststudyfiles.ucdavis.edu/uploads/cs_public/84/4f/844f0de3-7c1c-4bd3-89ac-893e3388c140/cherry2017_nsjv_final_draft4.pdf

University of California Agriculture and Natural Resources Cooperative Extension. (2017b). *Sample costs for processing peaches*. Retrieved from https://coststudyfiles.ucdavis.edu/uploads/cs_public/65/6c/656ce5fa-f4f9-4e95-99fe-a01c1ea8a7a4/2017peachsvsjv-ecling-final_draft2.pdf

University of California Agriculture and Natural Resources Cooperative Extension. (2018). *Sample costs to produce pears*. Retrieved from https://coststudyfiles.ucdavis.edu/uploads/cs_public/3d/01/3d01cbc0-8ca3-482d-a6d6-3780700bcb96/18pearssv-7218.pdf

University of California Cooperative Extension. (2014). *Sample costs to produce processing apples*. Retrieved from https://coststudyfiles.ucdavis.edu/uploads/cs_public/9c/bf/9cbfcb79-c037-47fe-af0b-79ec2edb4031/apples-santa-cruz-conv-processing.pdf

University of California Cooperative Extension. (2015). *Sample costs to establish an orange orchard and produce oranges*. Retrieved from https://coststudyfiles.ucdavis.edu/uploads/cs_public/19/d4/19d4f1bb-408a-443e-a759-36fd53a2948f/oranges_vs_2015.pdf

Upadhyaya, S. K., Cooke, J. R., & Rand, R. H. (1981). Dynamics of fruit tree trunk impact. *Transactions of the ASAE, 24*(4), 846–855.

USDA-NASS. (2018). *Non-citrus fruits and nuts: 2017 summary*. United States Department of Agriculture. Retrieved August 31, 2020, from https://usda.mannlib.cornell.edu/usda/current/NoncFruiNu/NoncFruiNu-06-26-2018.pdf

USDA-NASS. (2019a). *Citrus fruits 2019 summary*. Retrieved from https://www.nass.usda.gov/Publications/Todays_Reports/reports/cfrt0819.pdf

USDA-NASS. (2019b). *Noncitrus fruits and nuts 2018 summary*. Retrieved from https://www.nass.usda.gov/Publications/Todays_Reports/reports/ncit0619.pdf

Van Henten, E. J., Hemming, J., Van Tuijl, B. A. J., Kornet, J. G., & Bontsema, J. (2003). Collision-free motion planning for a cucumber picking robot. *Biosystems Engineering, 86*(2), 135–144.

Vicencio, K., Davis, B., & Gentilini, I. (2014). Multi-goal path planning based on the generalized traveling salesman problem with neighborhoods. In *2014 IEEE/RSJ International Conference on Intelligent Robots and Systems* (pp. 2985–2990). IEEE.

Wachs, J. P., Stern, H. I., Burks, T., & Alchanatis, V. (2010). Low and high-level visual feature-based apple detection from multi-modal images. *Precision Agriculture, 11*(6), 717–735.

Wan, S., & Goudos, S. (2020). Faster R-CNN for multi-class fruit detection using a robotic vision system. *Computer Networks, 168*, 107036.

Wang, Q., Nuske, S., Bergerman, M., & Singh, S. (2013). Automated crop yield estimation for apple orchards. In *Experimental robotics* (pp. 745–758). Springer.

Washington State Employment Security Department. (2013). *Agriculture workforce report*. Retrieved from https://fortress.wa.gov/esd/employmentdata/docs/industry-reports/agricultural-workforce-repo rt-2013.pdf

Whittaker, D., Miles, G. E., Mitchell, O. R., & Gaultney, L. D. (1987). Fruit location in a partially occluded image. *Transactions of the ASAE, 30*(3), 591–0596.

Xu, F., Liu, G., Zou, X., Chen, Z., & Xu, C. (2011). The virtual prototype design and simulation of litchi fruit flexible picking manipulator. In *International Conference on Computer Distributed Control and Intelligent Environmental Monitoring, CDCIEM 2011* (pp. 1919–1921). https://doi.org/10.1109/CDCIEM.2011.383

Yaguchi, H., Nagahama, K., Hasegawa, T., & Inaba, M. (2016). Development of an autonomous tomato harvesting robot with rotational plucking gripper. In *2016 IEEE/RSJ International Conference on Intelligent Robots and Systems (IROS)* (pp. 652–657). https://doi.org/10.1109/IROS.2016.7759122

Yau, W.-Y., & Wang, H. A. N. (1996). Robust hand-eye coordination. *Advanced Robotics, 11*(1), 57–73. https://doi.org/10.1163/156855397X00047

Yu, Y., Sun, Z., Zhao, X., Bian, J., & Hui, X. (2018). Design and implementation of an automatic peach-harvesting robot system. In *Proceedings - 2018 10th International Conference on Advanced Computational Intelligence, ICACI 2018* (pp. 700–705). https://doi.org/10.1109/ICACI.2018.8377546

Yuan, Y., Zhang, X., & Zhao, H. (2009). Apple harvesting robot picking path planning and simulation. *2009 International Conference on Information Engineering and Computer Science*, 1–4. https://doi.org/10.1109/ICIECS.2009.5366245

Zahid, A., He, L., Zeng, L., Choi, D., Schupp, J., & Heinemann, P. (2020a). Development of a robotic end-effector for apple tree pruning. *Transactions of the ASABE, 63*(4), 847–856. https://doi.org/10.13031/trans.13729

Zahid, A., Mahmud, M. S., He, L., Choi, D., Heinemann, P., & Schupp, J. (2020b). Development of an integrated 3R end-effector with a cartesian manipulator for pruning apple trees. *Computers and Electronics in Agriculture, 179*, 105837. https://doi.org/10.1016/j.compag.2020.105837

Zahid, A., He, L., Choi, D. D., Schupp, J., & Heinemann, P. (2020c). *Collision free path planning of a robotic manipulator for pruning apple trees*. ASABE Paper No. 200439. ASABE. https://doi.org/10.13031/aim.202000439.

Zhang, Z., Heinemann, P. H., Liu, J., Schupp, J. R., & Baugher, T. A. (2016). Design and field test of a low-cost apple harvest-assist unit. *Transactions of the ASABE, 59*(5), 1149–1156.

Zhang, Q., Karkee, M., & Tabb, A. (2019). The use of agricultural robots in orchard management. In J. Billingsley (Ed.), *Robotics and automation for improving agriculture* (pp. 187–214). Burleigh Dodds Science. https://doi.org/10.19103/as.2019.0056.14

Zhang, Z., Igathinathane, C., Li, J., Cen, H., Lu, Y., & Flores, P. (2020). Technology progress in mechanical harvest of fresh market apples. *Computers and Electronics in Agriculture, 175*, 105606.

Zhao, J., Tow, J., & Katupitiya, J. (2005). On-tree fruit recognition using texture properties and color data. In *2005 IEEE/RSJ International Conference on Intelligent Robots and Systems* (pp. 263–268). IEEE.

Zhao, D., Lv, J., Ji, W., Zhang, Y., & Chen, Y. (2011). Design and control of an apple harvesting robot. *Biosystems Engineering, 110*(2), 112–122. https://doi.org/10.1016/j.biosystemseng.2011.07.005

Zhao, Y., Gong, L., Huang, Y., & Liu, C. (2016). A review of key techniques of vision-based control for harvesting robot. *Computers and Electronics in Agriculture, 127*, 311–323. https://doi.org/10.1016/j.compag.2016.06.022

Zheng, L., Ji, R., Liao, W., & Li, M. (2018). A positioning method for apple fruits based on image processing and information fusion. *IFAC-PapersOnLine, 51*(17), 764–769.

The manufacturer's authorised representative in the EU is Springer
Nature Customer Service Centre GmbH, Europaplatz 3, 69115 Heidelberg,
Germany. If you have any concerns regarding our products, please
contact ProductSafety@springernature.com

Printed and bound by CPI Group (UK) Ltd, Croydon, CR0 4YY
30/09/2026
02225496-0011